WITHDRAWN
UTSA LIBRARIES
D0154444
RENEWALS: 691-4574

Design with Advanced Composite Materials

Design with Advanced Composite Materials

Edited by
Leslie N Phillips

The Design Council
London

Springer-Verlag
Berlin Heidelberg New York
London Paris Tokyo

Design with Advanced Composite Materials
edited by Leslie N Phillips

First published 1989 in the United Kingdom by
The Design Council
28 Haymarket
London SW1Y 4SU

Typeset by Santype International Limited, Salisbury, Wiltshire

Printed and bound in Great Britain by
Biddles Ltd, Guildford and King's Lynn

All rights reserved. No part of this publication may be reproduced, stored in a retrieval system or transmitted, in any form or by any means now known or hereafter invented, electronic, mechanical, photocopying, recording or otherwise, without the prior written permission of the Design Council.

© 1989 in the collection: the Design Council
Contributors retain copyright within their individual chapters

British Library Cataloguing in Publication Data

Advanced composite materials
1. Composite materials
I. Phillips, Leslie N.
620.1′18

ISBN 0-85072-238-1
ISBN 3-540-51800-2
ISBN 0-387-51800-2

Library
University of Texas
at San Antonio

Contents

Preface

This book is about composites, that somewhat exotic and mysterious group of materials which are of great interest to scientists, engineers, businessmen and designers.

It is the purpose of this book to dispel the mystery by explaining what composites are: how they are made; what properties they possess; how they can be employed in many different fields.

Quite simply, composites are materials comprising exceptionally strong, stiff fibres, embedded in a surrounding continuous phase or matrix. The emphasis in this book is on that branch wherein the second phase is a high polymer or plastics.

To understand composites, the services of physicists, polymer chemists and technologists, engineers and designers are required. Together, the authors represent a modest 'gang of eight'.

The subject falls neatly into ten main areas which are dealt with separately in chapters as follows:

Chapter 1 introduces formal definitions of fibre, matrix and composite, by way of a short history explaining the efforts to provide the enormous number of novel materials with special properties which are required by modern industry. The many different types of fibre (their preparation, properties and uses) are explained in detail, for without such fibres there would be no advanced composites. The other, necessary, partner is the matrix resin, which must adhere to the fibre surface. Indeed, the composite works, as a structural system, because the matrix transfers stress to the fibre by shear at the interface between them.

Chapter 2 deals with the art and science of fabrication, or how the mix of resin and fibre is shaped to final dimensions. A number of moulding processes, introduced over the past 40 years or so, depend critically on the flow of the matrix under different conditions of time, temperature and pressure. Note that the production of shapes by machining from blocks of material is rarely practised with composites. The production engineer is more likely to worry about staff time in moulding than in machining.

The importance of rheology, or flow, is an integral part of moulding technology, depending directly on the detailed chemistry of the matrix polymer. In *Chapter 3* George Green, of Ciba-Geigy Plastics, Cambridge, lists the many families which yield useful matrices, and describes their synthesis and properties in detail. In this field, it is very much a question of 'horses for courses': different fibres and specific surface treatments react best to particular matrix resins.

It is the performance of the composites themselves, their mechanical, thermal and electrical properties, which determine the most appropriate end-use by industry. These composite properties are recorded in numerous graphs and tables by James Quinn of Fibreforce Composites Ltd, Cheshire, in *Chapter 4*. A short introduction on design centres around the engineer's task when using rod, tube and plank composite products made by the pultrusion process. A brief recapitulation of the properties of neat resin and of elementary fibre underline the lesson that the prediction of composite properties from those of its constituents is not yet an exact science.

Frank Matthews, Director of the Centre for Composite Materials at Imperial College, London, explains in *Chapter 5* that their low bearing strength (relative to steel and light alloy) means that the use of mechanical fasteners, such as rivets or nuts and bolts, is less widespread than in engineering with metals. What is required is careful design around bolt holes, with the generous use of doubler strips and $\pm 45°$ orientation of fibres, together with adhesive-bonding techniques, in order to produce safe and reliable methods of joining composites, both to themselves and to metals.

Most of the materials and products used by industry are bought and sold subject to passing appropriate specifications as to composition, quality, or performance. This is particularly true for the defence industries, where the highest standards apply. Thus *Chapter 6*, by Jim Methven, an independent consultant and lecturer at The University of Manchester Institute of Science and Technology (UMIST), has made a complete record of the relevant specifications for fibres, resins and composites, listing both civil and military, and foreign as well as British Standards.

Having absorbed the considerable body of information given in the first six chapters, the way is open to deal with the wide variety of applications described in *Chapter 7*. Anyone is free to suggest novel uses for a material, but it takes a knowledge of fabrication, testing procedures and design experience, as well as an entrepreneurial streak, to bring daydreams to successful reality. Many applications, from aeronautics to sports equipment, are described in this section.

Once a component has been moulded, it must be inspected. There is a requirement for methods of inspection which are non-destructive, and it is fortunate that such methods have been developed at exactly the moment when composites have reached a mass-market. Paul Teagle, of Qualcorp Air–Space Systems in California, brings a wealth of experience to this field, and in *Chapter 8* explains the physical basis of NDT inspection, and how different fibre types, different orientations and different manufacturing defects affect the results and their interpretation.

The final two chapters deal with the mathematical aspects of designing with composites. It is sad but true that the engineers' formulae for stress and strain, the bending of plates and beams etc, are more complex than is the case with isotropic materials such as the metals. This is understandable, since fibres in a lay-up may run in several different directions. For this reason, the contributions of Len Hollaway and Peter Beaumont (professor and lecturer in the Universities of Surrey and Cambridge, respectively) are grouped together, and may be studied together with advantage. First, the three-dimensional analysis is set out by Hollaway, then the similar treatment of fracture processes by Beaumont, which is very different from the 'fracture mechanics' approach normally offered to engineers studying the behaviour of metals.

General readers, with limited mathematics, can derive benefit from considering the first eight chapters only; readers with sufficient mathematics to understand *Chapters 9 and 10* will considerably improve their grasp of the concepts underpinning the design process.

Leslie N Phillips
Farnborough, April 1989

Acknowledgements

Figures 1.6, 2.1, 2.5, 2.6, 2.14 and 7.22 are reproduced with the permission of the Controller of Her Majesty's Stationery Office.

Glossary

AEW	anhydride equivalent weight
aramid	aromatic polyamide
ARP	aramid-fibre reinforced plastics
AV	amine value
−C	continuous reinforcement
C(F)RP	carbon-fibre reinforced plastics
EEW	epoxy equivalent weight
FEP	fluorinated ethylene-propylene
G(F)RP	glass-fibre reinforced plastics
HM	high-modulus
HP	high-performance
HS	high-strain
HTS	high tensile stress
ILSS	interlaminar shear strength
IM	intermediate-modulus
Kevlar	an aramid manufactured by Du Pont
LEFM	linear elastic fracture mechanics
MPP	mesophase pitch
Nomex	an aramid manufactured by Du Pont
PAN	polyacrylonitrile
PEEK	polyether-ether ketone
PEK	polyether ketone
PES	polyether sulphone
pultrudate	pultruded laminate
−S	surface-treated
size	protective surface coating
SMC	sheet moulding compound
XA	high-strength

Conversion Factors

$1\ \text{MN/m} = 5709.92\ \text{lbf/inch}$
$(1\ \text{kN/mm})$

$1\ \text{MN/m}^{3/2} = 910.0142\ \text{lbf/in}^{3/2}$
$(1\ \text{MPa}\sqrt{\text{m}})$

$1\ \text{MPa} = 145\ \text{lbf/in}^2$
$(1\ \text{MN/m}^2)$

$1\ \text{kN/cm}^2 = 1450\ \text{lbf/in}^2$

$1\ \text{J/m} = 0.0728\ \text{cal/ft}$

$1\ \text{J/m}^2 = 0.0222\ \text{cal/ft}^2$

$1\ \text{g/cm}^3 = 0.5780\ \text{oz/in}^3$
$(1\ \text{Mg/m}^3)$

$1\ \text{J/g} = 6.77\ \text{cal/oz}$

$1\ \text{Å} = 10^{-10}\ \text{m}$

$1\ \text{kg} = 2.205\ \text{lb}$

$1\ \text{g} = 15.43236\ \text{grain}$

$1\ \text{t} = 0.9842\ \text{ton}$

$1\ \text{mm} = 0.03937\ \text{inch}$

$1\ \text{m} = 3.281\ \text{feet}$

$1\ \text{m}^{1/2} = 1.8113\ \text{ft}^{1/2}$

$1\ \text{km} = 0.6214\ \text{mile}$

$^\circ\text{F} = (^\circ\text{C} \times 1.8) + 32$

1 Introduction

Leslie N Phillips OBE FRIC CChem FPRI

formerly Head of Section,
Plastics Technology, Materials Department,
Royal Aircraft Establishment, Farnborough

and

Professor (ret.) in Department of Civil Engineering,
University of Surrey

1.1 THE IMPORTANCE OF MATERIALS

This book is about new materials. A material is any substance employed in making some useful thing or artefact. Thus a song, or a mathematical formula, or a drop of water are not materials, although when an Eskimo saws frozen snow into blocks in order to build an igloo, he is quite certainly using ice as a material.

Evidently materials are used because of their properties or attributes, one of which, as in the above example, is mechanical strength. A supply of many different materials, each possessing a different mix of properties, is needed for the efficiency, comfort and convenience of modern life. Should readers need persuading of the truth of this, the variety of materials used in their nearest town centre would reward study.

In the oldest parts of town there remain several traditional buildings made of natural stone or brick, with much use of wood for doors, window frames and roof beams. There may be bronze statues in the parks and squares, often with their cast-iron railings still intact. The pavements are of split stone and the narrow alleyways are cobbled underfoot. The whole scene has an early Victorian flavour, though it has remained unchanged for a couple of centuries. We can imagine the rumble of wooden wheels and the rattle of hansom cabs, as hundreds of horses provide the motive power to carry goods and people about their business.

When we move on to the modern town centre the picture is different. The tarmac streets swarm with cars, trucks, buses and vans in many shapes and sizes. Most of these roll by on pneumatic rubber tyres. The principal buildings are made of concrete and enormous quantities of glass. On new building sites, where construction is in full swing, the raw skeletons of structural steel and the precise grid of reinforcing rods are ready to receive their placing of concrete.

The decorative facing of the buildings and their surrounding paved precincts are also generally of precast concrete units. Thus, in a little over 100 years, the external appearance of the city has been changed by the introduction of just two materials, steel and concrete. It is easy to forget that cheap, mass-produced steel was the invention of the Bessemer convertor in 1856. The use of reinforced concrete in construction (Winter 1970) was pioneered by Hennebique (1843–1921), and the Flatiron building in New York in 1903 was the first American example in that city of skyscrapers.

However, we have to look inside the larger buildings before we can understand the full extent of the transformation. Within the office blocks, factories, hospitals, shopping arcades and sports centres there runs a network of pipes, wires and cables for various services.

We can identify tubes and ducts for central heating and air-conditioning. Electric cables supply power for pumps, elevators and escalators. They drive the machines and welders in the factories as well as the office machines and computers. There are cables for telephones, public address systems, closed-circuit television, fire detectors, alarms and sprinklers. All of these systems require a variety of conductors, semiconductors and insulators.

1.2 THE RISE OF POLYMERS

Up to the end of the 19th century, the metal and ceramic industries provided the conductors and the refractory materials. Insulation was based on natural products such as wood, paper, shellac and gutta-percha (Ridding 1964). Since the early 1900s, one of the most fruitful developments in materials has been the discovery and exploitation of synthetic high polymers.

These are substances, produced originally in chemical laboratories, which consist of long chains or networks, built up by the repeated linkage of small reactive molecules. Each of the simple units is called a mer; hence a polymer is simply the result of the joining together of many identical units. When the chains get long enough (say a molecular weight above 5,000) new properties appear such as hardness, stiffness and mechanical strength, which have no relation to the original liquids or gases which comprise their starting molecules.

Depending on the detailed chemistry and the spatial arrangement (architecture) of the chains and their cross-links, the following can be produced:

- plastics
- rubbers
- adhesives
- coatings
- fibres

There may be many hundreds of possible substances in each of these classes (Kaufmann 1967). Not all of them possess sufficiently useful properties, and graduate to becoming materials of commerce. The remainder stay on file.

It is true to say that, as urban life becomes more complex and technical, there is an unending demand by engineers for materials with novel or improved properties. Generally speaking this demand has been met, over the past eighty years or so, by the combined efforts of metallurgists, ceramic technologists and polymer chemists, but that is not the end of the story.

With a clearer understanding of the chemistry and physics of these materials (metals, ceramics and polymers) it has become possible to combine them with fibres in order to produce an enormous range of hitherto unknown substances which are loosely referred to as 'advanced composites'.

Although these bear a family resemblance to the composite materials produced by nature, of which wood is the best known example, it is a mistake to push the analogy too far. For example, wood is constructed from the natural fibre cellulose, which is hygroscopic. Thus wooden articles are not dimensionally stable – they 'move' with the weather and become weaker when saturated. They are also subject to biological attack – anything from termites to fungal rot – which destroys their integrity.

Modern man-made fibres and composites are not vulnerable in this way. Their stability, combined with outstanding physical properties, make them ideal engineering materials, able to outperform many of the metals for structural applications.

However, there are complications. Composites are fabricated by strange methods (see Chapter 2), quite outside those which are familiar in metal-working practice. Again, composites can be made with a pronounced grain, and are at their most efficient when constructed with the fibres running in certain directions. The engineer and the designer have a great deal to learn before they can rise to the challenge of advanced composites and exploit them fully.

1.3 THE NATURE OF COMPOSITES

We define a composite material as one comprising a large number of strong, stiff fibres, called the reinforcement, embedded in a continuous phase of a second material known as the matrix. This definition is very wide. It permits the fibre to be natural or man-made, metallic, inorganic or organic.

Likewise, the matrix can be a metal or metal alloy, an inorganic cement or glass, or a natural or synthetic high polymer (Dietz 1963).

In order to keep this volume to a reasonable size the discussion is restricted to those composites wherein the matrix is a synthetic polymer, ie a resin or plastics material.

The justification is that this constitutes the best developed branch, in terms both of volume and of value. There is now a recognisable industrial base in many countries, and a sizeable body of theory behind it.

Hopefully, the day will come when composites based on metal and ceramic matrices will rival in importance those based on polymers. Experience teaches that, since it has taken about one generation to succeed with the latter, it will take a roughly equal time with the former.

The scale of possibilities needs to be appreciated. There are many potential reinforcements, which are discussed in detail in the following Sections. There are also many potential matrix resins. If these two numbers are multiplied together we shall come close to the number of possible composites. It is possible to vary the amount of fibre which is

Table 1.1 Fibres, matrices and matching pairs

Fibre	Matrix	Matching pairs (good adhesion)
asbestos	phenol-formaldehyde	asbestos/phenolic
glass	melamine-formaldehyde	asbestos/melamine
carbon	polyester	glass/polyester
aramid	epoxy	glass/epoxy
boron	silicone	glass/silicone
silica	polyimide	aramid/epoxy
	polybenzimidazole (PBI)	carbon/epoxy
	furane	FC/silica
	Friedel-Crafts (FC)	boron/polyimide
		boron/PBI

added, as well as the orientation; it is also possible to change the resin/hardener combinations used for the matrix.

As each of these recipes gives a substance whose properties are unique, they may be considered different materials. When we employ the usual shorthand description of a composite (fibre first then matrix), for example a 'glass/polyester' or a 'carbon/epoxy' laminate, we are ignoring the fact that each is a family.

Table 1.1 gives a list of fibres, resins and suitable matching pairs.

1.4 HOW COMPOSITES WORK

An unimpregnated bundle of thin filaments is useless when considered as an engineering structure. It has no definite shape and no definite hard surface. It is virtually impossible to machine it accurately or to polish it to a smooth surface. Like a piece of string it resists tensile forces, but it is hopeless at compression, torsion, or bending like a beam.

When a bundle of filaments is dipped into a bath of resin, drained and allowed to harden, it has properties as a composite approaching that of steel and is invaluable to an engineer. It is able to resist enormous forces in tension, compression and bending. It has a definite shape, a durable surface and it can be machined with great accuracy.

Table 1.2 Mechanical properties of fibres and other materials

Fibre/material	Specific gravity density (g/cm^3)	Tensile strength		Young's modulus	
		(MPa)	($\times 10^3$ lbf/in^2)	(GPa)	($\times 10^6$ lbf/in^2)
E-glass	2.54	2410	349	69	10
S-glass	2.49	2620	380	87	12.6
carbon type 2	1.75	2410	349	241	35
aramid (high modulus)	1.44	3450	500	124	18
boron	2.63	2760	400	379	55
alumina	3.30	3000	435	297	43
asbestos (chrysotile)	2.40	1490	216	183	26.5
wood	0.50	69	10	7	0.95
light alloy	2.69	476	69	72	10.50
steel (structural)	7.85	413	60	207	30
titanium alloy	4.52	711	103	117	17

Considering the reinforcement more closely:

(i) The mechanical properties of the fibres are extremely high and often exceed the bulk properties of most engineering metals (Holister and Thomas 1966). Important fibre properties are tensile strength, tensile modulus (always called Young's modulus) and density or specific gravity. Some typical values are given in Table 1.2.

(ii) Fibres are necessarily long and thin, in the sense that their length is many times greater than their diameter. The ratio of length/diameter (l/d) is known as the aspect ratio; where this exceeds 100 or so we expect that a useful reinforcing action will be possible.

(iii) Because of the exceedingly small subdivisions of material, the superficial area of a fibre bundle is very large. Most commercial fibres are less than 20 μm (0.8×10^{-3} inch) in diameter, ie about a tenth the width of a human hair. A million fibres can be packed into a volume of 10 mm × 10 mm × 10 mm (0.4 inch cube). An actual (superficial) area of more than a square metre is available for adhesion to the matrix resin.

(iv) It should be noted that the mechanical properties of matrix resins are very modest. They are lower in strength and stiffness than reinforcing fibres by a whole order of magnitude and are rarely used by engineers in any long-term load-carrying capacity. They are, however, excellent

Figure 1.1 *Percentage strength against ratio of moduli $E_{f/Em}$, where E_f = modulus of reinforcement, and E_m = modulus of matrix. At high ratios of modulus, most of the total load is carried by the fibres*

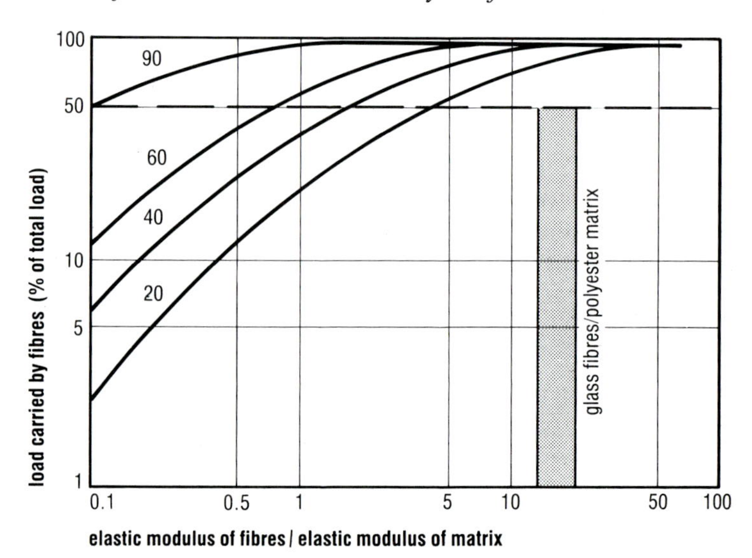

> adhesives and are well able to transfer forces to any reinforcing fibre by shear at the interface. According to Gordon (1978), shear is the type of force which is developed when one portion of a material slides past another.

A general picture of the relationship between fibre and matrix is given in Figure 1.1, where the percentage of the total load which is being carried by the reinforcement is plotted against the modulus ratio of fibre to matrix.

Different curves, of similar shape, are given by a family of composites which differ from each other in the volume of reinforcement present. It will be seen that as the volume fraction V_f of fibre rises, the proportion of total load taken by it rises beyond 90%.

It is evident that the essential role of the matrix is to transfer and distribute the incoming loads to the fibres which alone can sustain them. Both components of the composite are necessary for the system to function.

To look at shear in a more analytical manner, consider a single filament of circular cross-section embedded normally in a small disc of matrix, as shown in Figure 1.2 *a*.

If the filament is embedded for a small depth, say one or two diameters only, the adhesive grasp of resin on fibre is insufficient; on continued loading it is pulled bodily from the matrix, leaving an empty socket (Figure 1.2 *b*).

If the filament is submerged to a considerable depth in the matrix, as the tensile loading is increased more and more, the fibre eventually fails and it fractures outside the matrix, leaving a stump behind it (Figure 1.2 *c*).

Between these two extremes there must be a critical length l_c the size of which is related to the aspect ratio and the degree of adhesion. If the length is less than l_c there is pull-out: at and above l_c there is a tensile fracture of the reinforcement. At the critical length, therefore, the failing load is nicely balanced between shear and tension, so

(tensile strength of fibre) × (cross-sectional area of fibre)
= (area of adhesion) × (shear strength of interface)

Thus,
$$\sigma_f \pi r^2 = 2\pi r l_c \tau$$

where
σ_f = tensile strength of fibre
r = radius of fibre
l_c = critical length
l = embedded length
τ = shear strength

Figure 1.2 *Cross-section of single filament embedded in matrix. A well-bonded fibre will fail in tension*

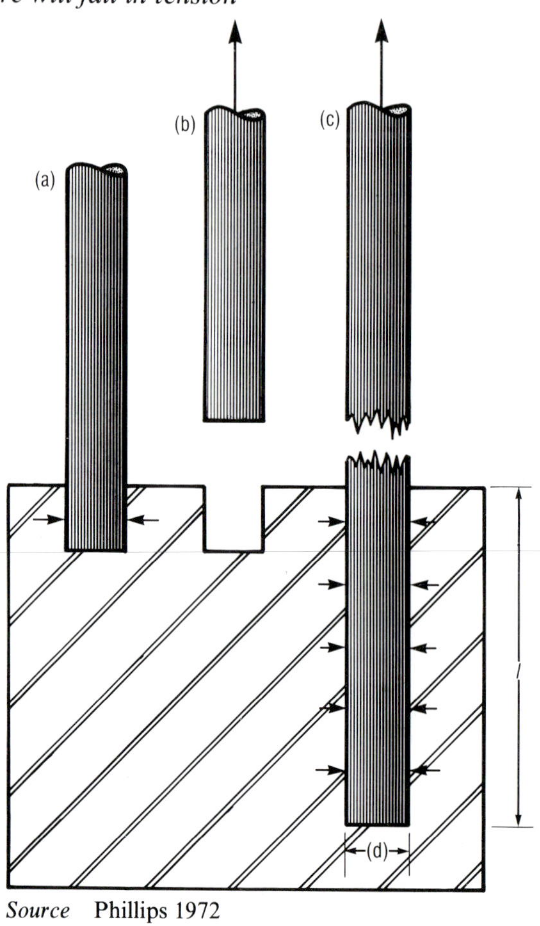

Source Phillips 1972

Simplifying, $\tau = \dfrac{\sigma_f r}{2l_c}$

The transfer of a load to the fibre is a gradual process. At the extreme end of the fibre, the tensile stress will be zero. Proceeding inboard, along the fibre, the tensile stress rises to a constant (maximum) once l_c has been passed. Conversely, shear stress is greatest at the fibre end, falling inboard towards zero.

1.5 THE REINFORCEMENT

There are many potential fibres, some found in nature, others man-made. All are distinguished by high tensile strength and many have great stiffness, ie quite simply, they stretch very little under load. The more valuable ones also have a low density, for it is their specific strength

$$\frac{\sigma_f}{\rho}$$

and specific stiffness

$$\frac{E_f}{\rho}$$

which are important criteria when comparing one engineering material with another (ρ = fibre density).

1.6 PRINCIPAL FIBRES AND THEIR INDUSTRIAL IMPORTANCE

Normally, the introduction of novel materials into engineering practice follows a curve of exponential growth. This curve is characterised by a slow initial rise of some years' duration – while the material emerges from the laboratory and struggles for acceptance – followed by a rapid, almost explosive, rate of growth when mass markets are found and commercial production is established.

There is no way of accurately predicting when this massive upturn will occur; what is certain is that, after two or three decades of active research and development, the composites industry has entered its period of rapid growth. For example, between 1983 and 1988 the European market demand for carbon fibre grew from about 254 tonnes (250 tons) to over 915 tonnes (900 tons), a 25% increase per annum; while the demand for aramid fibre grew from 132 tonnes (130 tons) to some 406 tonnes (400 tons) over the same period – an annual growth rate of 18%. Similar expansions took place in the USA and the Far East (Japan, Taiwan etc).

The overall world demand in 1987 was for some 11,792 tonnes (11,607 tons) of fabricated composite, valued at 3 billion US dollars. If the market continues to expand at the present rate of 11% per annum, it would produce some 31,750 tonnes (31,250 tons) of composite by 1997, valued at over 9 million US dollars.[1]

Approximately 98% of the present industrial market for composites is satisfied by the reinforcing fibres: carbon, glass and aramid, while the oldest fibre – asbestos – is increasingly of less importance for reasons which are noted in section 1.6.7.

However, it should not be forgotten that the fabrication techniques of high and low pressure moulding date from the pioneering days of asbestos and glass. They have been adapted for both carbon and aramid, and could readily be extended to other reinforcements – as and when these are introduced.

1.6.1 Glass fibres

Glass is the common name given to a number of mutually soluble oxides which can be supercooled, ie cooled below their true melting points, without crystallising (devitrification). They are all clear amorphous solids when cold and break with typical conchoidal fracture surfaces.

The key oxide appears to be silica in the form of silica sand (SiO_2), while other oxides such as calcium, sodium, aluminium etc reduce the melting temperature and hinder crystallisation.

1 Grades

Several grades of glass fibre are produced commercially, of which the most important are:

- E-glass – the main formula for normal glass fibre
- S-glass – a stronger stiffer glass originally developed for military hardware such as rocket bodies
- R-glass – a civil version of S-glass, used for important structural applications
- C-glass – used for chemical resistance, mainly against acid attack
- Cemfil – specifically used for resisting the alkali in Portland cement, to make a range of glass-cement articles.

The chemical compositions of E-glass and S-glass are shown below:

Oxides used:	CaO	Al_2O_3	MgO	B_2O_3	SiO_2
for E glass	17.5	14.0	4.5	10	54
for S-glass	—	25	10	—	65

By pulling swiftly and continuously from the melt, glass can be drawn into very fine filaments. These range from 3 to 24 μm (0.1 to 0.9×10^{-3} inch) in commerical production, although between 10 and 13 μm (0.4 and 0.5×10^{-3} inch) is the more usual fibre range.[2]

The process is shown schematically in Figure 1.3. On the industrial scale the mix of silica sand and other oxides is weighed automatically from the storage bins and conveyed to a furnace where it is melted, stirred and maintained at a uniform temperature.

The base of the furnace is equipped with platinum alloy nozzles called ‘spinnerets’, which are electrically heated (because the molten glass is a conductor of electricity at these high temperatures). As the drops of

Figure 1.3 *Schematic diagram of glass fibre melt*

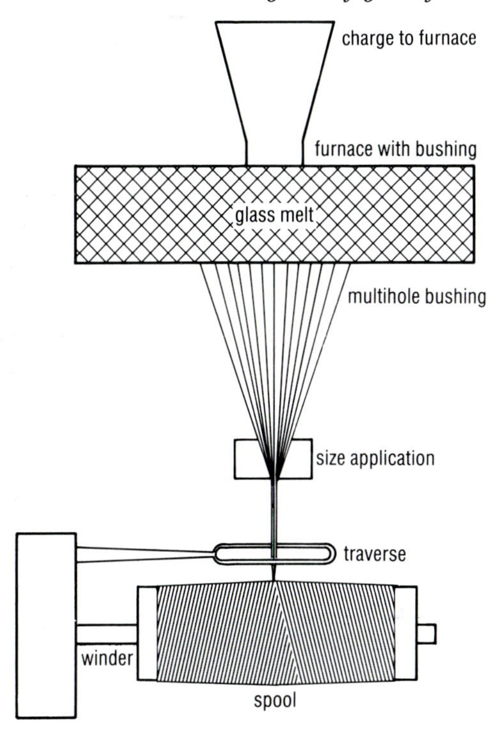

molten glass gather at the spinneret they are drawn by hand into a thread which is transferred to the surface of a rotating drum. Drawing speeds are high – 25 m/s (56 miles/hour) is quite common. There are 200 nozzles to a typical spinneret, and the individual filaments, having been sized, are drawn together into a strand or 'end'. Further strands may in turn be brought together into an untwisted bundle called a 'roving'.

Both ends and rovings can be woven into a variety of cloths, which have been virtually standardised. Rovings may be chopped and reassembled into chopped strand mat, which is a random array, or ends can be made into thin random veils which are used to produce resin-rich surfaces on composite mouldings.

Multiple-end rovings may also be used directly for the filament-winding technique (see Chapter 2).

2 Strength

The strength of glass is a complex subject, and the following facts are important to understanding (McCrum 1971):

- Glass fibres are much stronger than bulk material such as window glass or microscope slides.

- When microscope slides are etched they become much stronger – often by a factor of 10 – though never as strong as glass fibre.
- If carefully prepared glass in the form of fibres or rods is kept away from dust particles, other rods and the presence of water vapour, the strength remains high. A mere touch of the finger causes a drastic reduction in strength.
- Contrary to earlier expectation, thick glass fibres are just as strong as the thin ones, certainly over the wide range of immediate interest, say 5 to 50 μm (0.2 to 2.0 $\times$ 10^{-3} inch) in diameter.
- Glass is strongest at the very lowest temperatures, for example liquid nitrogen or liquid helium. Above −80°C (−112°F) there is some loss in strength, especially in moist air. Between −80°C and 200°C (−112°F and 392°F) in vacuum, glass fibre has a constant strength.
- The measured strength is a function of the length under test – what we call the gauge length. As the gauge length is increased, so the strength falls off. The strength of different samples of glass should always be compared at the same gauge length.
- When a glass fibre is held under constant load, at stresses well below the instantaneous static strength, it will sooner or later fail, so long as the stress is above some minimum value. This phenomenon is known as *static fatigue*, and its occurrence depends on atmospheric conditions, water vapour being especially serious.

All of the above facts are consistent with the theory that the surface of glass contains submicroscopic voids which act as stress concentrations. These voids are known as 'Griffiths cracks', and they must be some 15 Å (1.5 nm, 0.06 $\times$ 10^{-6} inch) in depth, with sharp radii at the bottom of the cracks.

Material ahead of the cracks is in a state of high stress, and the influence of corrosive water vapour is sufficient to propagate the crack until fracture occurs. (It must be remembered that moist air contains a small concentration of the weakly acidic carbon dioxide.) At very low temperatures the reaction with water is too slow to be significant, above the freezing point and up to, say, 200°C (392°F), adsorbed water is being lost from the glass surface.

For these several reasons it is important to protect commercial glass fibre from the mechanical damage necessarily incurred in gathering, spinning and weaving, as well as handling during the fabrication of composites.

Over the years a number of sizes have been developed. The earlier ones were based on starch, and later polyvinyl alcohol solutions were employed. Although these substances were effective at protecting and lubricating the fibres, they interfered with their subsequent adhesion to the matrix. Further, they seemed unable to prevent slow ingress of water to the interface, so that a significant decrease in strength appeared after prolonged immersion in water or on exposure to atmospheres of high humidity over a period, such as monsoon conditions.

At one time the problem was a serious one which threatened the future of glass-reinforced plastics for structural use in adverse conditions such as the marine environment. Some slight improvement was obtained by heating the loom-state glass cloth in an air-oven, in order partly to destroy the starch used for sizing. This process was known as 'caramelising'. The trick was to burn off the starch at a temperature which did not permanently weaken the glass itself. A temperature of 275–300°C (527–572°F) for 90 minutes to 2 hours was found suitable, at which point the glass had assumed a light golden colour.

Much greater benefit was obtained by treating the heat-cleaned glass with the chrome complex methacrylato-chromic chloride (Volan), which was standard good practice for several years.

Even greater benefit was obtained by using vinyl trichlorosilane and its many derivatives. The theory is that the chlorine atoms are hydrolysed by reaction with the hydroxyl groups which are present on the surface of glass and adhere tenaciously. The opposite end of the molecule, ie the vinyl group, is oriented outwards and is in the best position to react with any unsaturation present in the surrounding polyester resin matrix. This is shown diagrammatically in Figure 1.4.

Other derivatives of vinyl silane are readily synthesised, to place other groups such as amino in an outward-facing position relative to the glass surface, and these groups can promote adhesion to epoxy, phenolic and other matrix resins.

Figure 1.4 *Mode of reaction of vinyl silanes with glass surface and polyester resin. The double bond reacts with those of the polyester, while chlorine atoms of the silanes are hydrolysed by hydroxyl groups on the glass*

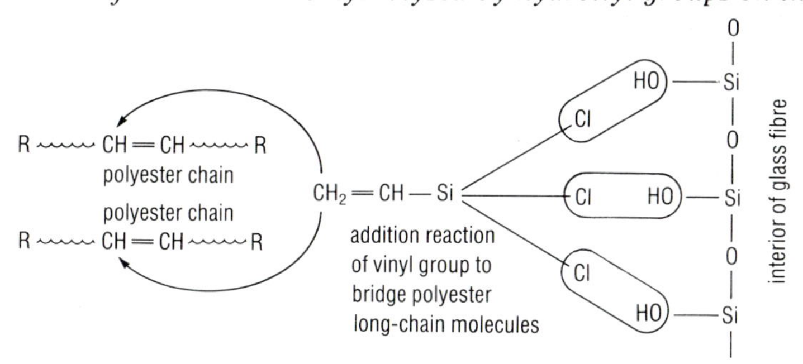

The consequence of these improvements is to reduce the loss in strength after the 4-hour boil test to between 5–10% of the original strength, depending on the matrix used.

1.6.2 Carbon fibres

Carbon fibres have a long history. They were first produced because of the known ability of carbon to conduct electricity.

Davy had invented the arc-lamp, using rods of carbon as the electrodes. Pioneers such as Crompton had improved the mechanical feed of the electrodes towards each other as they were slowly consumed at incandescence (Bowers 1971).

The early electric lamps produced by Edison used much finer carbon filaments, which were intended to last indefinitely in a vacuum. Edison experimented with the carbonisation of many different precursors, mainly of vegetable origin such as slivers of bamboo, and eventually concluded that carbonised cotton thread worked best (Clark 1977).

In England, Swann was concerned with improving control of precursor diameter, and preferred to extrude threads from a solution of regenerated cellulose, followed later by carbonisation of the washed and dried product. In the event, fibres from synthetic precursors proved the more uniform and reproducible; Edison and Swann soon came to a commercial agreement on exploitation.

Cellulose cloth can be carbonised, ie heated strongly in the absence of air to give a carbon cloth, but this is very much weaker than glass and has roughly the same stiffness, so was not used except for high-temperature applications such as rocket nozzles.

Looking at the general situation regarding reinforcing fibres in 1963, it was clear that the low specific stiffness of glass, as compared with the metals, was its main stumbling-block. A major increase in stiffness – three to five times better than that of glass – was the goal.

The ability of carbon to appear in many allotropic forms was intriguing, and especially intriguing was the reported Young's modulus of graphite 'whiskers' at 68 GPa (9.9×10^6 lbf/in^2).

A whisker is a single crystal, in this case a sheet of carbon atoms in a hexagonal array rolled up like a Swiss roll. It represents sheer atomic perfection, in sharp contrast to the random carbon arrangement in a piece of coke, for example.

Thus, orientation is the key and all steps should be taken to start with a highly oriented precursor and to maintain and preserve this orientation until the desired carbon fibre has been reached.

In the experiments planned with my colleagues Watt and Johnson in the Autumn of 1963, I had recommended the use of polyacrylonitrile (PAN) as the most suitable precursor (Watt, Phillips and Johnson 1966). Exceptionally, among the synthetic fibres which were then available, it neither shrank a lot nor melted on heating in air, both of which phenomena imply a considerable loss of orientation. On the contrary, it converted to another nitrogen-containing fibre form[3] which looked ideal for further carbonisation.

1 Production

- Untwisted tows of continuous polyacrylonitrile (PAN) fibres were wound on to sturdy frames (see Figure 1.5), and roasted in an air-oven at 220–230°C (428–446°F). The fibre, which is originally white in colour, turns slowly yellow, then brown and finally black.
- During this time oxygen is being absorbed, and in spite of the loss of volatile material there is an increase in weight. This volatile matter comprises some water, a little ammonia and a fair quantity of hydrogen cyanide. This is a poisonous gas, and due precautions must be taken to neutralise it.
- When the process of oxygen absorption is substantially over, the oxidised fibre can be cut from the frames and packed into graphite boats ready for carbonisation. This comprises heating in an inert atmosphere – 'white spot' grade of nitrogen is suitable – at temperatures from about 400°C up to 1600°C (752–2,912°F).
- Some tar and further gases are evolved during carbonisation. They include cyanogen (C_2N_2), which is another deadly poison and is treated in the same way as hydrogen cyanide. Also, at the higher end of the temperature scale, some grades of PAN produce a white crystalline deposit around the walls of the vessel, which is sodium cyanide (Watt and Johnson 1969).
- A final, optional stage is heating the boats and their charge of carbon fibres for an hour or so to even higher temperatures, in furnaces equipped with carbon resistance heating elements, at about 2,300°C (4,172°F) (Watt and Johnson 1969).

The effect of temperature on the mechanical properties of the resultant fibre can be seen in Figure 1.5. It will be noted that the modulus rises smoothly throughout the temperature range. The explanation is that the amount of crystallinity in the fibre increases at the expense of amorphous material. This is shown in Figure 1.6. On the other hand, the tensile strength, after going through a maximum at about 1,600°C (2,912°F) falls

Figure 1.5 *Preparation of carbon fibre from PAN (Royal Aircraft Establishment process)*

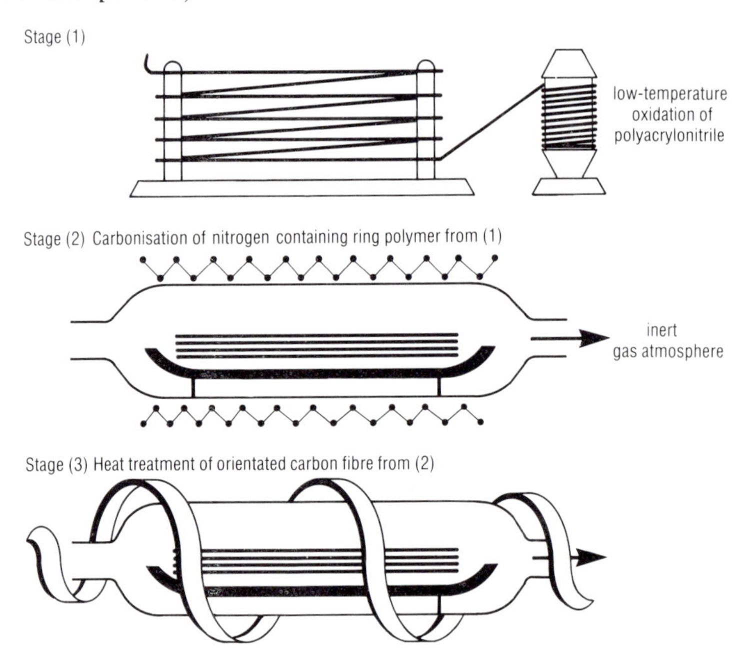

Figure 1.6 *Schematic diagram of carbon fibre structure. Bundles of oriented crystalline carbon are held in a matrix of amorphous carbon*

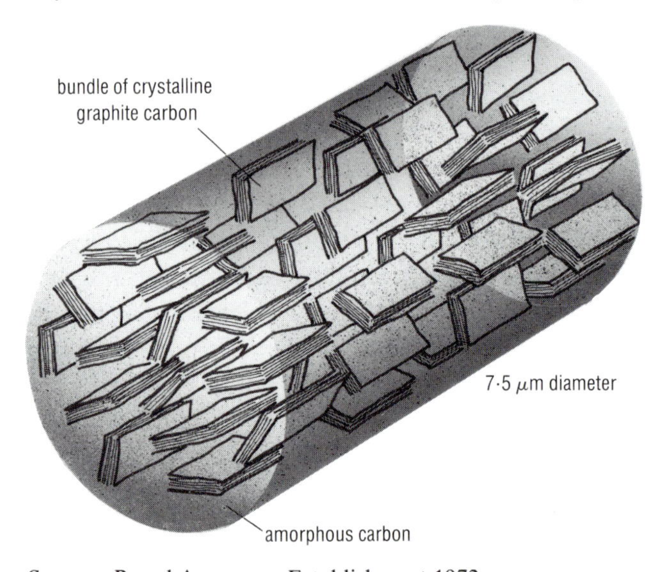

Source Royal Aerospace Establishment 1972

to a plateau which is virtually unchanged to the highest temperatures used.

From this behaviour, we were able to select three distinct and reproducible types of carbon, namely:

Type 1: This is the stiffest carbon and the one discovered first.

Type 2: This is the strongest fibre, carbonised at the temperature where the best tensile strength was recorded. As stiffness is lower than the first type and the strength higher, the failure strain (or elongation at break) is somewhat increased, a move much appreciated by aircraft engineers, makers of sports equipment, and others. Also at the optimum temperature of 1,600°C (2,912°F), a more durable platinum-wound furnace could be used as this lowers the costs of production (Figure 1.7).

Figure 1.7 *Properties of carbon fibre against temperature. There is a steady rise of fibre modulus with increasing temperature of treatment. However, the tensile strength shows an optimum at about 1600°C*

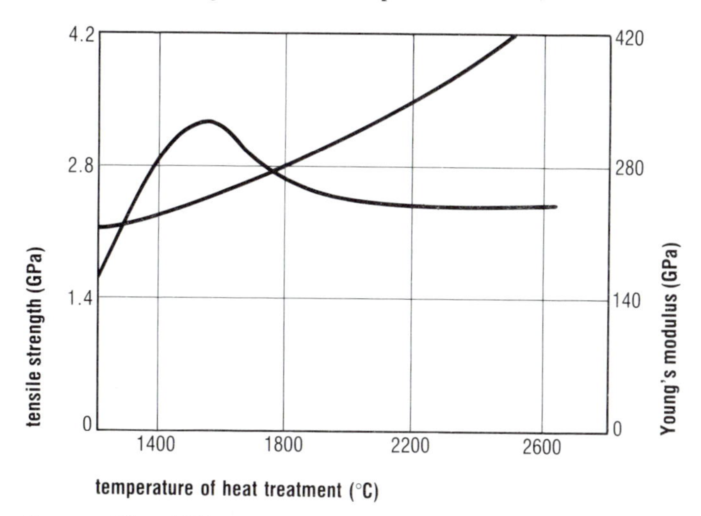

Source Watt 1970

Type 3: This denotes the cheapest fibre. Following the above theme, one can produce a perfectly acceptable fibre from a bank of normal laboratory furnaces equipped with standard Nichrome resistance heaters. The stiffness is lower than the previous grades but the strength is quite reasonably good.

Table 1.3 gives typical values for all three types of fibre, together with the unidirectional properties of composites made from them, all having a 60% volume fraction of reinforcement V_f.

Table 1.3 Typical fibre and laminate properties

Fibre value \ Fibre type	1		2		3	
Young's modulus, $lbf/in^2 \times 10^6$ (GPa)	50	(345)	35	(241)	29	(200)
ultimate tensile strength, lbf/in^2 (GPa)	290,000	(2.00)	350,000	(2.41)	320,000	(2.21)
specific gravity	1.92	—	1.75	—	1.70	—
Unidirectional composite ($V_f = 60\%$)						
Young's modulus, $lbf/in^2 \times 10^6$ (GPa)	26	(179)	17.5	(121)	14.5	(100)
ultimate tensile strength, lbf/in^2 (GPa)	145,000	(1.00)	200,000	(1.38)	175,000	(1.21)
specific gravity*	1.72	—	1.55	—	1.52	—

* Specific gravity of epoxy resin taken as 1.25

The essential groundwork for a batch process having been established, the next step was manufacture on a continuous basis. The main development work was done by Lloyd, MacMullen and their team of the Plastics Engineering Laboratory, Farnborough. Two complete lines were laid down, the first having variable controls, so that the optimum conditions for transit speed, temperature zones and gas flow could be clarified.

The second line was operated over a long period with 'locked' controls, in order to produce a succession of nominally identical batches of fibre, so that the basic scatter in mechanical properties could be established.

2 Strength

The strength of carbon fibre can be found unequivocally by a direct tensile test on individual filaments. Details are given by Watt (1970). Normally a batch of 50 filaments is taken. Later tests on the whole tow (eg 6000 filaments) were used for routine control.

The recorded tensile strength depends markedly on the gauge length; long gauge lengths appear to be weaker and short gauge lengths appear stronger. Engineers and purchasing officers must be very aware of this phenomenon, as the strength quoted by manufacturers may vary from 20 mm gauge up to 100 mm gauge (0.8 to 4 inches).

The analogy with glass fibre is a strong one. The strength of glass, as we have seen, depends on gauge length also, and the 'Griffiths crack' theory is used to explain it. The cracks are not directly visible. In carbon the flaws may be seen under the microscope during suitable conditions.

The difficulty lies in identifying the original fracture, since the filament fails suddenly and explosively because of the large amount of strain energy stored within. The trick is to carry out the test under glycerol, or similar viscous fluid, when the two broken ends are easily recovered.

On examination under high power, the conchoidal fracture surfaces initiating from faults or specific inclusions in or below the surface of the filament can be seen. These facts led to the experiments of Moreton, on the spinning of PAN precursor dope under 'clean room' conditions. This produced fibres of improved strength.

Such improved working practices led to commercial fibres of better performance, such as those from Courtaulds 'Apollo' process which depends on meticulous filtration, and the T400 and T800 grades from Toray Industries who operate especially clean premises.

According to the survey of Lovell, at the 3rd International Conference on Carbon Fibres in 1985, there are now 17 manufacturers who between them are offering 74 different grades of material (Lovell 1986).

3 Sizing and surface treatment

It should be understood that the continuous carbon fibre produced for sale has been sized, ie given a protective surface coating (Welling 1981). This is because raw fibre straight from a carbonising furnace is apt to pick up surface charge in rubbing contact with guides, rollers etc and is then awkward to handle. The electrostatic charges are clearly seen on passing a comb through a bundle of the tows, since the fibres fly apart. Winding or unwinding from a spool, passage through guides, weaving and other textile processes all lead to snagging, breakages and fine floating particles which can short electrical machinery.

In ascending order of susceptibility to damage, weaving is more abrasive than simple filament-winding, while braiding machines inflict the most severe abrasion. The fibre intended for braiding needs a heavier coating of size.

Many different polymer solutions have been tried, but the general consensus is that a low-molecular-weight epoxy resin (or a blend of liquid and solid for braiding) is the most satisfactory. This is normally applied without added hardener.

In addition to a size, carbon fibre may need a surface treatment to improve adhesion to the matrix. Surface treatment is essentially an oxidative process. The idea is not simply to clean the surface but also to produce an etching or pitting of the surface and to provide reactive oxide sites by chemical reaction.

An early wet oxidation, first introduced by AERE Harwell, was sodium hypochlorite solution. The 'short-beam' shear test, a good measure of interlaminar adhesion, demonstrated a useful increase; however, the tensile strength and the impact strength both decreased considerably. When continuous carbon fibre became freely available, the hypochlorite method was abandoned in favour of the electrolytic process (Fitzer 1986).

Since carbon fibre is a good conductor of electricity, a section of tow passing between a pair of separated conducting rollers can be made into the anode of an electrolytic cell. A number of anions are theoretically available, for example sulphate, phosphate, chloride and permanganate, which on discharge at the anode produce nascent oxygen, which is an effective oxidising agent. In practice, salts are chosen to yield the minimum permanent residue.

By varying temperature, concentrations of electrolyte, current density and speed through the bath, surface attack on the fibre can be accurately controlled. Moreover the speed is convenient to interpose the electrolytic process between the last carbonisation furnace and the sizing bath.

Another method of surface treatment which leaves no residue is thermal oxidation. It is usually carried out at 300–450°C (572–842°F) in air, although atmospheres of other gases have been used to speed up the process.

It should be explained that the reactivity of carbon fibre towards both electrolytic and thermal oxidation varies with the temperature of original carbonisation. The least reactive surfaces are those which are most crystalline, ie those which were exposed to the highest temperatures during formation.[4]

1.6.3 Aramid fibres

When an aromatic diamine is reacted with an aromatic dicarboxylic acid, an aromatic polyamide is produced. The situation may be compared with that of nylon, where both the amine and acid are aliphatic and an aliphatic polyamide is produced.

If, as an extra restriction, both the amino groups are directly attached to the aromatic ring in the para position (as in paraphenylene diamine) while both of the carboxyl groups are attached to their ring in the para position also (as in terephthalic acid), then the resulting polymer is polyparaphenylene terephthalamide.

If this polymer is spun and drawn into fibre form, then the name generally adopted as a short descriptive term is an 'aramid' fibre (see Table 1.4):

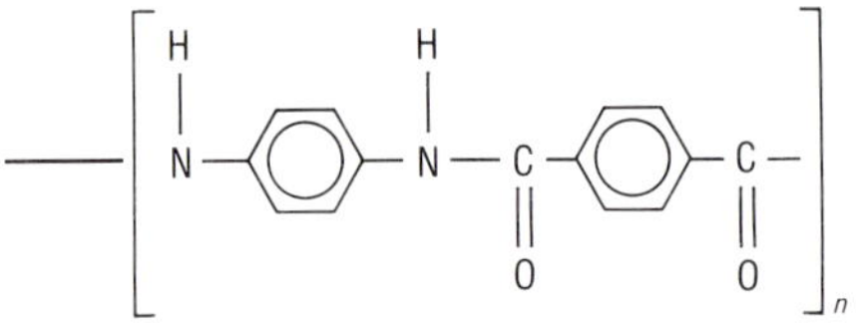

$$-\left[\mathrm{N(H)}-\mathrm{C_6H_4}-\mathrm{N(H)}-\mathrm{C(=O)}-\mathrm{C_6H_4}-\mathrm{C(=O)}- \right]_n$$

Table 1.4 Typical fibre properties of aramid

	Tensile strength		Young's modulus		Specific gravity	Elongation of break
	(MPa)	($\times 10^3$ lbf/in^2)	(GPa)	($\times 10^6$ lbf/in^2)		(%)
lower-modulus fibre	2,650	384	59	8.6	1.44	4
higher-modulus fibre	2,650	384	127	18.4	1.45	2.4

The combination of a regular repeating pattern of rigid benzene rings which are always paralinked, with hydrogen bonding in the transverse direction, leads to a strongly crystalline polymer on which a high degree of orientation can be imposed during drawing.

The result is not only a textile fibre of high strength but also one which has an exceptionally high stiffness. It is much in excess of glass fibre and approaches that of type 3 carbon.

The world is now well supplied with aramid fibre. In Japan the Teijin Company are producers. In Europe the Dutch textile group AKZO make 'Twaron', while in the USA and in Northern Ireland, Du Pont are major manufacturers under the name 'Kevlar'.

The polymer is dissolved in sulphuric acid and wet-spun into an aqueous bath. The design of plant and the development of procedures to handle this aggressive solvent in safety calls for very considerable skill in chemical engineering.

Normally the plant produces two types of aramid, having a higher or lower Young's modulus. The stiffer material has a lower elongation at break, while the less stiff material has the higher elongation.

The lower-modulus aramid is more useful in purely textile applications and is recommended for ropes and cables where the ability to withstand abrasion and the severe mechanical handling of cabling machines is important. Since aramid fibre is degraded by strong ultra-violet light, it is usual for ropes and cables to be jacketed.

The rather low adhesion of many polymers to the aramid surface is less important in ropes where much flexing occurs; waxes and oils are often used as impregnants and lubricants.

High-modulus aramid fibre is the choice for rigid laminates of maximum stiffness, and here the poor adhesion of fibre to matrix is a disadvantage. However, in impact-resistant applications, such as helmets and body-armour, delamination is a useful mechanism for absorbing energy.

High-modulus aramid composites have gained wide acceptance in the boat industry, for sports-car bodies and for many secondary structures in the aircraft industry.[5,6]

In the case of primary aircraft structures, ie those in which failure is likely to lead to the immediate loss of the aircraft, aramid composites suffer from the disadvantage of low compressive strength. Compression causes a crease or buckling on the inner surface of the fibre at a comparatively low load; it is a fundamental and inherent fault as far as all beams are concerned.

At the expense of increased structure weight, it is possible to get over this difficulty – what engineers delight in calling 'designing round the problem':

- one method is simply to increase the cross-sectional area of the compression face of the beam. Apart from a large weight penalty, the main drawback is uncertainty about which is the compression side. Sometimes, as when an aircraft suffers a very heavy landing, the tension side momentarily becomes compressive and vice versa.
- another method is to use a hybrid construction of carbon and aramid, usually in a simple 50:50 ratio. This brings the overall compressive strength to an acceptable value without too much change in the modulus.

On the whole, it is better to avoid any uncertainty and use aramid where the strength is known to be sufficient but where the stiffness of the structure needs increasing. This occurs most often where glass fibre has been used in the past as the sole reinforcement. By substituting, in a hitherto all-glass laminate, successive layers of aramid cloth to replace the layers of glass cloth which have been removed, a number of prototypes can be created with both decreased weight and increased stiffness. The optimum lay-up can then be selected.

Of course, to avoid possible difficulties in compression, aramid fibre can be used purely in tension. In this loading mode it is highly effective, having the best specific strength of any reinforcement. Pressure vessels are a good example of such an application.

1.6.4 Boron fibres

Boron is the name given to the strong, stiff fibre obtained by high-temperature reduction of boron trichloride vapour on to a tungsten or carbon substrate.

In the original process, tungsten wire is passed through a narrow chamber fitted at either end with a mercury seal. This is connected to an electrical supply, so that the length of wire between the electrodes is raised to a high temperature (1,200°C, 2,192°F) by resistance heating.

Two similar chambers are linked together in this process. In the first, the tungsten filament is outgassed and cleaned of any lubricant left from the original wire-drawing process. At this stage there is an atmosphere of hydrogen. In the second chamber the main reduction and deposition of boron take place. The atmosphere is a mixture of hydrogen and boron trichloride; the gases are recirculated and replenished to maintain the supply of boron to the substrate (Figure 1.8).

Figure 1.8 *Preparation of boron fibre on tungsten substrate. Boron trichloride is reduced to the element and deposited on the tungsten wire*

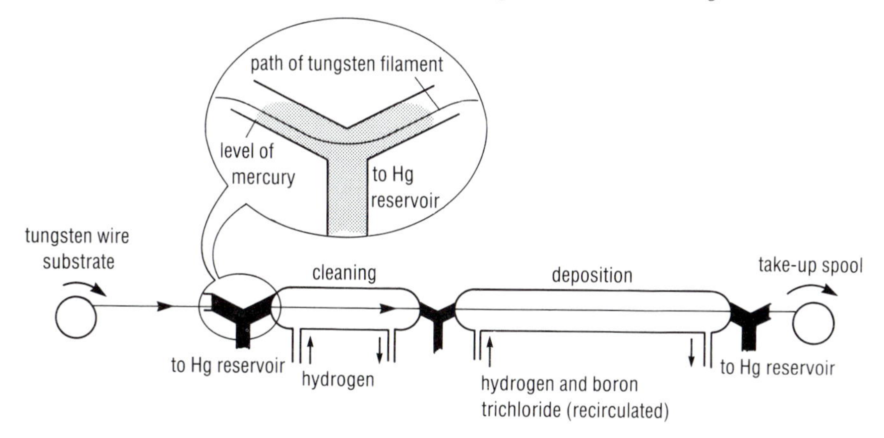

The equation for the reaction is:

$$2BCl_3 + 3H_2 \rightarrow 2B + 6HCl$$

Compared with glass, carbon and aramid, the boron is a 'jumbo' fibre, the more popular sizes being 50 μm and 140 μm in diameter (2 and 5.5×10^{-3} inch).

Because of its large size and high stiffness, it is not possible to carry out normal textile processes such as weaving. For this reason, boron is processed in the form of single-thickness, parallel-laid, pre-impregnated sheets or narrow continuous tapes, which require autoclave or high-pressure moulding (see Chapter 2).

Epoxy resins are the usual choice for the matrix, although polyimide pre-pregs have been used for performance at high temperatures, in excess of 200°C (392°F).

Boron fibres are heavy. The commercial filament has a specific gravity of 2.63, compounded of the extremely dense tungsten substrate (specific gravity of 19.3) and the surrounding coat of pure boron, with a specific gravity of 2.3.

For this reason, combined with the difficulty of machining the composite and the slow, expensive processing, there is a move away from a tungsten substrate towards the use of a carbon-fibre substrate.

For this purpose, the use of a pitch-based carbon is more suitable than PAN-based fibre because the former is more highly graphitised and has a more suitable electrical resistance. The strength of the boron fibre with a carbon precursor is a little lower than with a tungsten core, but the difference is marginal. What matters with boron is that the modulus is very high, combined with good strength, whatever the inner substrate used.

Fibre properties are typically as follows:

ultimate tensile strength	(GPa)	2.76 – 3.45	0.4–0.5 × 10^6 lbf/in^2
Young's modulus	(GPa)	380 – 400	55–58 × 10^6 lbf/in^2
specific gravity		2.63	
specific strength		1,100 – 1,300	
specific modulus		145 – 150	

Boron in an epoxy matrix has been used with success in many pioneering aircraft structures, in sailing (for spinnaker poles), as superior fly-fishing rods, as tail-rotor blades on helicopters etc. Nevertheless it is made by an inherently more expensive process than carbon fibre; after considerable competition in the 1970s and 1980s it appears that the grand future predicted for boron, as a reinforcement for a polymer matrix, will not be realised. On the contrary, it can be argued that boron has more promise for reinforcing a metal matrix.

Detailed consideration is outside the scope of this book; two examples must suffice:

(i) Magnesium is a suitable matrix metal, since it melts at 700°C (1,292°F) and boron can withstand a short exposure at this temperature. Thus a continuous casting process has been devised for feeding both fibre and molten metal simultaneously into a heated die, wherein infiltration and cooling are both very rapid. I-beams, Z-stringers, continuous profiled tapes and sections can all be obtained (Schwartz 1984).

(ii) In the case of aluminium, where reaction between boron and the metal is very considerable at 700°C (1,292°F), it is necessary to protect the fibre. The problem has been sufficiently overcome by the vapour deposition of a thin coating of silicon carbide. The coated material, known as 'Borsic', has been used successfully with an aluminium matrix for the fan rotors in jet engines, and similar aircraft uses are being investigated.

A comprehensive review of boron fibre, including economics and testing procedures for composites, was given by Parker (1974). After comparing boron and carbon, she concluded, 'which fibre will finally be chosen will depend on ease of manufacture of components, on performance, and above all on relative cost!' Now, in 1989, the choice in favour of carbon, at least in so far as polymer matrices are concerned, is clearly established.

1.6.5 Alumina fibres

Although the potential of alumina (aluminium oxide, Al_2O_3) as a reinforcement has been advocated many times before, it has taken 25 years to proceed from possessing the basic scientific information to actual commercial production.

As Kelly pointed out in his book *Strong Solids*, fibres of a polycrystalline alumina can be made by first extruding a thickened mix of fine alumina powder suspended in an alginate binder and then sintering the resultant fibrous mass at high temperature (Kelly 1973).

The product has a tensile strength of 0.5 GPa (72,500 lbf/in^2) and a Young's modulus of 310 GPa (45×10^6 lbf/in^2). These properties should be contrasted with those of filaments made from a single crystal of sapphire, which have an average tensile strength of 2.07 GPa (300,150 lbf/in^2) and a Young's modulus of 460 GPa (66.7×10^6 lbf/in^2). This evidence – it existed in the shape of discrete fibres in haphazard diameters and lengths up to 30 m – was real enough but hardly constituted a production process.

Nowadays it is clearly recognised that alumina exists in different forms with different properties. There are three important forms:

(i) Because, in a polycrystalline alumina, a smaller crystal size leads to a stronger product, small additions are nowadays added to the basic extrusion mix in order to suppress crystal formation. As an example, Du Pont produce a polycrystalline-grade FP. It analyses at 99% pure Al_2O_3 and is produced with a round cross-section to the filament. It is resistant to temperatures of 900–1,000°C (1,652–1,832°F).

The FP fibre is provided routinely with a thin coating of silica, 10 μm (0.4×10^{-3} inch) in thickness, which improves the strength somewhat when a polymer matrix is used. Epoxy, polyimide and bismaleimide are the resins preferred. When a metal such as aluminium or magnesium is used as the matrix, the silica coating virtually disappears, dissolving in, and reacting with, the metal. There is little doubt that the silica coating is helpful in allowing the metal to wet the fibre rapidly and completely.

A new semicontinuous fibre called 'Safimax' is available in development quantities from ICI Ltd. It is a polycrystalline material comprising 95% alumina and 5% silica.

Two grades are marketed, both having the same fibre diameter of 3 μm (0.1×10^{-3} inch). The low-density grade has a specific gravity of 2.0, a tensile strength of 2.0 GPa (290,000 lbf/in^2) and a Young's modulus of 200 GPa (29×10^6 lbf/in^2); its specific stiffness is very similar to the SD grade.

However, there is a big difference in the temperature resistance. The LD grade has a maximum service temperature of 900°C (1,652°F), while the SD grade can withstand 1,600°C (2,912°F).

(ii) In single-filament alumina, spun from a single crystal, the situation has greatly changed by virtue of the new technology used in industrial heating.

In the technique known as the 'floating-zone' method, a laser beam is focused on the tip of an alumina crystal until melting begins. A seed crystal is placed momentarily against the molten surface and withdrawn at a steady rate so that the amount of material melting at the tip is equal to the amount being withdrawn as filament. A crucible is no longer required.

The floating-zone technique is not restricted to alumina, but can produce single-crystal filaments of ceramic materials up to 4,000°C (7,232°F).

As lasers become smaller, cheaper and more powerful it will become possible to design plant capable of processing many fibres simultaneously, thus creating the required production capacity.

(iii) The final form of alumina is the single 'whisker'. This is a very small elongated crystal of alpha-alumina which has a smooth surface and can withstand a temperature of 1,650°C (3,002°F) for 2 hours without degradation. These conditions are sufficient to ensure survival in metal casting and in ceramic sintering and firing processes. We may see many new products in which such whiskers are used to reinforce polymer, metal and ceramic matrices.

The Vista Chemical Co of Houston claims to be able to produce such whiskers on a production scale as a mixture with diameters of

0.5 μm up to 30 μm (0.02 to 1.18 $\times 10^{-3}$ inch) and with an aspect ratio between 3 and 10. One imagines that a separation into more narrowly graded fractions and an increase in the aspect ratio will be needed before the potential benefits are properly realised.

1.6.6 Polyolefine fibres

An exciting discovery in the field of strong fibres is the development known as 'polyolefine fibres', the greatest interest being in polyethylene itself.

During the 1980s there has been an increased understanding of polymer flow phenomena, including the fact that elongational flow through orifices causes stretching of the macromolecules; also, entanglements between such molecules can cause sufficient orientation to produce crystallisation. Likewise, whereas solutions of polymer may be quite stable while at rest, at high rates of extrusion, the contained polymer may come out of solution as a gel phase. Both of these mechanisms are exploited to produce commercial products.

In the first method, scientists at DSM (Dutch State Mines) have used the gel spinning and drawing technique to make novel fibres which are now in production at Allied Fibres and Plastics under the 'Spectra' tradename. Excellent fibre properties are obtained. It should be explained that this process is expensive and demands an efficient solvent recovery.

In the second method, developed by Ward and his co-workers at the University of Leeds, normal grades of polyethylene are melt-spun and then subjected to 'super-drawing' at controlled speed, close to the T_g of the polymer. A bath of high-boiling non-solvent such as glycerol is used to maintain the polymer at optimum temperature. There is an astonishing difference in behaviour. Whereas in the normal way polyethylene has a draw ratio of 7:1, with an absolute maximum at 10:1, after being superdrawn it has a draw ratio of 30:1 with an enormous increase of both strength and stiffness. Ward describes his product as ultra-high-modulus polyethylene (UHMPE) (Ladizesky and Ward 1985, and Ferguson 1988).

It is clear that if the figures for strength and stiffness are taken in conjunction with the low density, then the specific strength and specific stiffness will bear comparison with well known structural reinforcements such as glass. Table 1.5 makes this point clear.

When it comes to producing composites, however, none of the usual matrix resins adheres well to the fibre, and it becomes essential to devise an efficient surface treatment.

Table 1.5 Specific properties of glass aramid and polyolefine fibre

Material	Specific gravity ρ	Tensile strength (GPa)	Young's modulus (GPa)	Specific modulus (GPa/ρ)	Specific strength (GPa/ρ)
E-glass	2.54	3.0	70	27.6	1.18
UHMPE	0.96	1.5	70	72.9	1.56
aramid	1.45	3.6	125	86.2	2.48

Using pull-out tests on individual filaments embedded at one end in discs of candidate matrix, two treatments were investigated:

(i) immersion in an acid solution, of which chromic acid was the best. This treatment worked well on polyethylene fibres of low draw ratio but was less effective with fibres of high draw ratio such as UHMPE
(ii) continuous plasma etching in the presence of oxygen. This gave a fibre reticulation or cellular structure to the surface of the filament, into which liquid resin penetrated easily. The hardened matrix produced an excellent mechanical bond, with an adhesive pull-out value ten times better than before.

From the filament attributes, some interesting properties can be predicted from the composite:

(i) Since the fibre density is extremely low at 0.96 cm^3 (0.55 oz/in^3), a composite having a high V_f within a matrix of normal density can have an overall specific gravity less than unity and will be buoyant in water and sea water.
(ii) UHMPE fibres have a high elongation at break, two or three times higher than glass or aramid. A good impact resistance could be predicted; this is borne out with a flat 'Charpy' test value of 120 kJ/m (8.73 kcal/ft) on first impact, falling to 85% of this value after three successive impacts. Both carbon and glass would break under these conditions.
(iii) UHMPE is a thermoplastic fibre with a maximum use temperature of 130°C (266°F). This is a limitation and in any case one would expect rather high creep from such a material under constant load. However, the creep resistance may be improved dramatically by cross-linking. Exposure to electron-beam irradiation is highly effective and creep is then virtually eliminated.

One can expect to find applications as lightweight textiles: in buoyancy and boats, for radar-transparent structures and as personal armour, including helmets of various sorts.

1.6.7 Asbestos

Finally, some detail of the asbestos fibres is necessary. While they are now used increasingly less, owing to known health hazards, they are none the less of historical importance.

Asbestos is the generic name for a group of crystalline silicate minerals which are mined in Canada, USA, South Africa, Russia and other countries. They occur in asbestos-containing rocks as thin veins of parallel-orientated fibre 25 mm–100 mm (1–4 inch) in depth. These veins are separated by crushing the surrounding rock and the pieces are cleaned and opened up by splitting along the fibres, which is relatively easy.

There are three main types of asbestos, and these are of two different molecular structures:

(i) The first contains sheets of atoms which are then rolled up like a Swiss roll. This formation is called serpentine; its sole example is the valuable chrysotile asbestos;
(ii) the second structure comprises chains of atoms called amphibole with two representatives, amosite and crocidolite. The chemical composition, Young's modulus and ultimate tensile strength of types of asbestos are given in Table 1.6.

Table 1.6 Composition and mechanical properties of types of asbestos

Asbestos type	Sample length (mm)	(inch)	Tensile strength (MPa)	($\times 10^3$ lbf/in^2)	Young's modulus (GPa)	($\times 10^6$ lbf/in^2)
amosite (brown) $1.5MgO.5.5FeO.8SiO_2.H_2O$	35	1.4	1020	148	163	24
chrysotile (white) $3MgO.2SiO_2.2H_2O$	35	1.4	1290	187	160	23
crocidolite (blue) $Na_2P.3FeO.Fe_2O_3.8SiO_2.H_2O$	35	1.4	3230	468	187	27

1 Amosite

This is the tan-to-brown coloured grade which is found in Russia and South Africa. The length of fibre is comparatively long, and veins 100 mm (4 inches) deep are not uncommon. The bundles of fibre are more difficult to break down than either blue or white asbestos, and amosite is not used for spinning. It is used mainly for asbestos–cement compositions.

2 Crocidolite

This is the well known blue asbestos from South Africa. It has good stiffness and strength and has been widely used in Europe for the lagging of industrial boilers, for pipe insulation and for asbestos–cement products. However, over the years it has been shown to be a considerable health hazard, responsible for severe respiratory illness, and occasional carcinoma. Recent EEC health regulations have virtually banned it in Europe, and the safe removal of old installations requires great care and expert supervision.

3 Chrysotile

Chrysotile asbestos is the fine, white, soft spinning grade, which can be split into finer and finer filaments. The ultimate fibrils are 100 mm (0.4 inch) long and only 100 μm (4×10^{-3} inch) in diameter.

It can be combed, twisted and woven into a variety of cloths, which are used in fire-resistant industrial clothing, for fire curtains in theatres, for chemical filtration (Gooch crucibles) and a number of asbestos/cement compositions. Chrysotile is an alkali-resistant fibre and is able to withstand attack from Portland cement.

Braids and felts, either impregnated with fluorinated plastics such as PTFE, or with oil-resistant synthetic rubbers, are used as seals and gaskets in chemical plant. In the form of large felts of roughly random orientation[7] impregnated with water-soluble phenol-formaldehyde resin, it has been used since the early 1940s as a candidate for structural composite in aircraft applications.

Under the name of 'Durestos'[8] it has been used for experimental aircraft fighter wings, fuel drop tanks, radar dishes, rocket nose-cones and rocket nozzles (venturi tubes).

Because the orientation is near random, the strength is fairly modest by today's high standards; efforts have been made by Gordon and his co-workers at ERDE, Waltham Abbey, to produce a more highly oriented material.

If a suspension of fine short fibres is made in a viscous medium such as glycerol or sodium alginate solution and the mixture is forced through a narrow slot or nozzle, the filaments are aligned along the direction of flow. A reciprocating head, bearing such a nozzle, can be run over a vacuum box fitted with a fine wire gauge. This builds up a felt comprising overlapping strips in a rectangular array. Alternatively, a similar head, centrally located, can be traversed automatically across the inside of a hollow rotating cylinder for the same purpose. After felt formation, any residual alginate can be washed away prior to drying.

Asbestos, along with glass, has been a seminal fibre during the early days of structural composites. However, the environmental hazards associated with its use lead to a doubtful prognosis for the future. Many important companies, such as Turner and Newall, have adopted a general policy of converting to non-asbestos products whenever possible.

1.6.8 Miscellaneous reinforcing fibres

There are several other fibres besides those already mentioned. At the moment (1988) they are expensive or difficult to produce, or insufficient is known about the reproducibility of properties or their reactions with different matrices to be clear on their commercial prospects.

Various methods of preparation have been mentioned, as follows:

- filaments are obtained by the melting and drawing of a single crystal. The melting, usually with a laser, follows the lines already given for alumina
- whiskers, ie small single crystals, are made from the vapour phase (see, for example, the experiments of Gordon and Evans (1964) on silicon nitride using the 'bran tub'
- chemical vapour decomposition on to a filamentary substrate such as tungsten or carbon
- thermal decomposition, with rearrangement of the atoms within a polymer fibre, which contains the chemical elements required in the final reinforcement, and
- precipitation of a strong orientated crystal – usually large numbers of them – packed together within a more random matrix during the cooling phase. This applies particularly to the metal matrix where casting techniques are very common; there is evidence for the formation of such distinct internal structures in liquid-crystal polymers.

Oxides, carbides, borides and nitrides of various elements have been examined as the possible source of strong fibres, among which should be mentioned: silicon nitride (Si_3N_4): zirconia (ZrO_2): tungsten carbide (WC) and titanium carbide (TiC).

Silicon carbide (SiC) is probably the best known of these candidates. It can be made by the decomposition of a silane or silicon tetrachloride on to an incandescent carbon filament. Depending on the conditions, one can aim for a thin carbide coating on an unchanged carbon fibre, or a more thorough conversion to SiC.

Another example is method (iv), used in the production of the Japanese fibre 'Nicalon'. A special carbo-silane polymer is spun into precursor

Table 1.7 Physical properties of experimental whiskers

Material	Tensile strength (GPa)	Max. value recorded ($\times 10^6$ lbf/in^2)	Young's modulus (GPa)	Young's modulus ($\times 10^6$ lbf/in^2)	Specific gravity p (g/cm^3)
graphite whisker	19.6	2.8	686	99	2.2
alumina Al_2O_3	15.4	2.2	532	77	4.0
silicon nitride Si_3N_4	14.0	2.0	385	56	3.1
silicon carbide SiC	21.0	3.0	700	102	3.2
silicon Si	7.0	1.0	182	26	2.3
titanium carbide TiC	20.0 (yield stress)	2.9	496	72	4.9
titanium nitride TiN	14.0 (yield stress)	2.0	345	50	5.4
beryllium oxide	7.0	1.0	350	51	3.3

fibre, which is then decomposed by a high-temperature 'bake' in an inert atmosphere.

Typical properties of several such fibres are given in Table 1.7. In general, the products are dense materials of a refractory nature; the end object is evidently the reinforcement of metals and ceramics.

1.7 THE POLYMER MATRIX

Fewer than a dozen strong, stiff materials have so far emerged as candidate reinforcing fibres for use in composites. The position is somewhat different in the case of candidate resin matrices.

While only a few of the twelve or so families of thermosetting polymers have become industrially important (with the various epoxy resins accounting for 80% of the total), several of the newer thermoplastics show considerable potential. The further possibility of mixing different classes of polymer to give improved properties – such as greater adhesion and toughness – creates hundreds of formulated blends which need detailed evaluation and selection.

This section deals with the matrix which surrounds, protects and supports the reinforcing fibre. A glance at Figure 1.9 shows the essential differences in the stress/strain properties of the two phases.

The fibre should be stiff, ie with the maximum slope to the stress/strain curve. It should also be strong, with the curve extended in an upward direction as far as possible. The matrix is less stiff (lower slope), and although weaker than the fibre, should have a larger extension at break,

Figure 1.9 *Stress/strain curves for fibrous reinforcement and matrix. In order to develop the best composite properties, the matrix should be able to strain more than the reinforcement*

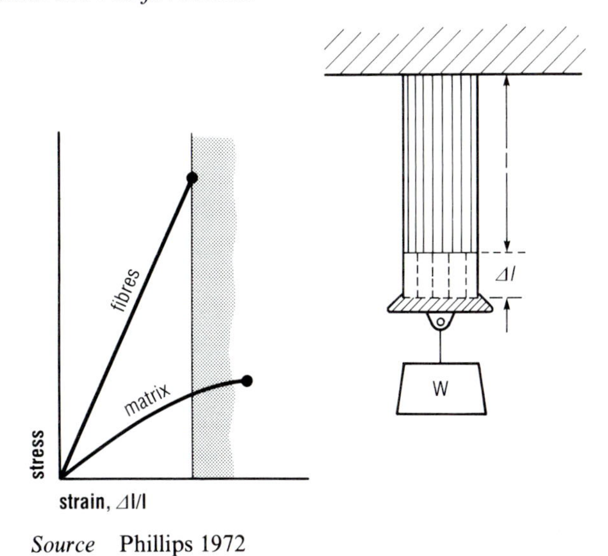

Source Phillips 1972

so that the fibre reaches its full extension and realises its full strength before the matrix fails.

Desirably, the energy stored at failure, which is the area under the stress/strain curve, is as large as possible, since this indicates a tough composite.

When it comes to fabrication processes it is the matrix which is the important partner. At some stage during fabrication the matrix must soften and flow before it hardens; it is the time, temperature and pressure causing these things to happen which are specific to the matrix.

Polymers may be grouped into two broad classes.

(i) *Thermosetting resins*: These harden by a process of chemical cross-linking, whereby resins of low molecular weight and good solubility grow into products of very high molecular weight and limited solubility. The cross-linking is an irreversible process.
(ii) *Thermoplastics*: These are already high-molecular-weight strong solids. They soften on heating, and on cooling regain their original mechanical properties. The process is reversible.

Among thermosetting materials the commonest are phenol-formaldehyde, melamine-formaldehyde, polyester, epoxy, silicone, furane and Friedel-Crafts. Polyimides, bis-maleimides and polybenzimidazoles are more expensive, more difficult to process and are reserved for high-temperature applications. The chemistry of thermosets is discussed in more detail in Chapter 3.

The choice among thermoplastics is quite restricted, although the class is a large one. Not only must a composite have some degree of solvent resistance, it must also have an adequate working temperature.

Thus nylon, polycarbonate, polysulphone, polyether sulphone, polyphenylene sulphide and polyether-ether ketone (PEEK) are the usual possibilities.

Where a polymer is considered a likely candidate, there are five stages to the selection process; an unsatisfactory performance at any stage is grounds for rejection:

Stage 1: It is simpler and cheaper to carry out preliminary work on the unfilled resin, as sheets of cast resin are cheaper to produce than composites. First of all, the strength and stiffness in flexure, tension and compression are determined at room temperature and repeated at 50°C or 25°C (90°F or 45°F) steps to determine the effect of elevated temperature on strength and stiffness.

At the same time, a tension test will give the elongation at break and the shape of the stress/strain curve. Immersion of small squares of matrix reveals the water absorption. The 24-hour immersion is used for comparative purposes but the test can be prolonged for 49 days in order to get the long-term pick-up at saturation.

A polymer 'finger-print' of fundamental importance is the glass transition temperature T_g. This can be obtained by a number of analytical devices for obtaining the thermomechanical response. Thus the automated torsion pendulum, or the machines produced by Du Pont or Stanton Instruments Ltd, will give the T_g with quite a small sample of material.

Stage 2: These are pre-preg trials. The solubility of the candidate matrix in various solvents and the ease of wetting of the reinforcement are examined, together with the stability of the system, known as its 'shelf-life'. Normally the curing characteristics and the flexibility of the pre-impregnated sheet begin to change soon after manufacture; time limits and recommended temperatures for room-temperature and refrigerated storage must be determined.

The moulding variables (time, temperature and pressure) and the degree of pre-cure (a low-temperature advancement of cure to control flow) will be investigated. When the first well bonded laminates are made, the next stage can begin.

Stage 3: Here the short-term mechanical properties are established, including the statistical treatment of data (standard deviation and coefficient of variation) in order to establish the '*A*' and '*B*' values of the aircraft designer.[9]

It cannot be too strongly emphasised that suppliers of material to the aerospace industries must demonstrate to the inspecting directorate that they have a system of quality control, laboratory testing and record-keeping to the highest standards.

Stage 4: Here the long-term effects are investigated. The behaviour of new materials such as composites with respect to fatigue (repeated cycles of stress), creep (strain during continuous loading at steady stress), heat-ageing (the continuous exposure to high-temperature) and the effect of environmental exposure (on the ground and in flight) must all be established before the engineer can have sufficient confidence to adopt them.

Stage 5: Finally, special tests are grouped according to particular industries. Examples are:

(i) Electrical tests, where a material has radar transparency and may be suitable for radomes and other dielectric windows. It will be tested for power loss (tan δ) and dielectric constant at different wavelengths. For printed circuit boards and similar insulation, the breakdown voltage, arcing and tracking resistance will be found
(ii) The resistance to fluids which are in common use in the aircraft industry includes fuels, lubricants and coolants[10]
(iii) In the chemical industry, apart from attack by particular organic solvents, there are acids, alkalis and salts at different temperatures and concentrations
(iv) Coefficients of fraction, wear rate, thermal conductivity and coefficients of expansion are of concern to bearing manufacturers and makers of precision equipment
(v) Lastly, and very important for many industries, ranging from furniture to aircraft designers of both military and civil aircraft, are the tests for fire-resistance, and the generation of smoke and its toxicity.

NOTES

1 Data from *Composites* (1988) 19(5)344; (6)426
2 Vetrotex. Product Directory, Vetrotex International, Geneva, Switzerland (1975)
3 This is the well known 'Black Orlon' experiment
4 A number of companies are developing processes for making carbon fibre from the liquid-crystal phase (mesophase) of various aromatic pitches. These can be oriented

by proper heat treatment and the shear forces which are generated during spinning and drawing. There is no difficulty in carbonising to high moduli and reasonable strength; the drawback is the low compressive strength obtained in production. For this reason the number of structural applications is limited to date, and applications are concentrated on the heat-resistant ones, eg carbon-carbon, aircraft brakes, rocket nozzles, nose tips for re-entry vehicles etc

5 *Properties and Uses of Kevlar.* Du Pont Information Bulletins (several), Geneva, Switzerland

6 *Tenax – Engineering With Fibres.* AKZO group ENKA AG, and Enka High Performance Fibres for Composites. Arnhem, Holland and Wuppertal, West Germany

7 The ratio of longitudinal to transverse strength is about 6:4

8 Durestos resinated asbestos moulding materials, Ref No SA6, Turner Bros, Rochdale, 1961

9 The so-called A and B values for acceptable design must fulfil the following conditions: (a) the design value shall be such that no more than 10% of the material has values which fall below that value; (b) the design value shall be such that not more than 0.1% of the material has values less than 0.9 times the design value. In statistical notation, and assuming the strength/frequency distribution approximates to the normal Gaussian curve,

$$f_{0.1} = \bar{x}(1 - 1.3v)$$

and

$$f_{0.001} = 1.11\bar{x}(1 - 3v)$$

where $v = \sigma/\bar{x}$, ie the coefficient of variation, and $f_{0.1}$ and $f_{0.001}$ are the design values fixed by the 0.1 and 0.001 proportions, respectively, of the weak specimens. The design value will be whichever is the less (Atkinson 1948)

10 A typical list of aircraft fluids will include methylethyl ketone, trichloroethane, kerosene (AVTUR50), methylene dichloride, mineral hydraulic oil (Castrol OM15), phosphate-ester-based hydraulic oil (SKYDROL B), synthetic lubricating oil (OX38) and a propylene glycol-based de-icing fluid (Kilfrost ABC)

2 Fabrication

Leslie N Phillips OBE FRIC CChem FPRI

formerly Head of Section,
Plastics Technology, Materials Department,
Royal Aircraft Establishment, Farnborough

and

Professor (ret.) in Department of Civil Engineering,
University of Surrey

2.1 THE BASICS ARE DIFFERENT

Not surprisingly, after a few thousand years of experience, the mechanical technology of the metals industry has reached a state of near-perfection. Processes may be grouped according to a common property of the metallic state, some examples being:

(i) Where the property is that of *fusibility* the art and science of the foundry have arisen, with the production of patterns, moulds, castings and die-castings on an enormous scale. Also depending on fusion is the welding of metal in all its forms – soldering, brazing, gas (oxyacetylene), electric resistance and arc-welding.

(ii) Where the properties are those of *malleability* and *ductility*, we have the operations of forging, stamping, spinning, rolling and wire-drawing.

(iii) The property of *machineability* depends on the ability of metals to part cleanly under a cutting edge without cracking, splitting or tearing lumps from the surface. The impressive array of apparatus in a modern machine shop includes mechanical saws and lathes as well as planers, millers, drillers and grinders. All of these are devoted to removing larger or smaller amounts of metal from the work in order to finish it to the required dimensions. Little of this machinery is directly relevant to the fabrication of composites; the endeavour rather is to mould to shape with a minimum of machining (Charnock 1942).

Although, as we shall see shortly, there are many ingenious variations on the theme of moulding, the basic steps are simply stated as follows:

(i) The entire surface of the reinforcement must be coated with matrix. A number of ways are available, including soaking, brushing or spraying, either with neat resin or a solution of polymer. Sometimes one can coat the polymer on to a film of release agent and transfer it to a reinforcement by passing both between heated rollers.
(ii) The fibres must be placed at their correct ratios in appropriate directions, to resist the worst stresses which are applied to the component. The orientation of the fibres to withstand stress must be taken seriously; detailed diagrams of the 'lay-up' are provided by the designer to the assembly teams.
(iii) The completed lay-up is shaped by forcing it against the surface of a suitable mould. Where an accurate external contour is needed a female mould is used; where accurate internal dimensions are required the mould will be male.
(iv) After a period of flow, the matrix is caused to harden or 'cure' and the composite is brought to its final shape. Cure is brought about by heat and pressure or by appropriate chemical agents which cause chemical cross-linking. In the case of thermoplastics, once heat and pressure have caused sufficient flow, cooling alone is enough to harden the matrix.

2.2 THE PROVISION OF FORMS

In any moulding process, it is an advantage to the fabricator[1] if the reinforcing fibre can be presented in a form which is convenient to handle. When a reinforcing fibre is available in continuous lengths, as in glass, carbon or aramid, it is possible to carry out many of the classical textile processes such as weaving, knitting or braiding (Weaver and Taylor 1985).

Machinery now exists to measure the length of continuous tows very precisely, so that a number of reels can be matched with respect to length. This makes it easy to produce an array of parallel-laid impregnated fibres in the form of a narrow tape, useful for tape-laying (Figure 2.1) or as a wide sheet of prepreg.

Matching lengths of fibre are also an advantage in the processes of filament-winding and of pultrusion. Continuous lengths can also be put through a chopper (25 mm–50 mm) and allowed to fall on to a screen in a random, overlapping manner to form a wide, resilient mat known as a 'chopped strand mat'.

Figure 2.1 *Tape-laying machine. The machine peels back the protective film, then lays the tape down and cuts it to the exact length before moving across a distance equal to the tape width. The sequence is automatically controlled*

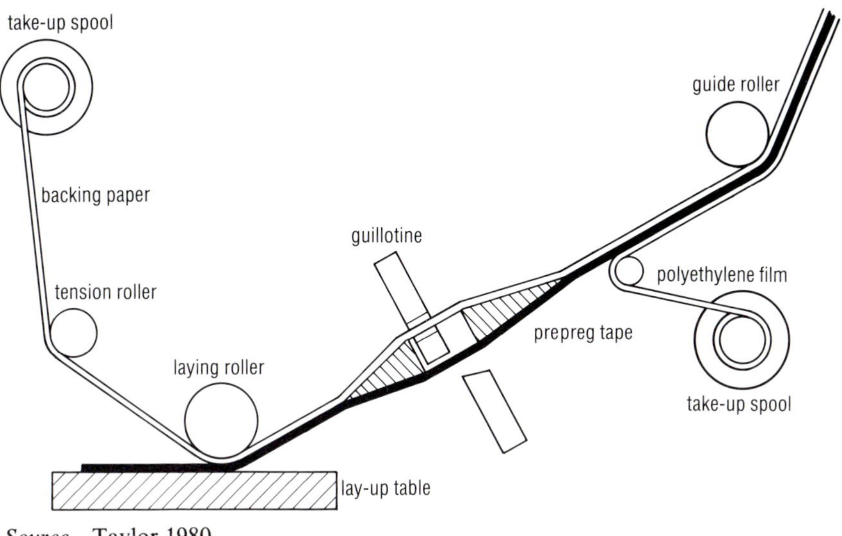

Source Taylor 1980

Finally, fibre can be chopped and separated even more finely in a machine such as a kitchen blender, using water and a small amount of wetting agent, to form a 'soup'. On the large scale this action is duplicated in a paper-maker's beater, to form the watery suspension corresponding to the 'stuff' of beaten cellulose fibres used in paper-making (Walker 1985). With a few modifications these materials can be handled on a Fourdrinier machine to produce novel papers of glass, carbon or hybrids.

In the case of fibres which are available as short natural lengths, such as various types of asbestos, other methods must be used, such as carding before conversion to felts, or spun before the resulting threads can be woven.

There is a need for each of these separate forms, and indeed many businesses have been created to meet the demands of particular users (Ballance 1985). For example, a prepreg manufacturer may supply impregnated sheet and tape, both woven and unwoven. There will be a range of thicknesses and resin contents with a selection of resin systems. The latter will be supplied by a resin-maker, while the cloths come from a weaver, who, in turn, obtains matched reels from a primary fibre producer.

2.3 FABRICATION TECHNIQUES

During the past 40 years, over a dozen moulding methods have been invented and absorbed into the skill of the industry. It is instructive to describe each in turn, starting with the simplest of all and proceeding to others which are more complex.

2.3.1 The 'leaky mould' technique (carbon fibre and cold-setting resin)

The apparatus consists of an open-ended metal trough with a loose-fitting top force which is T-shaped in cross-section. A clearance of about 0.05 mm (0.002 inches) is allowed between top and bottom forces.

Also required are a few grams of carbon fibre in the form of straight tow and about three times the quantity of low-viscosity liquid cold-setting resin. Both epoxy and polyester resins work well.

The mould having been lightly coated with stearate grease parting agent (see Section 2.5), a quantity of freshly catalysed resin with a pot-life of about 20 minutes (longer for an epoxy resin – say 50 minutes), is poured on to the bottom of the mould, and a combed and weighed amount of fibre is dropped into it. The resin gradually wets the fibre bundle, and bubbles of air can be seen coming away from the top. After about 10 minutes, the top-force of the mould is placed over the array and a heavy weight balanced on top. The apparatus now acts as a filter press, the fibre being trapped but the excess resin escaping from the ends of the trough and into the clearance between top and bottom forces (Phillips 1969). (See Figure 2.2.)

Figure 2.2 *The leaky mould technique. After the fibre bundle has been saturated with liquid resin, the mould is closed, to act as a simple filter press and remove excess matrix*

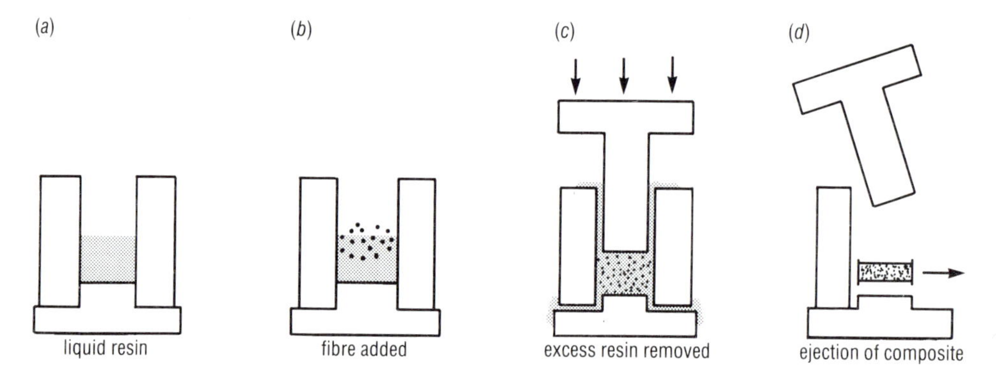

After the resin has hardened, the mould may be opened and the smooth parallel-sided rectangular specimen removed. It is easily trimmed to size and is suitable for mechanical tests.

Many simple variations on this theme are possible. For example, the mould may be placed on a laboratory hot-plate adjusted to a low setting or placed in an air-oven to accelerate cure. Another alternative is the use of a multiple trough mould. Where material is plentiful it is almost as easy to make a batch of eight specimens together as to make them singly. The weight of fibre in each cavity must be the same. If the thickness of specimens must be repeated for successive batches, the mould should be closed down to mechanical stops, one for each corner.

2.3.2 High-pressure compression moulding

This is the standard high-pressure technique used since the early days of phenolic resins, using tool-steel moulds in a hydraulic press.

Figure 2.3 *Design of compression moulds. The simple flash mould will remove excess material to a greater degree than the other designs, which require an accurately weighed charge*

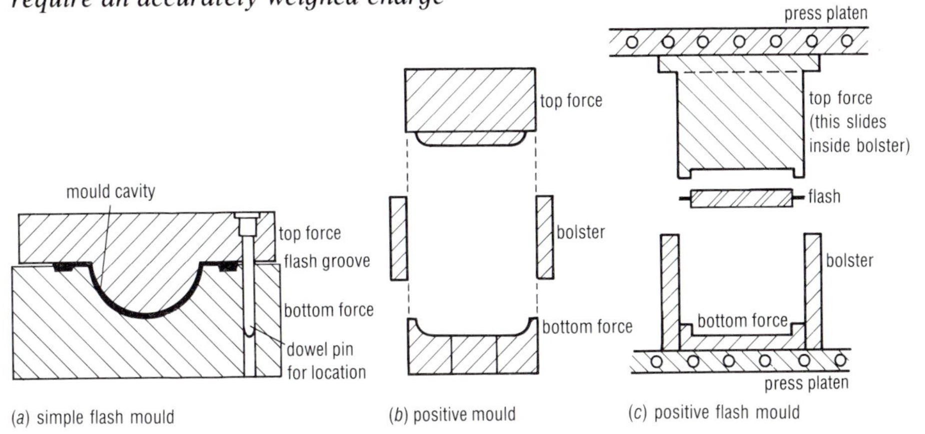

Three types of mould have evolved:

(i) *Flash mould*: This is the simplest type, consisting of two parts only – the top and bottom forces. Normally they are fixed to the top and bottom platens of the press and receive heat by conduction, though large components may have additional electrical heating. A slight excess of moulding material is placed in the lower force and the mould halves are brought together under high pressure – 14 MPa (2, 030 lbf/in^2) is common. Any excess material is squeezed out into a flash groove as a thin porous flash which is removed later (Figure 2.3 *a*).

(ii) *Positive mould*: This has a minimum of three parts. In addition to the top and bottom forces, there is a surrounding heavy bolster into which the two forces fit accurately. The bottom force is fixed within the bolster but the top force is able to slide downwards, causing the full force of the press to be exerted on the moulding charge, none of which can escape. It is therefore important to weigh the charge very precisely, otherwise there will be differences in thicknesses between one component and another (Figure 2.3 *b*).

(iii) *Positive flash mould*: The design combines features of the previous two. As in the flash mould, the top and bottom forces define the component, but both halves are located within a deep bolster so that the moulding charge receives the full pressure until the cavity has been filled completely, at which point there is provision for a small flash to escape into the clearance between the top force and the bolster. As before, and particularly when a multi-impression mould is produced, exactly the same weight of charge is placed in each cavity (Figure 2.3 *c*).

The shape of component and the moulding characteristics of each resin/fibre combination determine which type of mould design is the most appropriate.

For example, Rolls–Royce reported that, when engine motor blades for the RB162 booster engine were first made, a glass-fibre/epoxy prepreg was used. Epoxy resins are particularly free-flowing when hot, so provision was made to use a slight excess of material, with the excess flowing into generous flash grooves and moulding down to fixed mechanical stops.

There were difficulties in eliminating voids within the components and some fibre distortion with some rolls of prepreg. When it was decided that the sixth-stage rotor blades required improved high-temperature performance, the epoxy resin was replaced by the bis-maleimide resin Kerimid 601.

The prepreg was dried down to a low residual solvent content (which gives the best thermal stability), and this gave a dry and completely tack-free material. The prepreg was then cut out to 'pastry cutter' patterns heated to 125°C (257°F), assembled in two halves on shallow trays and, after bonding the two halves together with the minimum of resin, the assembly moulded at the comparatively high pressure of 6.9 MPa (1,000 lbf/in^2) in a positive mould held at 175°C (347°F). There were no fixed stops, and the moulds were positive along the faces, leading edge and trailing edge, which are important aerodynamically. There are small flash areas at the tip and root edges only (Johnson 1980).

By the use of very high pressure and restricting flow, fibre distortion is avoided, while voids due to traces of solvent are prevented by dissolving solvent back into the moulding. With high-pressure compression moulding the guiding principle is to produce the component in large numbers, which should emerge from the mould as near as possible finished and to the correct size. Considerable care and ingenuity is invested in the tools, replating when necessary to maintain dimensions. The moulding itself is a high-speed semi-skilled operation.

2.3.3 Autoclave moulding

For large components the autoclave method is almost universal where double curvatures and the highest-quality moulding are specified.

The autoclave is a cylindrical pressure vessel which can generate a pressure of several atmospheres. It is also equipped with the means of producing a vacuum within any airtight membranes placed within the vessel, so that volatile matter such as solvents or water vapour can be removed. Heating is closely controlled by electrical heaters which warm the atmosphere (usually nitrogen), and this transfers heat to the thermosetting lay-up by convection and conduction. A railway line transports moulds in and out of the apparatus.

A number of auxiliary materials are required when moulding in the autoclave; their relative positions are shown in Figure 2.4.

Figure 2.4 *Autoclave moulding. Care and expertise are required in autoclave moulding, in order to produce parts to aeronautical standards*

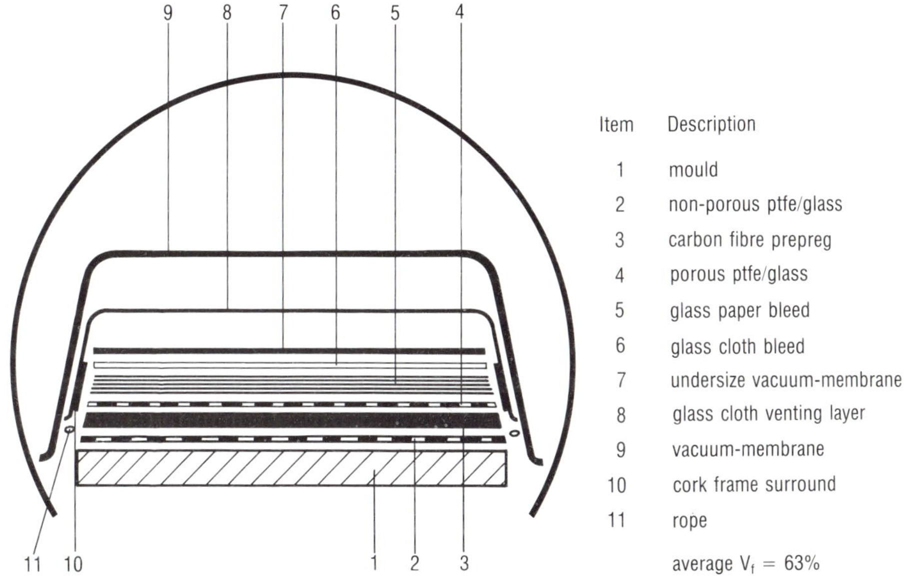

Item	Description
1	mould
2	non-porous ptfe/glass
3	carbon fibre prepreg
4	porous ptfe/glass
5	glass paper bleed
6	glass cloth bleed
7	undersize vacuum-membrane
8	glass cloth venting layer
9	vacuum-membrane
10	cork frame surround
11	rope
	average $V_f = 63\%$

The lay-up is placed against the mould surface and backed with a porous film parting agent and a 'bleeder layer' of glass-fibre cloth or glass-fibre paper. This allows excess resin which escapes from the moulding to be absorbed. Where, exceptionally, the resin is sufficient to fill all the voids between fibres without excess, the technique is known as zero-bleed.

Outside the 'bleeder-layer' is the pressure blanket which transfers pressure from the interior of the autoclave to the lay-up within. Surrounding the edge of the lay-up is a dam to stop it spreading sideways and to act as a vacuum channel.

It is a characteristic of epoxy resins that there is a sharp fall in viscosity when the temperature is first raised above, say, 40–50°C (104–122°F). From this point onwards, resin flow may occur and excess resin is expressed from the lay-up. Continued heating will cause an increase in viscosity followed by sudden gelation. There is then no further flow, only continued chemical cross-linking of the resin system.

Thus a balancing act takes place in an autoclave. Premature gelation will produce laminates which are too thick, have excess resin and therefore a low V_f, while excessive resin flow can lead to a resin-starved laminate with a high void content (Childs 1980).

Attempts to monitor changes in resin viscosity while the lay-up is inside the autoclave have not been very successful, although simulators can monitor such changes '*in vitro*' in the presence of temperature rises but in the absence of pressure. It is known that increasing the pressure within the autoclave increases heat flow to the laminate and speeds gelation of the matrix.

However, there have been so many practical experiments to evaluate and improve the quality of autoclaved parts that the best cure cycles are known; so long as there is careful monitoring of temperatures and pressures throughout and checks for voids, fibre content, thickness and weight of components, the permanent record can usefully assist and guarantee quality control.

A typical moulding cycle would be as follows:

(i) The vacuum is applied and the temperature allowed to rise.
(ii) There is a longer or shorter 'dwell' time at constant temperature to allow resin flow with little cure.
(iii) Pressure is applied to the autoclave and the temperature allowed to rise. Gelation occurs within the hour.
(iv) The temperature and pressure are maintained constant to the end of cure – approximately 2-3 hours.

(v) Heating is switched off and the moulding allowed to cool to room temperature. Some pressure is maintained during the cooling phase.

The total time taken to complete a floor-to-floor cure cycle is approximately 5 hours.

2.3.4 Vacuum-box
(carbon, glass, aramid with polyester and epoxy resins)

Because of the expense of autoclaves and the slow cycle times imposed by indirect heating, efforts have been made to simplify the procedure by placing the lay-up directly against electrically heated moulds and to mould by vacuum pressure alone (up to 100 MPa (14·5 lbf/in^2)). The apparatus developed at RAE, Farnborough, is shown in Figure 2.5.

Figure 2.5 *Vacuum-box moulding. Sometimes referred to as the 'poor man's autoclave', vacuum-box moulding gives good quality parts at less expense, but with a slightly higher resin content, than the autoclave*

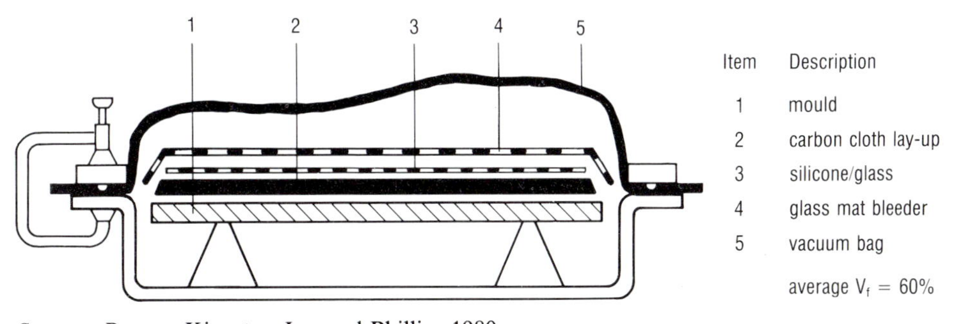

Source Rogers, Kingston-Lee and Phillips 1980

It consists of a deep tray within which is placed a hot-plate of the required area. The box is constructed from glass/polyester resin, stiffened with carbon fibre ribs across the base. The upper edge of the box is flanged and the flange carries a peripheral groove which takes a silicone rubber O-section ring to form a seal between the box and a covering vacuum blanket. Pressure is applied to the O-ring via a light-alloy retaining frame clamped over the blanket and against the flange, with quick-acting toggle clamps.

An electrical hot-plate uses a sheet of carbon-fibre cloth as the heating element. Entry and exit paths for current are provided by sandwiching the ends of the conducting layer between copper contact strips and placed between sheets of glass mat and cloth. The assembly is then hot-moulded in a hydraulic press, using the high-temperature epoxy system Epikote 828/methyl nadic anhydride/benzyl dimethyl amine.

After post-cure the hot-plate is completed by soldering copper terminal strips to the exposed edges of the contacts at each end of the panel. When pure tin is used as the solder, terminal temperatures up to 230°C (446°F) can be borne without damage.

The temperature distribution across the hot-plate is reasonably uniform. At 150°C (302°F) no point was more than ±5°C (9°F) different. The heat-up rate is a convenient 5°C per minute, and control is exercised by a step-down 100 A transformer driven through a single-phase mains and variable transformer.

Both prepreg and wet lay-up can be used in a vacuum-box. Above the lay-up are placed the normal auxiliary layers of parting cloth (porous) bleeder layer, vacuum-bleed cloth and the impervious vacuum blanket. A 50 mm (2 inches) thick layer of glass wool insulation is laid over the vacuum blanket to conserve heat. Excess resin, which is expelled from the wet lay-up is simply removed, as follows. The hot-plate is positioned at the same level as the flange by raising it on a supporting egg-box construction. A deliberate gap between the hot-plate and the flange is bridged by an inverted channel section, the top of which is perforated with drainage holes. Any resin not taken up by the bleeder layer flows directly into the drainage holes and so to the bottom of the box, where it hardens. Since internal surfaces have previously been coated with parting agent, these resinous residues are easily removed.

Double-curvature mouldings are accommodated by the use of conducting moulds cast with epoxy resin and filled with granular aluminium powder. The base of the casting is ground flat and placed against the hot-plate. The silicone-rubber vacuum blanket is made with a knitted cloth of nylon and polyester fibre. This has sufficient two-way stretch to vacuum-load the mould without tearing.

The vacuum-box system is a cheap and useful addition to the battery of methods for making components of excellent mechanical properties. It is used primarily for cloth reinforcement of polyester, epoxy or vinyl ester resins (Rogers *et al.* 1980). Note that a volume fraction of 60% can be obtained, whereas the autoclave method gives a volume fraction of 65%.

2.3.5 The vacuum moulding of Durestos

Whereas the high-pressure compression moulding and the autoclave moulding of a phenolic/asbestos composite are carried out by dissolving any steam back into the composite, under a pressure sufficient to cause steam to condense back to water, the process evolved at Farnborough for moulding phenolic/asbestos under vacuum works differently.

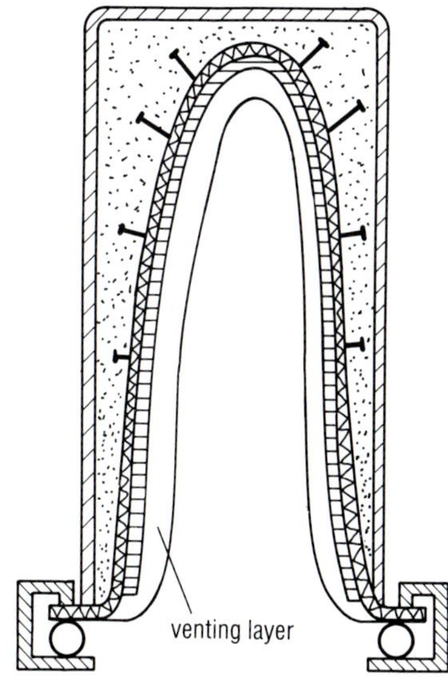

Source Phillips 1955

Figure 2.6 *Vacuum moulding of Durestos. A venting layer allows the back surface of the moulding to breathe. A large amount of water vapour is released during moulding as the phenolic matrix used with Durestos undergoes cure*

The water is encouraged to come away in the form of vapour as soon as it is formed by chemical reaction (condensation polymerisation). This implies a vacuum throughout the curing process, and since the entire back surface of the moulding is breathing out vapour, arrangements for venting must be thorough. Figure 2.6 shows the technique.

A Durestos mould, with embedded electric resistance heaters, is cured against a wood master and embedded in a leak-proof steel box. The space between the box and shell moulding is filled with concrete.

Prior to composite moulding, in this case an aircraft wing, a loading jig is placed below the mould and loaded with successive layers as follows:

(i) the rubberised fabric bag, coated with a heat-resistant rubber
(ii) an auxiliary heating mat, so that the moulding can be heated from both sides simultaneously
(iii) a layer of glass cloth with several thermocouple wires stitched in
(iv) a layer of glass cloth which acts as a venting path, or incompressible pipe, across the entire back of the moulding, and finally
(v) the lay-up, consisting of several layers of Durestos asbestos/phenolic felt, together with a honeycomb sandwich core. A brush coating is applied to the outermost surface of the lay-up, of an aqueous solution of methyl cellulose which acts as a parting agent, to prevent mould and moulding from welding together.

The loaded jig once offered into position within the mould, the bag is sealed round the rim of the steel box using rubber hose to apply pressure.

Vacuum is applied first, then the electrical heating switched on both to the mould and to the external heating blanket. A total time of 4 hours is allotted to the moulding cycle, comprising 1 hour warm-up, 2 hours cure at 160°C (320°F) and 1 hour cooling. Vacuum is maintained throughout. (A 50 kg (110 lb) moulding charge will produce about 5 kg (11 lb) of chemical water, so that a large capacity vacuum pump is required, with a suitable condenser between mould and pump, to trap all the water (Phillips 1953). Naturally, all such mouldings will be microporous.)

Figure 2.7 *Spray-up process using Mavrick high pressure airless spray gun. The stream of chopped fibres is seen entering the cloud of catalysed resin created by the two jets below*

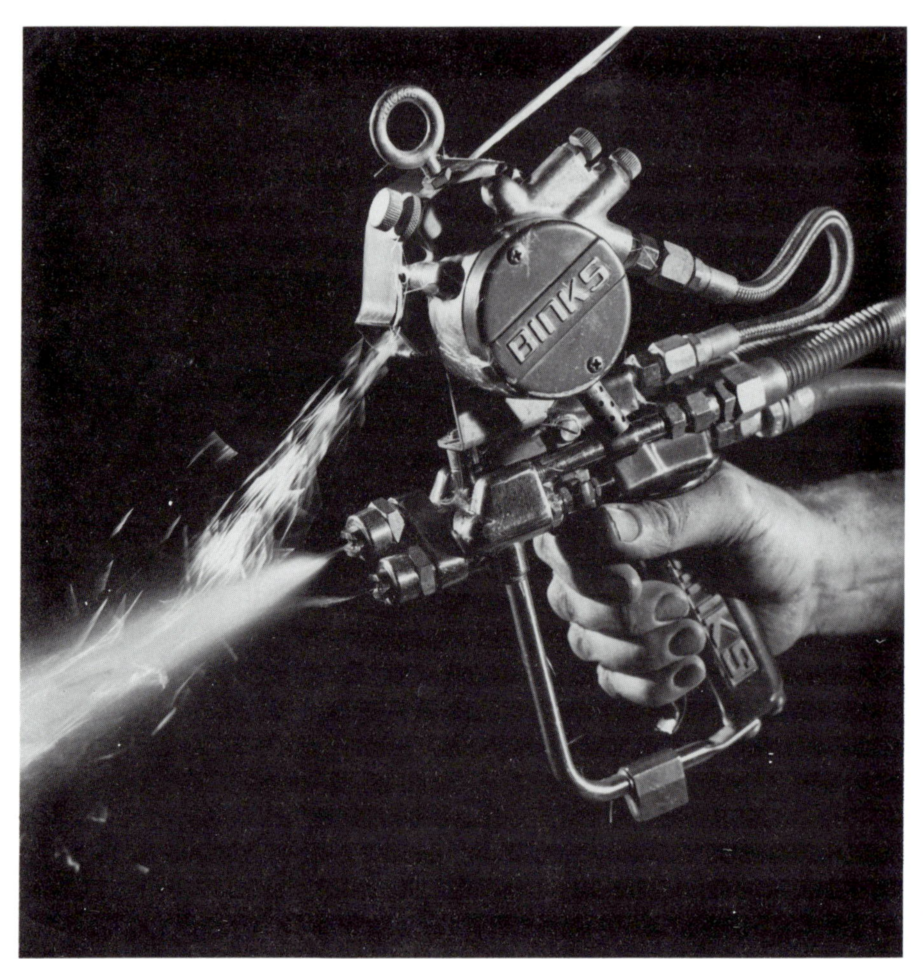

COURTESY BINKS-BULLOWS LTD

2.3.6 Spray-up (chopped glass or carbon, with polyester matrix)

An excellent method of delivering a considerable weight of fibre and resin simultaneously to a mould surface is the spray-up technique. It is applicable to continuous glass and carbon but is of little use for aramids, owing to the difficulty of chopping this tough fibre cleanly.

A fibre chopper consists of a guide, receiving several tows of fibre from a creel and feeding them between two rollers, one coated with a tough polyurethane rubber and the other grooved at intervals around the perimeter in a direction parallel to the control axis, in order to receive a number of hardened steel cutter blades (Figure 2.7).

By varying the position and number of blades, the length of the cut can be altered between 25 mm and 50 mm (1 and 2 inches). The spray of chopped fibres passes through a cloud of freshly catalysed resin, created by twin spray-heads positioned alongside the chopper delivery and angled so that the two spray-cones of resin and hardener solution mix together about 300 mm (12 inches) ahead.[2] The cut fibres, drenched in resin, impinge on a prepared mould surface placed about 1 m (39 inches) ahead.

A build-up of randomly oriented reinforcement takes place quickly. A good operator can deposit a mix of fibre and resin at 3 kg to 6 kg (6.6 to 13.2 lb) per minute. Because there is an excess of resin, along with numerous air bubbles, it is necessary to go over the work surface, carefully by hand with grooved rollers, in order to consolidate the spray-up and remove as many voids as possible. The laminates which are produced by this method are of variable thickness and fibre content, and have a larger number of voids than other techniques.

Nevertheless, for some objects with a large surface area and modest mechanical requirements, such as caravan bodies, architectural units, advertising models and displays, storage tanks and even some small boats and pedaloes, it has been proved as the most economical method.

2.3.7 Filament winding (glass, carbon, aramid with polyester or epoxy resin)

This technique can be applied to any continuous fibre bonded with either a cold- or hot-setting resin. Most installations use a liquid resin applied to the fibre seconds before it is wound on to the receiving mandrel. Whether it is allowed to harden without further heating, or heated *in situ*, or transferred to a separate oven for hot-cure, depends on size and the numbers required.

Less often, the fibre, having been pre-impregnated with a suitable hot-setting resin system, is wound in the dry state, then heated later to fuse,

Figure 2.8 *The difference between (a) circumferential and (b) polar filament winding. The latter is particularly suited to forming completely circumferential ends. α is the specified unit winding angle*

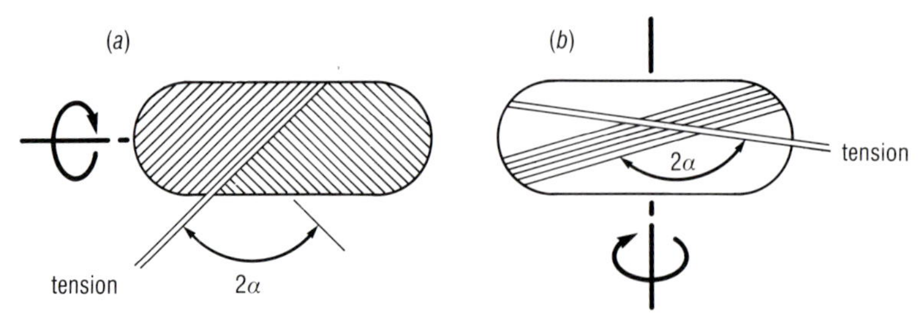

Source Tarnopol'ski 1983

then cure. Figure 2.8 shows the difference between circumferential and polar filament winding.

In circumferential winding, tows from a creel are gathered together, tensioned and passed through a small resin bath, and mounted on a carriage equipped with a collimator and swivelling comb. In this manner, an aligned continuous narrow ribbon of impregnated fibre is drawn on to the mandrel, which takes the place of the workpiece in the lathe.

The mandrel and the carriage are geared together so that the translation of the carriage delivers fibre at any desired angle to the rotating mandrel. Movements may be programmed so that the carriage is reversed automatically at the end of each travel. Arrangements can also be made to 'dwell' either at a specified place along the length, or at an end, in order to build up thickness with hoop fibres (Rosato and Grove 1964, Newling 1968, Kelly and Mileiko 1983, Ruegg 1981 and Cook 1980). This facility is useful when winding in an insert such as a locating ring or metal end-fitting.

The alternative winding pattern is polar. In this case the fibre is laid from end to end, so the ends can be closed, thus giving, say, a hemispherical shape. The creel and the resin bath are mounted on a common carriage which is carried round an elliptical 'race track' surrounding the mandrel. The latter is placed so that its axis is at a slight angle to the horizontal within the long horizontal axis of the race track. Once again, the indexed rotation of the mandrel is geared to the translation of the carriage around the track.

Typical of modern machines is that designed by Pultrex (Figure 2.9) in which the fibre tension, impregnation level and all linear and rotary movements of the winding unit are controlled by the CNC unit shown between the computer console and the filament winding machine. Once

Figure 2.9 *Pultrex five-axis filament winding machine, shown winding an aerofoil section*

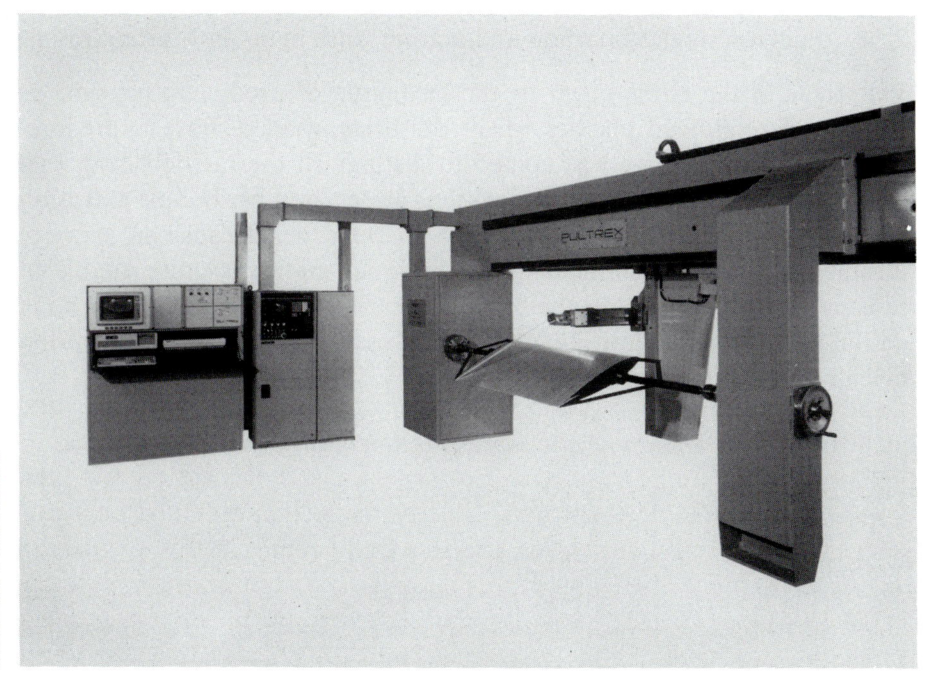

COURTESY PULTREX LTD

winding is complete, the winding programme and other selected parameters, such as the amount of reinforcement used etc, can be down-loaded to the microcomputer as a time- and date-referenced record. This information can be sent to another computer or saved either on disc or as a printout. Such information is a complete manufacturing record and can be used as such to give complete traceability in a recognised BS 5750 or other quality control recording system.

There are various ways to remove the filament-wound components from the internal mandrel:

(i) The mandrel is collapsible ie it is made from several parts which can be dismantled, in reverse order to their assembly, and withdrawn one piece at a time through an end opening.
(ii) The mandrel is made from a soluble plaster, which can be soaked in warm water to soften it, then washed away with a powerful jet of water.
(iii) The mandrel is made from a thin shell of metal, usually light alloy, which can be dissolved later in a dilute solution of warm alkali. The method is useful where the matrix resin is an epoxy type, since these are more resistant to hydrolysis than polyester.

(iv) The mandrel is made from a thin shell of electroplated nickel which acts as a permanent liner to which the composite is firmly bonded.

2.3.8 Pultrusion (glass, carbon and aramid, with many hot-setting resins)

Pultrusion is the name given to the technique of producing continuous sections of reinforced plastics where the orientation of fibre is predominately axial. The name was coined to distinguish the method from that used in metals technology where sections are created by forcible extrusion of metal in the plastic state through a shaping die. Because an array of unsupported fibres impregnated with a liquid matrix is quite unable to resist comprehensive loads, the fibres are transported through the forming die under tension. Figure 2.10 shows the essential parts of a pultrusion machine, or pultruder.

Figure 2.10 *Diagrammatic representation of the pultrusion machine*

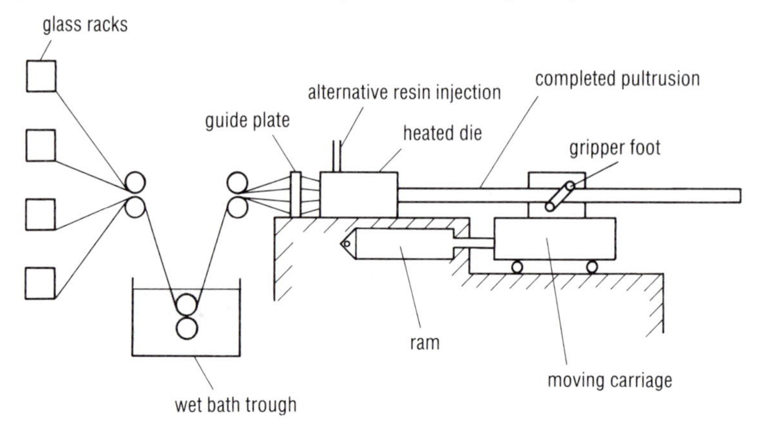

Fibre tows from a creel in matched lengths, together with sufficient mat or biaxial fabric to give the section adequate transverse strength and stiffness, are led through a resin bath and saturated with catalysed liquid resin. On leaving the bath, the reinforcement passes through a series of wiper rings, which progressively remove excess polymer, and through a spider to distribute and collimate the fibre before entering the die (Iddon and Blundell 1980).

Depending on the size and complexity of the component, the die is heated in one or two temperature zones. In the latter case, the first zone at lower temperature reduces viscosity and assists impregnation and compaction of the fibre bundle; the second zone gels and cures the matrix at the higher temperature. On leaving the die and cooling, the section is sufficiently hard and strong to be grasped by rubber pads, which pull it hand-over-hand to the cut-off wheel, or to a large-diameter collecting drum.

It will be seen that, so long as matched lengths of continuous fibre are fed to the resin bath, and so long as the latter can be topped-up with a constant supply of ready-mixed freshly catalysed resin, pultrusion is theoretically a non-stop process.

In practice it takes considerable organisation and quality control to create pultruded stock for electrical and optical cable in the lengths of 2 km (1.25 miles) which are now being demanded.

All the continuous reinforcing fibres can be pultruded, so that hybrid constructions can also be readily produced. Moreover, planks of plywood or of cellular plastics can be fed into the die, with fibre on either side, so that sandwich panels can be produced in one operation.

Design of the profile is very flexible, ie solid rods and tubular section can be made circular, square, oval, rectangular, wedge-shaped etc.

We may summarise by saying that pultrusion is a production method which is capable of giving a large throughput of material, with a high volume fraction of fibre in the composite.

The ready availability of standard rods and tubes gives the designer and structural engineer wide scope in many industries, from chemical plant to aeronautics and to civil engineering structures. Jim Quinn (Chapter 3) explores this fascinating subject.

2.3.9 Resin injection (glass, carbon and aramid)

The technique of resin injection was first introduced by Dr Muskat in the USA, who produced small boats in glass/polyester materials during the late 1940s.

He positioned a dry glass lay-up between a substantial male mould of aluminium and an outer glass/polyester shell, then sucked liquid resin upwards from a surrounding 'moat' or recess in the male mould.

The female mould was sufficiently thin that the progressive wetting of the lay-up could be followed by a visible darkening of the zone of impregnated glass which was evident on the translucent shell. The latter dipped down below the level of resin in the moat, making a seal. The resin suction was always from the topmost level in the inverted position, usually at about mid-keel.

Where the reinforcement was most tightly packed, the vacuum tended to be sealed off; resin would often by-pass such areas. Conducting channels of porous glass mat were inserted to improve vacuum distribution, not always with success.

In the modern process of resin injection, both halves of the mould are made from rigid non-porous materials and catalysed resin is injected under considerable pressure into the dry lay-up (Jones and Johnson 1980).

The matrix resin is frequently a hot-setting epoxy system with a reasonable pot life. Thus the epoxy resin, warmed sufficiently to attain a minimum viscosity, can be pumped through the lay-up and out through an upper sight-glass. Air bubbles are driven from the reinforcement and the resin flow is continued (recirculating if necessary) until the resin runs clear and bubble-free. At this point, the flow is cut off, the temperature is increased, the pressure maintained, and gelation and cure take place.

In Figure 2.11, which shows the mould for the Concorde radome, the reinforcement comprises a matched set of knitted glass stockings, each successive stocking being of slightly larger size.

The moulds used by Dowty-Rotol Ltd for their new family of propeller blades are themselves made from cast epoxy resin filled with aluminium powder to increase the conduction of heat. The blade lay-up is a hybrid of glass and carbon fibre oriented to achieve the design requirements in bending, tension and torsion.

Resin injection is a valuable process for making void-free components of medium size, where a high volume fraction of reinforcement and accurate dimensions are required.

2.3.10 Hand lay-up processes (fibrous sheet materials)

There are three distinct processes which go under the general heading of 'hand lay-up'. The term is used to indicate that special apparatus is not required to provide pressure during the moulding stage. As explained in the following sections, pressure is an essential part of consolidating a fibrous array, since all such arrays in the dry state have a certain springiness or resilience. The use of small hand tools – squeegees, smooth rollers, grooved rollers and even rounded glass rods – is sufficient in many cases to reduce the initial thickness and give an acceptable volume fraction of fibre. We deal in turn with the three processes:

1 No-pressure process for Durestos (asbestos/phenolic felt)

This method was developed in the period up to 1946, some time before cold-setting polyester resins became available in the UK. Durestos was introduced as a high-pressure moulding material, and the water, which is produced chemically during cure, was dissolved back in the laminate, at 150–160°C (302–320°F).

Since Durestos is bonded with a water-soluble phenolic resin, it pos-

Figure 2.11 *Concorde radome. Matched mould prior to closure and resin injection, showing reinforcement in position*

COURTESY BRITISH AEROSPACE

sesses the useful property of softening in warm water applied by a damp sponge; it thus becomes very pliable and can be stretched and moulded to considerable double-curvatures with little difficulty.

The problems of high curing temperature and 'spring-back' of the resilient felts as they dried out were solved by painting the outermost surface of successive felts with a mixture of resorcinol-formaldehyde and furfuryl alcohol, 15% w/w. The mixture performed two functions:

(i) Being water soluble it penetrated completely through the felts and mixed with the phenolic resin already present. The latter was catalysed to cure at 80°C (176°F instead of 150–160°C (302–320°F). This enables the water produced to evaporate at a low vapour pressure, so that external pressure is not required.
(ii) The matrix contracts strongly during cure. This counteracts any tendency to recover in thickness as the felts dry out and harden.

Moulds for the no-pressure process could be both light and simple. They could be made from a lightweight volcanic rock, which was readily worked with hand tools.

The technique was an important demonstration that high pressure was not necessary in the fabrication of composites, and therefore that powerful and expensive hydraulic presses and strong steel moulds were not an essential piece of equipment for the professional moulder (Phillips 1955).

Many pioneering structures, such as the internal ribs for the RAE plastic wing programme, were made by this method. Also, artificial limbs, items of chemical plant, rocket-launch tubes and small boats were the humble forerunners of many business developments.

2 Ordinary hand lay-up

This process has become important ever since the invention of cold-setting polyester resins in 1947. They were given a further boost by the discovery that a small quantity of paraffin wax, previously dissolved in the resin, migrated to the surface during cure and prevented the sticky surface normally present in contact with air.

Wherever a reinforcement is available as sheet material, ie as cloth, mat or wide tape, and so long as a resin matrix can cure at room temperature without evolution of volatile matter, there will be a host of practitioners of hand lay-up.

In the simplest case a smooth flat surface such as a sheet of thick plate glass or a metal panel is coated with parting agent (see below for a more detailed discussion) and a first layer of catalysed resin is applied by brush, spray, or simply by flooding on a small pool which is pushed thinner and wider by a squeegee.

A layer of cloth or mat is placed over it and the resin is absorbed. A second layer of resin and a second layer of reinforcement is applied in the same way; the procedure is repeated with alternate layers of resin and fibre until the desired thickness is achieved. The completed lay-up is squeezed or rolled with grooved rollers (smooth rollers tend to skid across the wet surface) to consolidate the lay-up and remove many of the air bubbles. The moulding is allowed to gel and then harden, either at room-temperature or, after gelation, removed to an air-oven to speed up the hardening process.[2,3]

3 Void-free laminates

An important variation of hand lay-up is the so-called 'void-free' technique which is used to produce laminates of superior quality, made from cloth. Close observation of bubbles, which are always present in a hand lay-up, shows that there are two kinds normally present. First, there are small bubbles of air which are trapped within the interstices of the fabric itself; second, there are larger and flatter bubbles which are caught between layers, when one piece of wetted fabric is placed over another such piece.

Careful soaking of resin into the fabric from beneath allows the first type of bubble to be expelled from the fabric. The second type of bubble must be removed by sliding it out from between the retaining layers of cloth with an excess of resin using the following technique. A marked area having been selected as the assembly site for the lay-up, a shallow pool of catalysed resin is added, about 30–50 mm (1.2–2.0 inches) wide beyond the edge of reinforcing material and surrounding it completely.

After assembly, the lay-up and plate are then covered with a clear sheet of release film (Du Pont FEP film is best) so that the pool of resin forms a complete seal around and beyond the laminate (Figure 2.12).

Firm strokes are made by hand, using a rubber-edged squeegee, or a metal blade with a rounded edge, or a ruler with a bevelled edge, from the centre to the edge of the lay-up. It will be found that all bubbles, both large and small, are swept out in a surge of excess, while the laminate correspondingly decreases in thickness. About three or four strokes in each direction are required for expelling bubbles. Then a couple of final strokes from one edge to the other will even out any ridges or differences in thickness.

It is important that no fibres extend beyond the lay-up, through the seal of resin to the outside air. If this happens, air leaks back along the fibre and into the laminate which promptly expands to its original thickness.

Figure 2.12 *Seal in void-free laminating. The trick is to preserve an intact seal of resin around the lay-up. Fibres must not be allowed to bridge this seal, or air will enter the assembly*

Although this method sounds difficult, it looks quite simple when demonstrated by an expert. The advantage of the system is that it applies not simply to flat laminates, but also to female moulds and to vacuum bag moulding.

If the object is a radome, where a void-free outer surface has a much better resistance to rain-erosion than a surface containing voids, the void-free quality may in fact be specified. It would be as well for the fabricator to master the technique on flat plates first before tooling up for the vacuum method in production.

2.4 THE MOULDING OF REINFORCED THERMOPLASTICS

The mechanical properties of most thermoplastics are considerably improved by the addition of fibres. The processes used for fabrication depend on whether the added fibres are short and randomly oriented – in which case injection moulding is used – or whether the fibre is continuous and highly oriented.

2.4.1 Injection moulding

Dealing first with injection moulding, the material is presented in the form of injection-moulding granules. These are diced or granulated pellets in which up to 40% consists of chopped fibre, evenly dispersed (Trewin, Turner and Cluley 1980). Every effort is made to preserve fibre length and to ensure good adhesion between matrix and fibre. This is done with an appropriate size applied to the fibre while it is still continuous and before feeding into the usual screw extruder for compounding with the chosen thermoplastic.

A schematic diagram of an injection-moulding machine is given in Figure 2.13. Dry granules are fed from the hopper into the heating chamber, which is fitted with a plasticising screw and a reciprocating plunger, the stroke of which can be controlled closely, to give a specific weight of material at each forward stroke. As the granules move forward and heat through, they become increasingly soft and plastic. They are forced by the internal torpedo against the walls of the heating chamber and attain the optimum moulding temperature (Bader 1983).

Meanwhile, the halves of the mould are being clamped together with great force. The nozzle of the injection-moulding machine, which has a cut-off valve, is brought into position and the molten charge is ejected through the nozzle into the mould cavity, through the gate and runner provided, under high pressure.

Mould-locking forces range from 5 MN (1.12×10^6 lbf) for popular sizes of machine up to 25 MN (5.62×10^6 lbf) for the largest. Injection-

Figure 2.13 *Injection moulding machine. A rapid production method for moulding thermoplastic polymers having short-fibre reinforcement*

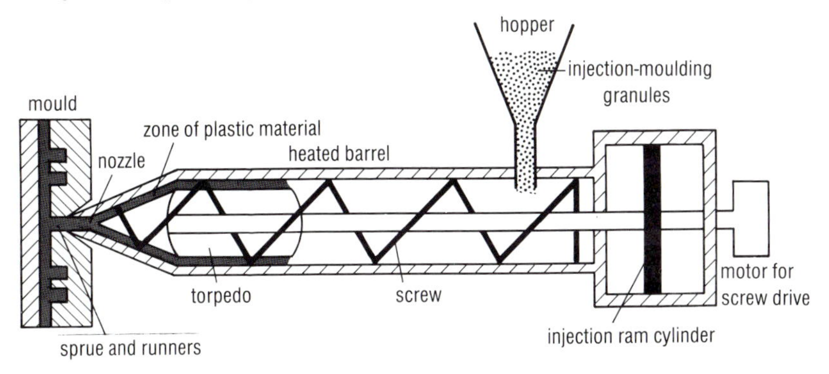

moulding pressures are correspondingly high up to 200 MPa (29,000 lbf/in^2) for filled materials.

Since both strength and stiffness are improved by fibre addition, one should be quite clear as to which effect is most needed. Where Young's modulus is satisfactory and a good increase in strength is desired, glass is the cheapest reinforcement, there being little point in adding carbon if strength alone is deficient. Where a dramatic increase in stiffness is called for, the more expensive carbon is the preferred fibre. Similarly, carbon gives lower shrinkage, greater conductivity of both heat and electricity, reduces the wear rate and has a lower coefficient of friction. Aramid fibre is less successful for reinforcing thermoplastics. It does not chop cleanly, as do glass and carbon and tends to separate at the runners and gates.

2.4.2 The film-stacking technique

The use of thermoplastics is relatively small despite obvious attractions:

- They produce stable prepregs which have constant moulding behaviour even after prolonged storage.
- Refrigeration is not necessary.
- Parts can be remoulded.
- Parts can be joined by hot-welding methods.
- Parts can be bent, twisted or otherwise hot-formed.
- Composites have greater impact resistance than thermosetting matrices.
- They are often quicker to mould than thermosets.

In order to understand the inherent difference between thermoplastics and thermosets, it is helpful to examine a piece of cloth impregnated with 45%–50% of a typical epoxy resin. The material is soft and flexible, with a carefully regulated amount of tack, so that successive plies adhere on press-contact during the lay-up operation.

In contrast, a similar cloth impregnated with the same amount of a thermoplastics such as polyether sulphone (PES) is hard and rigid without any stickiness or tack. It has a rough texture and will quickly break across if an attempt is made to bend and form it by hand at room temperature.

The explanation is that a single ply, impregnated with a sufficiency of high molecular-weight polymer is already a minilaminate of high stiffness.

One way to restore flexibility to such a prepreg is to reduce the amount of polymer present; it is found that a resin content of about 15% w/w gives a flexible product which can be draped, cut and readily handled.

The problem of bringing the final, overall volume fractions of fibre and resin to their correct levels is solved by adding additional matrix in the form of a thin film of pure polymer. Film-stacking therefore consists of alternating layers of fibre, impregnated with insufficient matrix, with polymer films of complementary mass to bring the overall laminate to the correct V_f and V_r, and then consolidating them under heat and pressure (Phillips 1980).

Among the common, soluble plastics, polymethyl-methacrylate and polycarbonate are simple to use and bond well to fibres, their main drawback being a low softening point.

Both polysulphone and polyethersulphone 'Victrex' have high glass-transition temperatures and make laminates which appear equally suitable for structural use. However, polysulphone is more readily attacked by substances which are common in aircraft use such as trichloroethylene ('Genclean') and polyether sulphone is much more suitable.

Other fluid-resistant matrices are polyether-ether ketone (PEEK of ICI), polyphenylene sulphide ('Ryton' of Phillips Petroleum) and polyimides.

In order to produce a laminate of high quality, ie one which is substantially void-free and of high fibre content, there must be adequate resin flow, not simply between cloth layers but also within the individual tows.

The best way to check on quality is examination of the cut and polished cross-section under the microscope. The appearance of poor flow and complete flow is easily spotted, as shown by Figure 2.14 *a* and *b*.

A pressure of 6 MPa to 12 MPa (870 to 1,740 lbf/in^2), temperatures from 275°C to 325°C (527 to 617°F) and dwell times of 20 to 30 minutes before cooling are suitable for the polysulphones and PEEK.

With regard to reinforcing fibres, glass, quartz, carbon and Kevlar have all been used successfully, glass and quartz being particularly useful in short-wave electrical applications such as radomes (Figure 2.15).

Fine metal mesh can be incorporated in the film-stack to give toughened laminates; stainless steel is the metal of choice. Aluminium mesh can be moulded into the surface of a carbon-fibre laminate to improve electrical bonding and to protect against lightning strike.

In a similar way, a different polymer can be used for outside protection, with a cheaper thermoplastic used within. Thus PEEK gives superior protection against rain-erosion, while polyether sulphone is used for the bulk of the laminate.

Last, but importantly, thermoplastic laminates can be produced in very long planks of materials, as follows. The films of polymer and films of prepreg are arranged in sequence so that they enter a heating zone stacked alternately. The pre-warmed material enters the press for the cycle of heating with pressure and cooling below T_g. The laminates can then be

Figure 2.14 *Film-stacking micrograph showing (left) faulty and (right) correct flow of the thermoplastic matrix into bundles of reinforcing fibre*

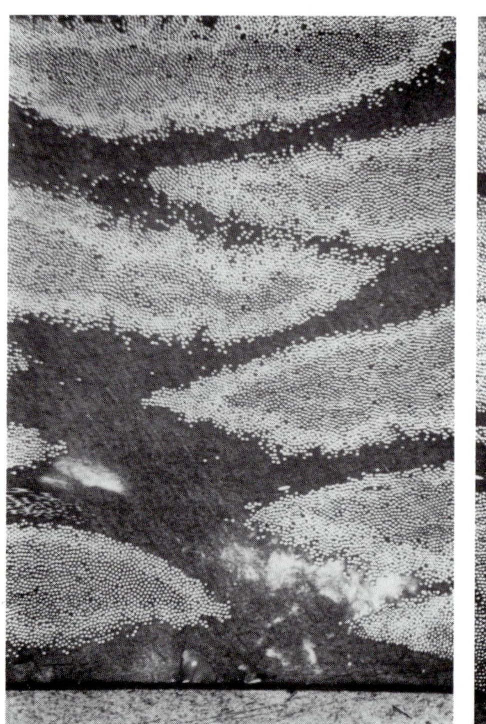

Source Phillips and Murphy 1980

drawn to a third zone for rapid cooling, after which the continuous plank can be coiled for packing and transport. Such planks, up to 50 m (150 feet) in length have been produced for the NASA beam-welding machine.

2.4.3 Prepreg materials with PEEK

ICI has developed a method of impregnating carbon fibre with PEEK, to produce continuous tow and, from it, sheets of prepreg. It is expected that woven impregnated cloth will shortly be available in commercial quantity materials).

The moulding conditions have already been given and it will be seen that they are conventional for high-pressure high-temperature moulding of thermosets, except that there is no allowance for curing time.

It is hoped that the APC materials can be adapted to normal autoclave moulding, when the absence of curing time will bring a substantial reduction in overall length of the moulding cycle.

Hot-forming operations on laminated sheet, such as stamping, rolling

Figure 2.15 *Belly radome of Hercules transport aircraft. The thermoplastic matrix is better than epoxy in impact resistance. It was introduced to withstand the stones on the Port Stanley runway in the Falklands*

and flanging, offer promise of rapid production methods for sheet materials based on thermoplastic matrices.

2.5 PARTING AGENTS

There are few things more disconcerting, when a prototype component has been laid-up in a well designed mould for the very first time, to find that the part cannot be extracted without serious damage because it has stuck. There are so many different resin systems in use for making laminates, and so many different materials used in mould construction, that serious problems of adhesion will doubtless arise several times in the working life of every moulder.

It is hoped that the following notes, based on years of experience, will decrease the annoyance, waste of time and money of the great stick-up!

2.5.1 Film-formers from solution

One of the best parting agents for polyester resins is an aqueous solution of polyvinyl alcohol. The more viscous solutions are better for clinging to vertical surfaces and undercuts without draining away to the bottom of the mould.

At the same time, these films take several hours to dry (they must be completely dry) and the addition of a small proportion of isopropyl alcohol and a current of warm air from an industrial air-heater will speed things considerably.

Sodium alginate is another excellent film-former which uses water as the solvent. There is so little adhesion that films tend to bridge across corners and undercuts as they dry out. When this happens the addition of 10% of glycerol gives the films sufficient flexibility to prevent it.

The third water-based film-former is methyl cellulose. Like sodium alginate it is a useful parting agent for reinforced phenolic resins, which can adhere so tenaciously to metal that large patches of fibre and resin can be torn from the surface.

Because of the long drying time for water-based solutions, a parting agent was developed based on cellulose acetate, which uses a blend of volatile organic solvents. By formulating lacquers for brush or spray application, drying time was reduced from an average of four hours with polyvinyl alcohol to ten or fifteen minutes only.[4]

Where the mould material is porous and absorbent, as in the case of wood or plaster of Paris, it is advisable to seal the surface first, for example with stearate grease (see Section 2.5.2), before applying the cellulose acetate.

For high-temperature thermosetting resins, the acetate has too low a softening point. In such cases it is better to use a lacquer based on cellulose-acetate-butyrate (Eastman Kodak grade EAB-171-2), useful for epoxy resins or cellulose triacetate (Bayer's Triafol grade) for phenolic resins.

All the above film-formers can be dyed so as to provide a visible indicator as to whether an area of mould surface has accidentally been left untreated or whether the parting agent has not covered a screw-thread.

Similarly, if a moulded surface has to be prepared for adhesive bonding, removal of a surface by washing or sanding is made obvious by the colour change.

2.5.2 Film-formers from oils, soaps and waxes

Many high boiling hydrocarbons such as paraffin oil, if smeared on to a non-porous surface such as a metal sheet, act as parting agents, but the mould surface is thereafter contaminated and the contamination is extremely persistent. This is the main trouble with silicone oils, where the contamination is transferred to the moulded surface and is subsequently responsible for many coating and bonding problems.

Metallic soaps, such as zinc stearate, are good parting agents. They exist in the form of a fine powder which can be incorporated into thermosetting moulding components, from which they migrate to the hot surface of the mould during the moulding process. In order to present such soaps in a form suitable for the laminator, the 'transient grease' has been devised.

A grease is normally made by thickening a soap with a non-volatile petroleum fraction such as lubricating oil. In the case of aluminium stearate the soap is dissolved in a volatile solvent such as toluene. A small amount of bentonite (a special montmorillonite clay) is used to stabilise the grease to prevent separation on storage.

The stearate grease spreads smoothly to a thin coherent film over the mould surface. In a minute or two the solvent evaporates and the mould is again ready for use. With porous moulds it is always better to seal the surface with stearate grease before using one of the film-formers.

Polishes are useful on smooth non-porous surfaces such as metal moulds, or moulds made from GRP. When polishes contain a hard wax such as carnanba, they can be buffed to a mirror finish. A good shine eases parting considerably. Polishes contain natural or synthetic waxes (which are fatty esters of monohydric alcohols) dissolved in solvents such as turpentine.

Unfortunately many manufacturers now add silicone oils routinely to every polish; it may be very difficult to find a product which is guaranteed to be silicone-free.

2.5.3 Polymer moulds and polymer films

Thermoplastics materials, either in bulk form or as calendered films, show little tendency to stick to other resin systems which polymerise on contact.

For example, laboratory beakers made of nylon, polyethylene or polypropylene, which are used for measuring and mixing polyester or epoxy resins of the cold-setting variety, will not adhere. If a small residue is put aside after mixing, to cure undisturbed, with a wood stirrer propped against the wall of the vessel, the residue, once set, can be lifted from the beaker as a solid block, with the stirrer still embedded.[5] The limitation is the melting point of the polymer. If a large quantity of quick-setting resin is left to set, the resulting exotherm may distort or even melt the beaker, with considerable adhesion. PTFE and nylon are satisfactory, but polyethylene less so.

In a similar way, thin films such as nylon, polypropylene or PVC can be 'patch-tested' by pouring a pool of catalysed resin on to the centre of a flat sheet. The pool sets to a roughly circular disc of solid resin from which the plastic film can be peeled. PVC and co-polymer sheet (chloride-acetate) make excellent bags for the vacuum moulding of cold-setting matrices, but the drawback is that the plasticisers used in the formulation of sheet may be soluble in the matrix resin. The area previously wetted in the patch test should be examined carefully for any signs of swelling or softening or stickiness. Any such signs are cause for rejection.

One of the most effective polymer films for use as a parting agent is fluorinated ethylene-propylene (Du Pont's FEP film). It has a high melting point, is unaffected by many liquid resins and leaves a glossy surface on the laminate. It can be stretched almost like a rubber without tearing. Some rubbers act as parting agents, although when sulphur or carbon black is included in the rubber compound there may be inhibition of the resin.

Last, but not least, a fluorocarbon telomer (a low-molecular weight polymer) is produced by Du Pont as a fine emulsion in a volatile fluorocarbon. This is an excellent parting agent for non-porous surfaces such as steel or glass.

2.6 FINISHING OF COMPOSITES

In general it is desirable that a component made from a composite material should emerge from the mould complete – apart from the removal of superfluous flash – with respect to both specified dimensions and surface condition. The reinforced components being moulded, a gel-coat finish is initially considered.

In this method, the finish is applied to the mould immediately after the parting agent and allowed to cure, up to the condition of a firm gel. Then the lay-up is applied directly to the back of the gel coat in the normal way and cured completely. On breaking the component from the mould, it will be found that the finish has transferred itself to the outer surface of the component (Mohr *et al.* 1973).

Both polyester and epoxy resins can be readily formulated so as to give durable gel-coat finishes. They are normally filled and coloured and rendered thixotropic with a special fumed silica, so that there is no tendency to flow from vertical surfaces and sag before setting.

Since the gel-coat is exposed to the weather (ultra-violet and visible radiation, water vapour etc.) it must be tough and resilient, rather than rigid. Thus a blend of rigid and flexible polymers may be specified.

However, there are sometimes particular requirements of colour schemes, military camouflage, rain-erosion, reflectance or lightning strike protection which require special finishes; these are applied after the moulding is made.

In order for finishes to perform correctly, they must be well bonded to the substrate surface, and it is common knowledge that the 'as-moulded' surface is normally contaminated with traces of one or more of the parting agents used previously.

Sometimes the residue is clearly visible, as with scraps of coloured film formers. At other times the traces of silicone oil or fluorocarbon are invisible to the naked eye and sophisticated chemical analysis is called for to confirm the presence of Si or F atoms on the surface.

Thus, in a thorough study, Parker and Waghorne (1982) examined by X-ray photoelectron spectroscopy (XPS) the contamination of CFRP surfaces which had been moulded against release cloth, or sheets of silicone rubber, or metal plates coated with a number of parting agents.

Contamination was measured not only by chemical analysis, but also by attempting to bond to the surfaces, with a number of commercial adhesive formulations applied as pastes or as hot-cure film adhesives.

There was a strong correlation between the level of chemical contamination and the amount of adhesion, both before and after attempts at cleaning.

Cleaning was carried out with one of three methods:

(i) hand abrasion with a sanding block using 240 or 320 grit silicon-carbide paper
(ii) abrasion with an abrasive cloth (Scotchbrite®) until an appreciable amount of debris had accumulated on the cloth
(iii) grit-blasting with a dry alumina grit of 280 grade. Usually three passes of the gun were required.

Abrasion with silicon-carbide paper, or grit blasting, were the most effective, not only in removing chemical evidence of contamination, but in restoring adhesion to a high level (Parker and Waghorne 1982).

It should be noted that after abrasion all surfaces should be wiped with a lint-free cloth damped in acetone until all traces of black (carbon swarf) are removed.

With a clean surface, the usual kinds of surface finish may be applied with confidence:

(i) Normal aircraft or automobile finishes based on acrylic, alkyd, epoxy or polyurethane systems.
(ii) Special elastomeric polyurethanes or neoprenes with good abrasion resistance for leading edges and radomes subject to high-speed rain erosion (Phillips *et al.* 1964).
(iii) Sprayed metal (usually zinc or cadmium) in order to give a reflecting surface to radar aerials.
(iv) Sprayed copper or bonded copper or light alloy mesh, as protection against lightning strike.

NOTES

1 A fabricator (or laminator) combines fibres with matrix resins in order to create composite components. Specialisation may be by process (eg an autoclave moulder) or by industry (eg a small-boat moulder).
2 *Polyester Handbook* (1985) 2nd edition. Scott Bader & Co, Wellingborough, pp 25, 32.
3 Hand lay-up laminates with Kevlar 49 aramid fibre. Du Pont de Nemours. Brochure E.39608, Geneva (1983).
4 The most suitable lacquer, named T.10 after the particular grade of cellulose acetate (supplied by British Celanese), has been in use for over 30 years on a variety of surfaces.
5 Small indentations such as makers' names and volumetric marks are reproduced with great fidelity.

3 Properties of Thermoset Polymer Composites and Design of Pultrusions

J A Quinn CEng MIProdE

Technical Manager, Fibreforce Composites Ltd
Runcorn, Cheshire

3.1 INTRODUCTION

Thermoset polymer composites are defined as materials in which fibres with high elastic modulus, high strength and high aspect ratio are combined with a compatible polymeric matrix. This definition omits those composites which utilise the fibre primarily as a crack-stopping device, such as moulding compounds, DMC etc.

It is convenient to classify composites into families according to the type of fibre used as the reinforcement. Thus the properties and attributes which characterise the different groups may be compared. This considerably reduces the bewildering array of options available and allows global judgements to be made in the search for the optimum material for a particular application. The most important composites in order of market volume are glass fibre (GRP), carbon fibre (CFRP) and aramid fibre reinforced plastics (ARP).

3.1.1 Scope

The total possible range of composites is enormous, but the most important variants from which composites may be produced are fibre type, volume fraction, fibre geometry and matrix type:

Fibre type is restricted here to glass, carbon and aramid fibres. Although there are other fibre types available these three categories dominate the composites field.

Volume fraction of fibres: This is dependent on two main factors, the fibre geometry and the process being used to produce the composite.

Fibre geometry: This relates to the directionality of the fibres, of which there is an infinite variety but three main categories:

(i) Unidirectional. All the fibres lie in one direction.
(ii) Bidirectional. The fibres lie at 90 degrees to one another. This is achieved either by the use of woven fabric or by the use of separate layers of fibres, each unidirectional but successively laid at 90 degrees.
(iii) Random. The fibres are randomly distributed and are in-plane.

Matrix type: The family of polymer composites splits into those which use thermosetting resins and those which use thermoplastics. Of these the thermosets represent most of the current usage but the thermoplastics composites are becoming increasingly more important. The thermosets which are used as the matrix resins in composites are shown, together with the other options available, in Table 3.1.

Table 3.1 Range of composite options

Fibre	Fibre geometry	Fibre volume fraction %	Matrix
			orthophthalic polyester
glass			isophthalic polyester
	unidirectional	50–70	urethane methacrylate
aramid	bidirectional	30–55	vinylester
	random in-plane	15–35	het acid
carbon			bisphenol A
			epoxy
			phenolic
			bismaleimide
			polyimide

3.1.2 Mechanical characteristics of composites

It is important to realise that, unlike conventional materials, composites are not homogeneous. The properties vary and are dependent on position and the angle under consideration.

Composites are generally completely elastic up to failure and exhibit no yield point or region of plasticity. It can be seen from the stress/strain curve of composites (Figure 3.1) that they tend to have low strain to failure. The resulting area under the stress/strain curve representing the work done to failure is relatively small when compared to many metals. This would appear at first sight to be highly disadvantageous to engineers used to working with materials which have a substantial amount of 'yield' available to accommodate inadequate design. The designer of com-

Figure 3.1 *Stress/strain curves of typical composites and mild steel*

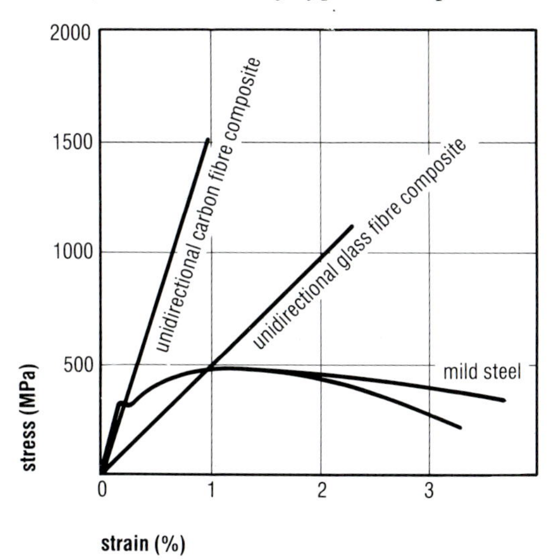

posite materials must ensure that he has taken into account all circumstances. He cannot ignore obscure local stresses, which in steel design would simply cause local yielding and thus dissipate the stress.

Even though the steel designer may have 10 or 15% strain to failure available it is rare that he is able to utilise it. Normally the design would be restricted to probably less than 0.5% strain.

3.1.3 Effect of fibre volume fraction and geometry

The properties of composites are, of course, dependent on the properties of both the fibre and the matrix, the relative quantity of each in the composite and the geometry of the fibres. If all the fibres are aligned in one direction then the composite will be relatively stiff and strong in that direction, but in the transverse direction it will have low modulus and it will have low strength.

Unidirectional composites tested at a small angle off the fibre axis show a pronounced reduction in strength. A similar but less pronounced effect is seen in the curve of laminate tensile modulus. Figure 3.2 illustrates this for a unidirectional E-glass fibre laminate.

If that same amount of fibre were divided equally into the longitudinal and transverse directions (a bidirectional composite) then the two directions would have equal strength and stiffness. However, neither would be as high as in the unidirectional case. If that same amount of fibre were randomly laid (in-plane) then the resulting composite would have equal

Figure 3.2 *Tensile modulus against fibre direction for uni-directional glass fibre composite*

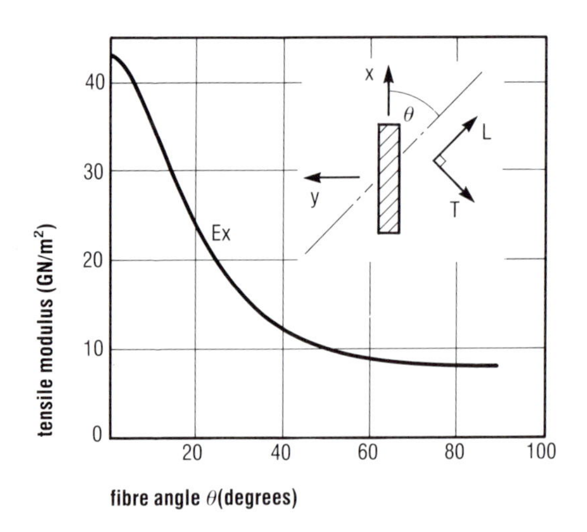

strength and stiffness in all directions (in-plane) but at a lower value than for the bidirectional case (in the direction of the fibres).

If only the fibre directionality is taken into account the approximate relationship of unidirectional, bidirectional and random (in-plane) modulus and strength is 1, 1/2 and 3/8, respectively. However, the amount of fibre which can be packed into a composite is a function of the degree of alignment. A unidirectional composite may have a fibre volume fraction perhaps as high as 75%. This can only be achieved if all the fibres are highly aligned and closely packed. A more typical fibre volume fraction for unidirectional composites is 65%. If the geometry is changed to put fibres in other directions then the maximum fibre packing is reduced further. A typical fibre volume fraction for bidirectional reinforcement (woven fibre) is 50% and a typical volume fraction for random in-plane reinforcement (chopped strand mat) is 20%.

Thus the mechanical properties of a unidirectional laminate are very different from those of a random laminate even if the same fibre type and resin type are used in each. There would be a factor of 1 : 3/8 to account for the directionality effect and a factor of 1 : 0.2 to account for the difference in the volume fraction of fibres. This would give an overall difference of 1 : 0.075, ie a 13-fold difference in modulus.

Some of the properties of the composite may be predicted from known fibre and matrix properties using the simple rule-of-mixtures equation, most notably the tensile modulus.

Table 3.2 Typical mechanical properties of composites

	UD E-glass/ epoxy	UD XAS carbon/ epoxy	UD Kevlar 49/ epoxy	0.90 woven E-glass/ epoxy	±45 woven E-glass/ epoxy	0.90 woven XAS carbon/ epoxy	±45 woven XAS carbon/ epoxy	0.90 woven Kevlar 49/ epoxy	CSM E-glass/ polyester
fibre volume, %	53	57	60	33	33	50	50	50	19
UTS 0°, MPa	1190	2040	1379	360	185	625	240	517	108
UTS 90°, MPa	73	90	30	360	185	625	240	517	108
UCS 0°, MPa	1001	1000	276	240	122	500	200	172	148
UCS 90°, MPa	159	148	138	205	122	500	200	172	148
USS, MPa	67	49	60	98	137	130	—	110	85
E0°, Gpa	39	134	76	17	10	70	18	31	8
E90°, GPa	15	11	5	17	10	70	18	31	8
G, GPa	4	5	2	5	8	5	27	2	2.75
ILSS, MPa	90	94	83	60	48	57	57	70	—
Poisson's ratio	—	0.263	0.34	0.24	0.7	—	—	—	0.32
density, g/cm³	1.92	1.57	1.38	1.92	1.92	1.53	1.53	1.33	1.45

UD = unidirectional

The tensile modulus is simple to predict from the rule of mixtures and the use of a factor to account for the effect of the geometry of the reinforcement (ie the directionality of the fibres). The equation is simply:

$$E_k = E_f V_f B + E_m V_m$$

where E_k = composite modulus
E_f = fibre modulus
E_m = matrix modulus
V_f = volume fraction of fibres
V_m = volume fraction of matrix
B = 1, 1/2, 3/8 for unidirectional, bidirectional or random (in-plane), respectively.
(Krenchel 1964)

This is not an accurate analysis but it does give a reasonable approximation.

The rule of mixtures may be used to predict composite properties such as density and Poisson's ratio, as well as tensile modulus. However, it is a poor model of the tensile strength behaviour of composite materials, although it could be said to represent a maximum upper limit.

Typical mechanical properties of a variety of composites are given in Table 3.2.

3.2 PROPERTIES OF MATRICES

Thermosetting resin matrices are isotropic materials which allow load transfer between the fibres, but there are several other duties. The matrix protects notch-sensitive fibres from abrasion and forms a protective barrier between the fibres and the environment, thus preventing attack from moisture, chemicals and oxidation. It also plays the dominant role in providing shear, transverse tensile and compression properties. The thermomechanical performance of the composite is also governed by the matrix performance.

The processability of a resin system determines to what extent it can be used in a variety of processes. Within this field are such factors as viscosity, shelf life, cure regime etc. The processability and the thermomechanical performance of a resin system are the two factors which most effectively characterise resin systems used in composites. The most commonly used thermoset resins in composites are unsaturated polyester, urethane methacrylate, vinylester, epoxy, phenolic, bismaleimide (Steiner *et al.* 1985) and polyimide. The properties of these resins are given in Table 3.3.

Table 3.3 Properties of thermoset resins

Property	Units	Isopthalic polyester 40-6034	Vinyl-ester 470-36	Phenolic	Epoxy MY720	Bismaleimide	Polyimide
density	Mg/m^3	1.23	1.15	1.25	1.15–1.25	1.32	
tensile strength	MPa	53	73		85		110
tensile modulus	GPa	3.7	3.5		3.3		3.6
compressive strength	MPa		127				
compressive modulus	GPa		2.2				
flexural strength	MPa	103	132		80	210	
flexural modulus	GPa	4.2	3.7		3.5	5.02	
hardness Barcol 939/1		45	40				
impact strength	J	1.0					
elongation at break	%	1.25	3–4		1.8	3.2	
water absorption	mg	25			25		
linear expansion coefficient	$10^{-6}/K$		53		70		
Poisson's ratio					0.35		
glass transition temp	°C			150	240	270	280–300

3.3 PROPERTIES OF GLASS FIBRES AND THEIR COMPOSITES

3.3.1 Introduction

Glass fibres can be categorised broadly into two sets: those with a modulus around 70 GPa (10.15×10^6 lbf/in^2) and with low to medium strength (ie E, A, C, ECR), and those with a modulus around 85 GPa (12.33×10^6 lbf/in^2) with higher strength (ie R and S2-glass) (Figure 3.3).

Figure 3.3 *Stress/strain curves for glass and aramid fibres*

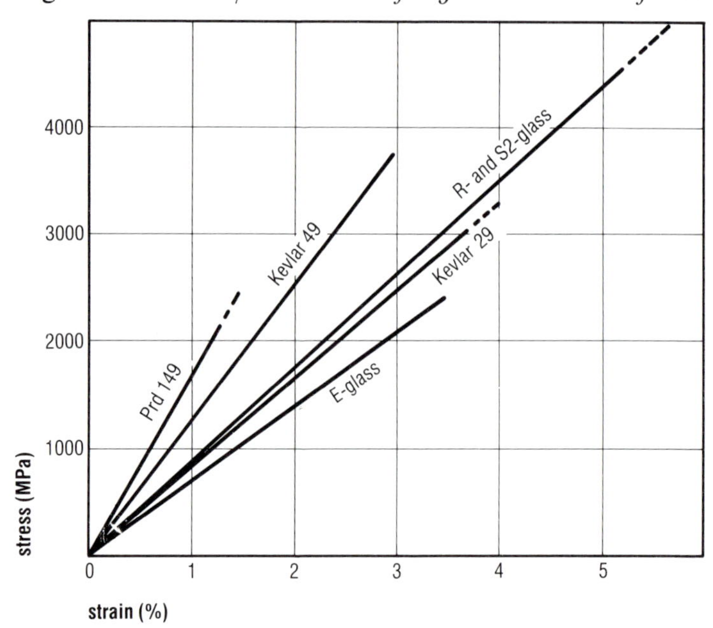

Although it is usual to consider glass fibres as very strong materials, this is not always the case. E-glass fibres are intrinsically reasonably strong. However, they are very susceptible to damage, both physical and environmental, which reduces their strength. This damage can occur during both fibre production and the composite manufacturing stage. Glass fibres which have been chopped for use in SMC, DMC etc are particularly affected. Fortunately this is not critical as in these cases its main purpose is to provide stiffness and to act as a crack-stopping device. The strength that it does have is quite adequate for the purpose.

S2- and R-glasses are both intrinsically high-strength and every effort is made to ensure that the fibres are not degraded before incorporation into the composite.

Table 3.4 Fibre properties

		Property				
Family	Fibre	Elastic modulus GPa	Tensile strength MPa	Ultimate strain %	Density g/cm^3	Max temp °C
Glass	E	72.4	2400	3.5	2.55	250
	A		3030		2.5	
	S2	88	4600	5.7	2.47	
	R	86	4400		2.55	300
	ECR					
	C	69	3030	4.8	2.49	
Aramid	49	125	2750	2.4	1.44	180?
	29	83	2750	4.0	1.44	180?
	PRD149	165	2400	1.3	1.48	
Teijin	Technora	70	3000	4.4	1.39	
Polyethylene						
spectra	900	117	2590	4–5	0.97	130
Carbon						
Grafil	XAS HS	235	3850	1.64?	1.79	600
	HP	235	3450		1.79	
	IM–S	290	3100		1.76	
	HM–S6K	370	2750		1.86	
Apollo IM	43–750	300	5500	1.80	1.77	
HS	38–750	270	5500	2.0	1.8	
HM	LF58	400	5200	1.3		
Toray	T300	230	3530	1.5	1.77	
	T800H	294	5590	1.9	1.81	
	T1000	294	7060	2.4	1.82	
	M40J	390	4200			
	M46J	450	4100			
Ceramic						
Nextel	440	190	2000		3.05	1400

The properties of the range of reinforcing fibres are shown in Table 3.4. The strength and strain figures quoted for fibres should be taken as the maximum upper limit available. The fibre performance is often degraded in processing and at the composite manufacturing stage. The fibre modulus, however, is not degraded in processing, which makes the composite modulus highly predictable.

The density of glass fibre is about 2.5 g/cm^3 (1.4 oz/in^3), which is high in comparison to other reinforcing fibres but by metallic standards very low (aluminium has a density of about 2.8 and steel 7.8).

Glass fibres are the most commonly used reinforcing fibres and there are several very good reasons for this. They have good properties, both absolute and with respect to weight. They have very good process characteristics and they are inexpensive.

The processing characteristics of particular types of glass fibre have been modified and optimised over many years to achieve the required performance, such as choppability, low static build-up, conformity to complex shape etc, and resin compatibility requirements such as fast wet out, good fibre/matrix adhesion etc. Thus glasses are available with characteristics which are designed to make them suitable for all the composites processes and compatible with the matrix resin.

E-glass fibres with a modulus of about 70 GPa (10.15×10^6 lbf/in^2) can only produce composites with modest moduli. As an absolute limit, assuming unidirectional fibres and the highest feasible fibre volume fraction of, say, 0.7, the stiffest E-glass fibre composite has a modulus of 50 GPa (7.25×10^6 lbf/in^2).

At right angles to this, in the transverse direction, the modulus approaches that of the resin itself at about 4 GPa (0.58×10^6 lbf/in^2).

An E-glass laminate made from random glass fibres with a more modest fibre volume fraction of, say, 0.2 would have a modulus of only 5 GPa (0.725×10^6 lbf/in^2). Hence E-glass fibre laminates are, generally, relatively flexible. Typical stress-strain curves for different geometries are shown in Figure 3.4.

The use of S2- or R-glass improves the composite modulus to about 60 GPa (8.70×10^6 lbf/in^2) for unidirectional and 20 GPa (2.90×10^6 lbf/in^2) for woven fabric (bidirectional) constructions. This is at some monetary disadvantage. They are both more expensive than E-glass and they

Figure 3.4 *Effect of fibre geometry on stress/strain curves – glass-fibre laminates*

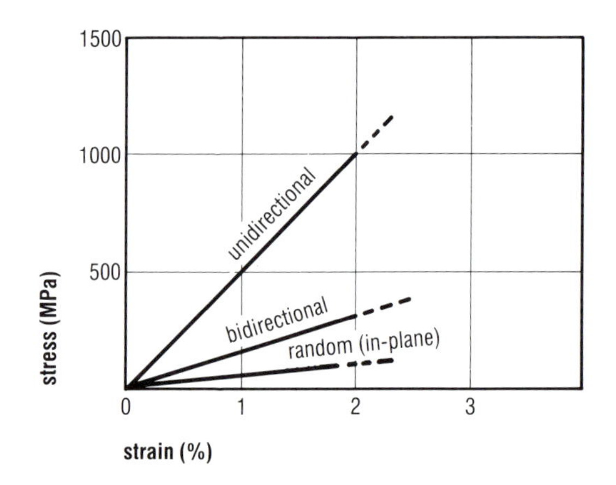

are only available in a fairly limited range of material types and resin compatibilities. Probably the most important virtue of S2- and R-glasses is their high strength, which is considerably higher than that of E-glass.

3.3.2 Physical properties

Glass fibres devitrify at about 675°C (1,247°F) for E-glass and about 780°C (1,436°F) for R-glass. They are therefore reasonably resistant to the effects of temperature and have an excess of performance available when used with polymeric matrices. These degrade at much lower temperatures and therefore have the limiting temperature performance.

Glass fibres are electrical insulators – hence their considerable use in laminates for electrical insulation applications. They are transparent to radio-frequency radiation and hence have found extensive use in radar antenna applications.

Since glass fibres are insulators of both electricity and of heat, their composites exhibit very good electrical and thermal insulation properties.

3.3.3 Chemical properties

E-glass is highly resistant to most chemicals but it is attacked by both mild acids and mild alkalis. The extensive use of glass-fibre reinforcement in chemical plant is reliant upon the corrosion resistance of the polymer matrix and its ability to ensure that the glass fibres do not become exposed to the environment (see Section 3.7.8, 'stress corrosion').

Other glass fibres, notably ECR-glass, shows an improvement over E-glass in a corrosive environment which may gain access to it. Bare E-glass in distilled water retains about 65% of its short-term ultimate tensile strength after 100 days. This compares with R-glass, which has a strength retention under the same conditions of 75% after 100 days.[1]

3.3.4 Fatigue

E-glass GRPs are generally more sensitive to fatigue than those composites with fibres of higher modulus. Because of their fairly low modulus they work at relatively high strains which approach the cracking strain of the matrix, thus allowing a fatigue process to occur and resulting in a reduced fatigue life. S/N curves for R- and E-glass are shown in Figure 3.5.[1]

The fatigue performance of E-glass/epoxy in a (90, 0, 90, 0, 90 degree) configuration is characterised in Mandell and Meier (1975) for zero-tension cyclic loading as monotonic strength of 419 MPa (60,755 lbf/in^2) and slope of the curve of 42.5 MPa (6,162.5 lbf/in^2) per decade of cycles.

Figure 3.5 *Tension-tension fatigue of unidirectional glass/epoxy at 65% volume fraction (Vetrotex data sheet – see note 1)*

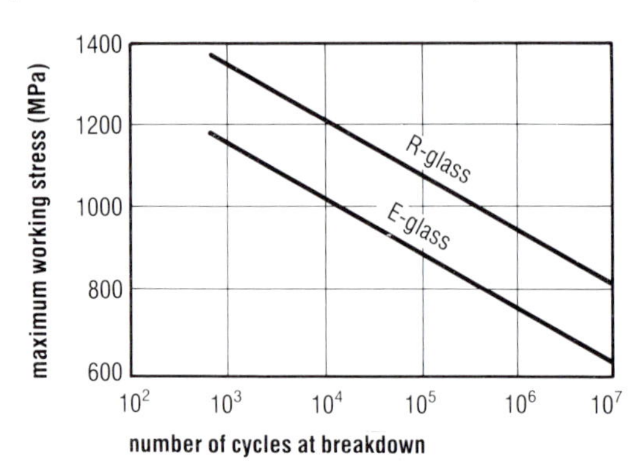

3.3.5 Creep

As would be expected, unidirectional reinforcement is the most creep-resistant construction. This is followed by bidirectional woven construction which has the disadvantage that the fibres tend to straighten out, thus increasing creep. The least resistant construction to creep is chopped strand mat (random in-plane), which suffers from relatively short length fibres (Holmes and Just 1983), thus making the composite creep performance more dependent on the creep resistance of the matrix.

Holmes and Just suggest an empirical model for the creep strain, beyond the primary creep range as follows:

$$E = m(t/t_0)^n$$

where E = creep strain
t = time (hours)
t_0 = 1 hour (unit time)
n = gradient of curve
m = value of strain after 1 hour.

Values of n for glass fibre composites have been determined as follows:

Composite	n
chopped strand mat laminates	0.08–0.1
bidirectional (woven) laminates	0.03–0.04
unidirectional laminates	0.008–0.01

3.4 PROPERTIES OF CARBON FIBRES AND THEIR COMPOSITES

3.4.1 Introduction

Carbon fibres are the predominant reinforcement used to achieve high stiffness and high strength. The term carbon fibre covers a whole family of materials which encompass a large range of strengths and stiffnesses.

Carbon fibre is most commonly produced from a precursor of polyacrylonitrile (PAN) fibre which is processed by first stretching it to achieve a high degree of molecular orientation. It is then stabilised in an oxidising atmosphere while held under tension. The fibres are then subjected to a carbonising regime at a temperature in the range 1,000 to 3,500°C (1,832 to 6,332°F). The degree of carbonisation determines such properties as elastic modulus, density and electrical conductivity. As an alternative to the use of PAN, routes via the use of pitch and Rayon have been successfully utilised, and such fibres are commercially available. These fibres tend to be of lower performance than PAN-based fibres. They also tend to be of lower cost due to their use of a lower-cost precursor.

3.4.2 Mechanical properties

The elastic modulus E, strength, strain and density of a range of carbon fibres commercially available is shown in Table 3.4. Their stress/strain curves are shown graphically in Figure 3.6.

The 'work horses' in the carbon fibre field are typified by T300 fibre from Toray and XAS high strain from Hysol Grafil. These fibres have very respectable values of modulus, strength and strain to failure, both absolute and relative to their cost. This allows them access to many market areas which are strongly driven by commercial considerations as well as technical requirements. They have tensile moduli of about 230 GPa (33.35×10^6 lbf/in^2) a tensile strength of 3,200–3,500 MPa (0.464–0.508×10^6 lbf/in^2) and a strain to failure of 1.5%. Thus unidirectional composites can be produced from them which will have typical properties as follows:

longitudinal tensile modulus (GPa)	125–135
longitudinal tensile strength (MPa)	1,700–1,800

At the other end of the performance scale, carbon fibres have been developed which exhibit strengths and moduli which can only be described as remarkable. Toray has developed T1000 fibre which has a tensile strength

Figure 3.6 *Stress/strain curves for a selection of carbon fibres*

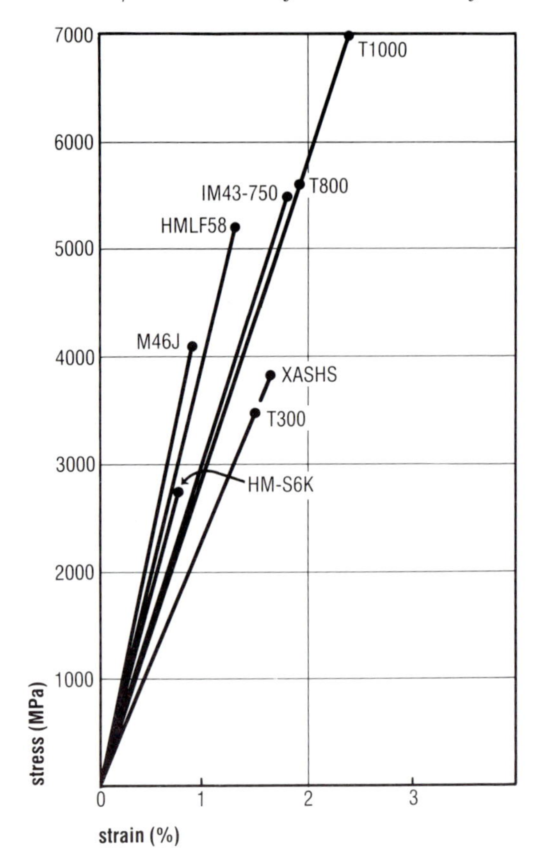

of around 7,000 MPa (1.015×10^6 lbf/in^2) together with a very respectable modulus of 294 GPa (42.63×10^6 lbf/in^2) (Tuhru Hiramatsu *et al.* 1987). Apollo HM LF58 from Hysol Grafil has a modulus of 400 GPa (58.00×10^6 lbf/in^2) together with a tensile strength of 5,200 MPa (754×10^3 lbf/in^2).

The primary characteristics of carbon fibre composites is without doubt their very high specific stiffness (ie the ratio of elastic modulus to density). The specific stiffness of a number of carbon composites is compared with other materials in Figure 3.7. In fact the elastic modulus of CFRPs can be very high, even in the absolute sense. Unidirectional CFRP composites using high-modulus fibre can be of the same order or even exceed the modulus of steel.

IM-S fibre from Hysol Grafil in a unidirectional composite can have a modulus of 190 GPa (27.55 lbf/in^2). Torayca M46 fibre, again in a unidirectional format, can have a modulus of 255 GPa (36.98×10^6 lbf/in^2).

Figure 3.7 *Specific stiffness of carbon composites compared with glass-fibre composites and metals*

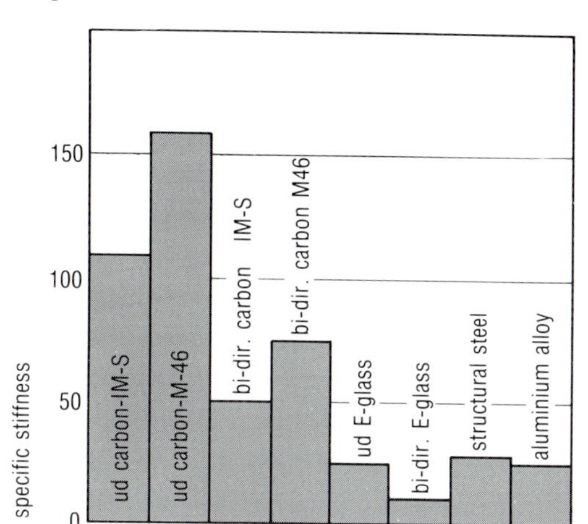

These compare with the modulus of steel of about 210 GPa (30.45×10^6 lbf/in^2). The composite densities of the two fibres is about 1.6 Mg/m^3 (0.92 oz/in^3), giving specific moduli of 118 for IM-S composite and 160 for the M46 composite. The comparative quantity for steel with a density of 7.8 Mg/m^3 (4.5 oz/in^3) would be 27. Therefore the IM-S composite is four times better and the M46 composite six times better.

It is generally the case that fibres must be placed in more than one direction to accommodate and sustain the complex loading systems that are applied to most structures. Thus it is rare that the composites designer is able to utilise these properties to the fullest extent. Even so, a very high specific modulus may still be achieved.

Although the modulus is the primary reason for the use of CFRP, it can have superb strength performance but not simultaneously with the highest modulus performance. Generally the designer must choose either high modulus or high strength or a compromise of reasonably high strength and reasonably high modulus.

3.4.3 Fatigue

CFRP composites exhibit outstanding fatigue performance when compared to metals and other composites. A typical set of S/N curves is shown in Figure 3.8, where the characteristically flatter curve for CFRP is in evidence. CFRP is relatively insensitive to tension fatigue damage when loaded in the fibre direction, even at very high stress levels (Walton and Yeung 1987).

Figure 3.8 *Fatigue property comparison S/N curves*

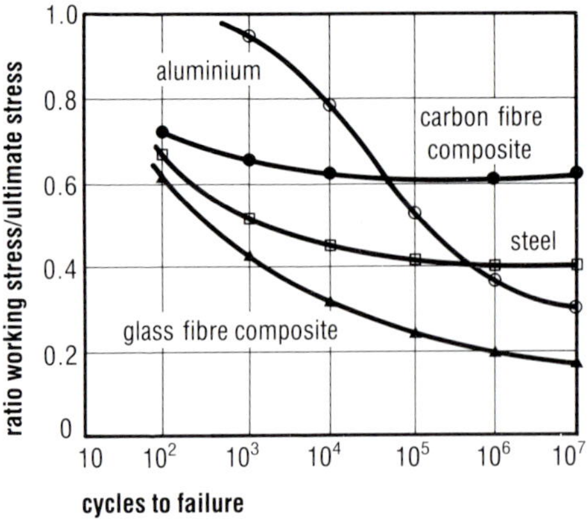

Source Courtesy Courtaulds Grafil

It should be stressed that the excellent performance of CFRP is associated with unidirectional composites loaded in the fibre direction. It should not be assumed that this applies to all configurations. If the matrix performance has a more dominant role, which is the case with ±45 degree configurations or when stressed transverse to the fibre direction or in compression, then the fatigue life is substantially reduced. However, even quasi-isotropic CFRP laminates can show benefit in fatigue resistance over steel and aluminium of 2- to 4-fold.

The ratio, after cycling, of tensile strength to short-term ultimate tensile strength for CFRP/epoxy after 10^7 tension cycles is in the range 53–58%. This applies to the configurations 0 degree, ±45 degree, quasi-isotropic and 90 degree configurations and compares with 2024 T3 aluminium at 28% and 4130 steel at 44% (Lubin 1982, p. 523).

3.4.4 Creep

The creep performance of CFRP in the direction of the fibres is remarkably good and compares very favourably with 'low-relaxation' steel and is significantly better than 'standard' steel (Anderson 1987). This is illustrated in Figure 3.9.

When subjected to tensile load, carbon composites are considered to have a major advantage in their ability to resist long-term creep. Here the totally elastic performance of the fibres dominates the behaviour of the composite. However, when under compression or 'off-axis', then the matrix properties become more significant. Polymers are viscoelastic materials and they deflect continuously with time under load.

Figure 3.9 *Creep data results comparing E-glass, Kevlar 49, carbon and steel* (*British Ropes Report R17/87012, 1987*)

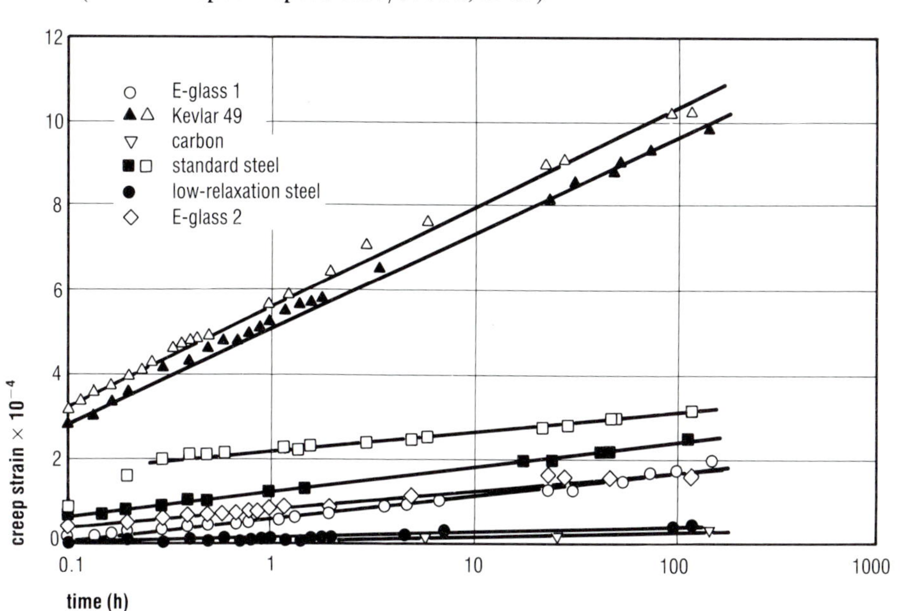

Under these circumstances the creep performance of the composite requires qualification.

3.4.5 Impact

CFRP composites do not generally exhibit good impact performance. They have low strain to failure – 1.5% is a typical value – and they behave totally elastically up to failure. Therefore their ability to absorb impact energy is limited. To improve the impact performance, fibres with high strain to failure are used. These may be high-strain carbon or glass or aramid fibres. Both glass and aramid have higher strain to failure than carbon.

3.5 PROPERTIES OF ARAMID FIBRES AND THEIR COMPOSITES

3.5.1 Introduction

Aromatic ether amide or aramid fibres are organic, man-made fibres which have found application in the composites field. There are various forms available: Dupont produces Kevlar (R) aramid in several versions;

Enka produces under the trade name Twaron, while the Teijin company has an aramid which is marketed under the trade name of Technora.

Aramid fibres are generally characterised as having reasonably high strength, medium modulus and a very low density. Their composites fit well into a gap in the range of stress/strain curves left by the family of carbon fibres at one extreme and glass fibres at the other.

Aramid fibres are fire-resistant and perform well at high temperatures. They are insulators of both electricity and heat. They are resistant to organic solvents, fuels and lubricants. A major distinction of aramid fibres is that they are highly tenacious in the non-composite form and do not behave in a brittle manner as do both carbon and glass fibres.

3.5.2 Mechanical properties

The tensile stress/strain curve of aramid fibres is essentially linear to failure (as shown in Figure 3.3). Aramid fibres have two distinct categories: those in which their elastic modulus is about the same as glass fibre, typically 60–70 GPa ($8.70–10.15 \times 10^6$ lbf/in^2), and those with a modulus at about twice this level. Kevlar 29 falls into the first category and Kevlar 49 into the higher-modulus category. It is generally the higher-modulus material which finds use in composites, but the lower-modulus aramids do have applications in composites in those circumstances where high strain to failure or high work to failure are required. The specific performance of aramids is their primary advantage, that is their strength/weight and stiffness/weight ratios. As the density of aramids is in the range 1.39–1.44 they show advantage over many carbon and glass fibres if either specific strength or specific stiffness is the selection criterion. Some aramids have a relatively very low compressive strength. Although a unidirectional ARP has a tensile strength of about 1,400 MPa (0.2×10^6 lbf/in^2), its compressive yield strength is about 1/6 of this at 230 MPa (33,350 lbf/in^2). This also results in poor flexural performance of about 300 MPa (43,500 lbf/in^2).

3.5.3 Fatigue

Unidirectional Kevlar 49 composites in tension/tension fatigue ($R = 0.1$, room temperature) show very good performance, superior to S-glass and E-glass composites and 2024-T3 aluminium. Only unidirectional carbon composites are superior. The picture is more complex for aramids in flexural fatigue. At a modest number of cycles they exhibit poor fatigue strength, inferior to E-glass. This is presumably owing to the poor static flexural performance of Kevlar 49. At a high level of cycles ($10^6 \simeq 10^7$), however, the excellent inherent fatigue performance of the aramid results in Kevlar 49 and E-glass having similar performance (Schaefgen and Wardle 1984).

3.5.4 Creep

In spite of their high inherent tensile strength, and even in unidirectional configurations, aramid fibre composites have creep rates generally very much higher than for glass or carbon composites.

3.5.5 Impact

An outstanding feature of aramid composites is their ability to withstand impact damage and particularly ballistic impact. They show better ballistic resistance than GRP of a comparable weight. They are more expensive than GRP, but as cost is rarely an influencing factor in ballistic performance aramids dominate the body armour market.

3.5.6 Chemical properties

Aramid fibres fall into two main categories with respect to their chemical resistance. PPTA fibres, of which Kevlar 49 is an example, are chemically quite stable and have high resistance to neutral chemicals. However, they are susceptible to attack, particularly by strong acids and also by bases. On the other hand Technora aramid fibre exhibits extremely high strength retention in both acids and alkalis.

Tests carried out in 40% aqueous solution H_2SO_4 at 95°C (203°F) for 100 hours showed 90% strength retention for Technora fibre compared to 20% for PPTA fibre (Imuro and Yoshida 1986), but to put this into context, E-glass fibre would have ceased to exist under these conditions.

3.6 PHYSICAL CONSIDERATIONS

3.6.1 Damping characteristics of composites

The damping characteristic of a material, that is its ability to reduce induced vibrations rapidly, is an important aspect of the material selection process in certain areas. Fishing rods, for instance, benefit from good damping properties because vibrations die out more rapidly, reducing frictional resistance of the line against the guides.

Mechanical equipment, in particular where subject to variable speeds, has a significant problem of vibration resonance which results in noise and reduction in performance. Such equipment benefits from the use of materials with good damping characteristics. Metals and ceramics have particularly low internal losses and hence have poor damping characteristics. Polymer composites, on the other hand, and particularly CFRP and ARP, have very good damping performance, which can be used to reduce vibration resonance.

3.6.2 Thermal expansion coefficients of composites

The thermal expansion of composites is dependent on several factors: type of reinforcement, type of matrix, geometry of reinforcement and volume fraction. Table 3.5 shows the thermal expansion coefficients for a variety of composites. The two negative values are a result of the negative fibre expansion coefficients which both carbon and aramid fibres exhibit.

Table 3.5 Typical thermal expansion coefficients

	Thermal expansion coefficient/degree C		
	UD comp. 0 deg	UD comp. 90 deg	bidir. comp. fibre dir.
E-glass polyester	8.6	14.1	9.8
carbon XAS HP epoxy	−0.3	28	1.9
aramid K49 epoxy	−4	79	3.2

Volume fraction of fibres = 65%

It is possible to tailor the expansion coefficient of a composite to a specific requirement, for example to match that of aluminium or even to be zero over a given temperature range.

3.6.3 Fire resistance

The factors by which fire performance is assessed include:

- surface spread of flame
- fire penetration
- ease of ignition
- fuel contribution
- oxygen index (ie the minimum oxygen content that supports combustion).

Each factor varies in significance depending on the particular circumstances. In short, there is no single simple test by which composites can be assessed and compared in terms of their fire resistance.

The fire performance of composites covers a wide spectrum, ranging from highly flammable to non-burning. The degree of flammability of a composite is governed by the following factors:

- matrix type
- quantity and type of fire retardant additives included
- quantity and type of fillers included
- reinforcement type, its volume fraction and its construction.

The dominant factor is, of course, the polymer matrix. Of the commonly used polymers in composites the approximate order of flammability without fire retardant additives is as follows:

polyester	burns readily
vinylester	↓
modar	
epoxy	
phenolic	excellent fire resistance

Polyesters, in spite of their relatively poor intrinsic fire resistance, can have very low flammability by the incorporation of suitable fire-retardant additives and a reasonably high glass fibre content. A very low-flammability polyester can have the following performance:

BS 476 Part 7:1971 Class 1	(surface spread of flame)
BS 476 Part 6:1968 Class 0	(fire propagation)
BS 476 Part 3:1958 Class AA	(fire penetration and spread of flame)
ASTM E84	flame spread of 25 (on a scale: asbestos cement board = 0, red oak flooring = 100)

However, this performance is very difficult to achieve with polyester, and it is expensive both in terms of cost and performance, particularly from weathering and corrosion resistance. Phenolic resins achieve this level of fire resistance without the inclusion of fire-retardant additives. They have extremely low smoke emission and virtually no toxic combustion products (Forsdyke 1984).

3.7 ENVIRONMENTAL PROPERTIES OF COMPOSITES

3.7.1 Temperature effects

As a general rule both the strength and the stiffness of composites are unimpaired by the effect of low temperature, and under some circumstances they may be enhanced. However, low temperatures tend to make polymers less flexible and therefore there may be a tendency towards damage by fatigue.

High temperatures are more problematic. As temperature is increased, all resins soften and reach a stage at which the polymer passes from a glass-like state to a rubbery state. This is the glass transition temperature T_g. Beyond T_g the properties of the polymer change dramatically. Hence this is a limit to the usable temperature range for all normal applications. However, the effect the fibres have is to improve the T_g of the composite above that of the matrix resin. Examples of glass transition temperatures for several polymer types are shown in Table 3.3.

3.7.2 Corrosion

1 Chemical corrosion

Composites are inherently corrosion-resistant and can show substantial cost benefits when used in environments which are extremely aggressive. Hence glass-reinforced plastics have been used extensively in chemical plant structures for many years. The polymers are selected on the basis of their ability to withstand the aggression of the chemical environment, be it strong acid, strong alkali or perhaps simply salt water.

In general terms the order in which polymers are corrosion-resistant is:

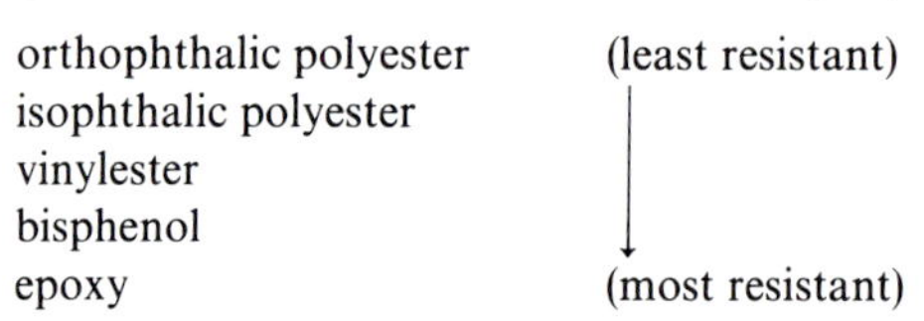

This is a very simplistic guide and does not apply in every case. There are certainly specific examples where the order can be reversed. Consequently the resin manufacturer's advice should be sought to determine the suitability of a resin system in a particular aggressive environment.

2 Galvanic corrosion

Glass fibre and aramid fibre composites do not suffer from this problem. However, carbon fibre is a conductor of electricity, and relative to various metals its composites are highly noble. Therefore if carbon composites are used in contact with metals, galvanic corrosion can take place. The severity of the corrosion depends upon distance between the composite and the metal, the extent of the polarisation and the efficiency of the electrolyte. If the conditions are favourable, then the metal, being more anodic, will corrode away. This would, of course, be disastrous if the two were to be bonded together as a structural joint. The situation is overcome simply by ensuring, by the application of a suitable coating, that the carbon composite and the metal do not come into contact.

3.7.3 Moisture effects

Moisture can penetrate all organic materials by a diffusion process which it will do until the equilibrium level is reached. All polymers are susceptible to its effects, which in general result in a reduction in mechanical properties and glass transition temperature. The loss of property is a function of the degree of moisture pick-up, and Figure 3.10 illustrates a typical

Figure 3.10 *Moisture absorption of 8 mm diameter glass-fibre (unidirectional)/polyester rods (ends sealed)*

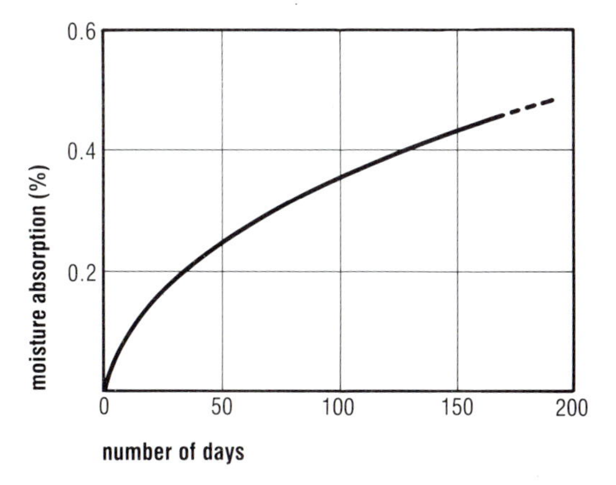

moisture pick-up curve for a totally immersed glass fibre/polyester composite. Resin systems can be selected which have excellent resistance to the effects of moisture (see Figure 3.11). However, the glass transition temperature can be reduced to 75% of its dry value by the effect of a moisture content of 4%. Flexural strength can be reduced to 50% of the dry value by the effects of 1.5% moisture content.

The designer must therefore take the effect of moisture into account at the design stage. If the composite is to be subjected to a wet environment then 'wet properties' must be used in the design process.

3.7.4 Thermal spike

It is known that carbon/epoxy laminates are degraded if they are heated when they have a high content of moisture. This is probably due to microcracking of the matrix and/or of the fibre/matrix interface. They can be severely degraded by the effect of a thermal spike, that is a sudden

Figure 3.11 *Effect of moisture absorption on flexural strength and modulus of 8 mm diameter glass-fibre/polyester unidirectional rod*

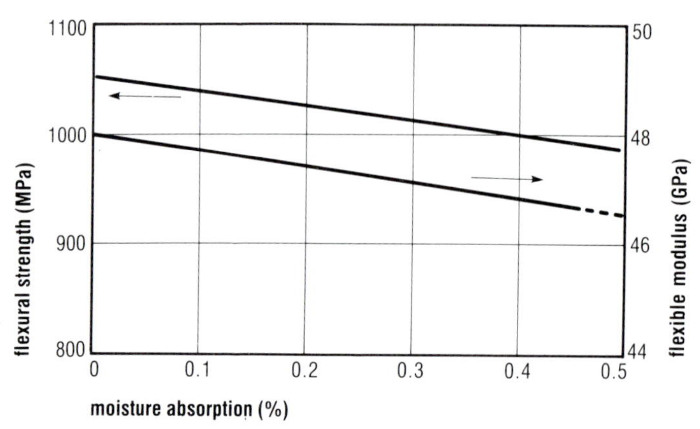

increase in temperature followed by sudden decrease. The degradation is partially reversible. However, voids and microcracks, created by the force of contracting moisture-filled free volume is irreversible degradation (Adamson 1983, McKague *et al.* 1975, and Ramani and Nelson, no date).

3.7.5 Weathering

The mechanical performance of composites is degraded by the effects of temperature, moisture, sunlight, wind, dust and acid rain. These effects come under the general heading of weathering. There are several mechanisms attributed to weathering, the most notable being: leaching from the resin of chemical constituents; sunlight attack on the resin, causing embrittlement and erosion of the resin due to its degradation; and the effect of wind and airborne particles. Thus the environment is allowed access to the fibres. However, for an epoxy/carbon composite in flying service, 'generally there is no indication of increasing degradation with length of time in service, the maximum deterioration occurs in the first year' (Jones 1985).

The most effective option to counter the effects of weathering is to use resins applicable to the environment, ie to incorporate ultraviolet stabilisers in the matrix and chemical-resistant fibres (C-glass, ECR-glass or polyester) in the surface of the composite.

3.7.6 Erosion

A significant aspect of weathering is the erosive effect of rain in rapidly moving air. This is the case, for example, in high-speed flight and with helicopter blades, in which case radomes, antenna covers, farings and the leading edge of helicopter blades etc are susceptible to erosion from rain, snow and ice. The rate of erosion is a function of speed, angle of impact, mass and frequency. For example, an unpainted edge cap on an aircraft rotodome which had been exposed for 19 years had a strength retention of 68% (Lubin 1982 and Blackford 1985). The most common method of countering its effects are by the use of rain-erosion-resistant finishes such as polyurethane. These can be so beneficial that rain erosion can be reduced from 'badly degraded' to 'no drop in tensile or flexural strength or modulus'. The advent of improved fibres has also significantly diminished the problem.

3.7.7 Lightning

Carbon-fibre-reinforced composites are more susceptible to damage from lightning than is aluminium. Hence, the use of CFRP in aircraft has prompted investigation into methods for their protection from lightning strike. It is well known that aircraft are at risk and have been lost as a result of damage from lightning strikes. The losses have been due both to structural damage to the aircraft and to electrical effects on systems and elements which are critical.

CFRP is a poor conductor of electricity. It therefore has neither the ability to dissipate electricity, as does aluminium, nor the insulating ability of GRP.

The most common method of protecting CFRP against 'arc root' damage from lightning is by the use of surface metallisation. This is achieved either by flame spray or by moulding-in a layer of aluminium mesh or tissue. The CFRP is insulated from the layer by a thin layer of GRP.

This is a very brief reference to a complex topic about which work is currently (1988) in progress. An excellent review is given in Payne *et al.* (1987).

3.7.8 Stress corrosion

The stress corrosion of glass fibre composites is a mechanism of failure where a laminate under stress in an acid or basic environment can fail catastrophically at very low stresses compared to the fracture strength in air.

The phenomenon is dominated by resistance to the aggressive environment of the glass fibre. The matrix has several roles: it must act as a chemical barrier, but more importantly, to avoid stress corrosion, it must be tough.

It is thought that the mechanism requires that the environment gains access to the glass fibre, either by diffusion or via surface cracks. Work has shown (Caddock, Evans and Hull 1986) that polyester resin is impermeable to hydrochloric acid and thus discounts the permeability mechanism for those particular laminates. This implies that the propagation of cracks between the fibre and the laminate surface is generally the route for environmental ingress. Having gained access, the environment may or may not degrade the fibre very rapidly, but it has been shown (Cochram and Scrimshaw 1980) that E-glass is susceptible to acid attack. In sulphuric acid E-glass strand suffers maximum strength loss in 1 to 6 N acid, and hydrochloric acid has a similar effect.

Hence the fibre is not able to perform one of its primary duties, to act as a crack-stopping device. Matrix cracks propagate, allowing the acid access to the next fibre. This is attacked by the acid, and fails, and so on.

An ECR-glass (a chemically resistant grade of glass) shows a much prolonged retention of strength in mild acids in comparison to E-glass.

There appear to be three types of specimen failure (Caddock, Evans and Hull 1986). At high loads failure is similar to dry failure under tension. At lower loads the failure surface is stepped, but at very low loads the fracture surfaces are very smooth.

A further complication is that some aggressive agents do not attack the fibre but severely degrade the polymer. Tests using 0.5 M orthophosphoric acid (Steard and Jones 1986) with E-glass/epoxy showed rapid micro-cracking of the laminate, but the fibres alone were found to be virtually crack-free. It is therefore dangerous to generalise in this field and necessary to refer to relevant data or to carry out accelerated stress corrosion trials.

Figure 3.12 (Hogg, Price and Hull 1984) shows log time to failure against log strain for E-glass and ECR-glass laminates with four resin systems:

Beetle 870 (equivalent to Hetron 197) het acid chemically resistant
Crystic 600 epoxy modified bisphenol
Crystic 272 isophthalic polyester
Crystic 272 + 30% Crystic 586 (flexible polyester).

It is apparent that the tougher, less chemically resistant, options perform better in a stress-corrosion situation.

3.7.9 Blistering

It has been observed that gel-coated polyester laminates reinforced with glass fibre may form unsightly blisters on their surface after lengthy

Figure 3.12 *Dependence of time to failure on applied strain for hoop wound pipe, tested by ring compression, in 0.6 M HCl at 20°C*

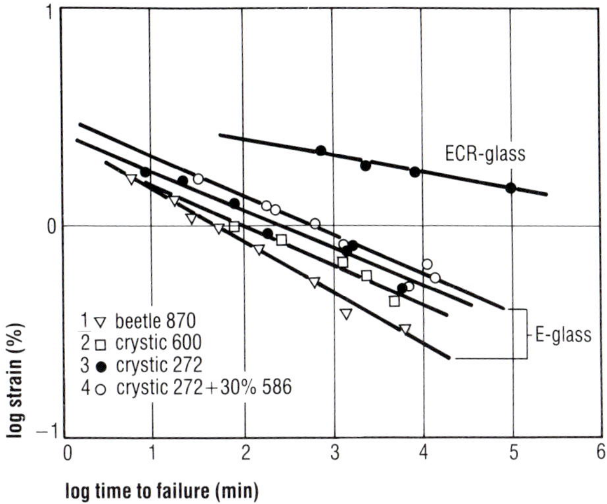

Source Hogg, Price and Hull 1984

contact with water. In particular, boat hulls and swimming pools have been affected.

The mechanism by which blisters occur is thought to be basically osmotic. The laminating resin and gel-coat resin are permeable to water molecules, which pass into the resin, albeit very slowly. Any voids in the laminate will pick up moisture. If the resin system is hydrolytically unstable (i.e. able to decompose by reaction with water), then salts may be present in the water to allow the gel coat to act as a semi-permeable membrane and thus create an osmotic cell. Sufficient pressure builds up eventually to create a blister.

Considerable experimentation has shown that to prevent blistering there are two major considerations: raw material selection and production techniques. The use of water-resistant resins, and preferably 'powder bound' chopped strand mat rather than 'emulsion bound' are the most important aspects of material selection. Production techniques must ensure thoroughly impregnated and wetted glass fibre and the correct resin/glass ratio (Birley *et al.* 1984 and Crump 1986).

3.8 DESIGN OF PULTRUSIONS

3.8.1 Introduction

The design process consists of several iterative stages, which result, hopefully, in the designer reaching a satisfactory solution to his problem. It is

Figure 3.13 *The design process*

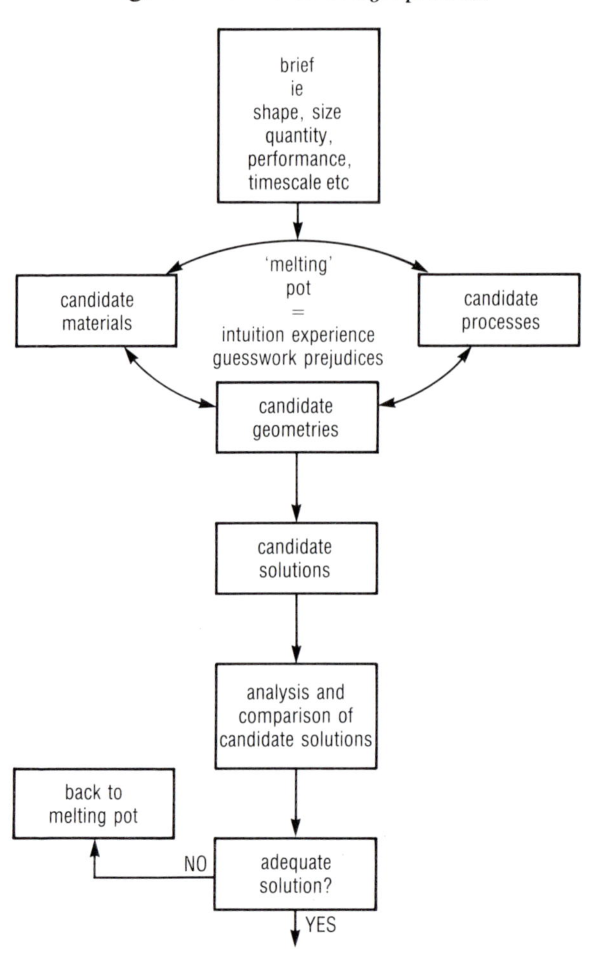

shown schematically in Figure 3.13. Although in principle the design process remains the same, in practice different activities dominate, depending whether the design process is primarily 'selective' or whether the intention is to produce an optimum, in which case the procedure has been iterated such that the lowest possible cost or weight has been achieved. There are techniques available which allow this to be carried out efficiently.

Selective design requires the same activities, but the design brief stipulates or implies a particular area of search. Once a solution has been found, very little or no attempt is made to optimise it. This is the case when designing with standard structural elements – 'I' beams, for example. Particular sizes are available from which to select; therefore there is no opportunity to refine or optimise. This provides rapid economic solutions to common engineering problems.

Composite structural elements are available as standard pultrusions in the form of box beams, angles and channels etc. The properties and performance are known, thus allowing structural analysis to be carried out. Hence designs may be made on a selective basis from these 'off-the-shelf' composites. On the other hand, composites in general and pultrusions in particular may be 'tailor-made' for a particular design brief, allowing an optimised design to be achieved.

In the case of optimum design of composites not only is the geometry (shape) designed but also the material itself. Material design and geometry cannot be solved in isolation. Candidate solutions are analysed and compared, the process being iterated until the optimum solution is found.

The analysis of composite materials is discussed in detail elsewhere and in various references, notably Tsai and Hahn (1980) which is concerned with classical laminate analysis. This is a powerful tool for the analysis of composite materials, including those which are pultruded. It allows materials to be modelled, consisting of multilayers of differing fibre reinforcements at any angle. The elastic properties may be predicted, and with the application of a failure criterion the failure can be predicted.

3.8.2 Component design

Performance criteria determine whether a standard structural element will be a suitable solution for a particular design brief. If not, then a custom shape must be designed. The geometrical shape of a pultrudate (pultruded laminate) is a function of the performance required in service and the restrictions imposed by the production process. The critical aspects of pultrudate geometry are thickness, radii, overall size, number of cavities and location of the reinforcement layers.

Thickness: Thick sections require a longer time for the heat of the die to penetrate and therefore run more slowly than thin sections. It is therefore desirable to keep wall thickness to a minimum, commensurate with achieving the required structural properties. The thickness range which is pultrudable is from 1 mm to 50 mm (0.04 to 2 inches) or more. However, a more practical range is from 2 mm to 20 mm (0.08 to 0.8 inch).

Radii: As a general rule corner radii should be as large as is feasible in order to avoid stress concentrations. This is not always possible, particularly if a tooling split line occurs at the radius, in which case a sharp corner cannot be avoided. Thickness should be kept constant throughout the radius, otherwise the reinforcement pack needs local modification. An acceptable minimum internal radius is 1.5 mm (0.06 inch); the external radius is wall thickness plus internal radius.

Overall size: The maximum size which can be pultruded is, of course, dictated by the dimensions and the pulling capacity of the machine available. The capacity of pultrusion machines varies from as small as 125 mm by 40 mm (4.92 by 1.57 inches) with a pulling force of 500 kg (1,102.5 lb), to a maximum of 1,000 mm by 165 mm (39.37 by 6.50 inches) with a pulling force of 15 t (14.76 ton). There is therefore considerable scope for the design of small solid sections at one end of the scale, to very large hollow sections at the other.

Cavities: Hollow profiles are possible and, in fact, common in pultrusion. They are very efficient structures but they do present production problems, particularly if there are multicavities. If hollow profiles can be avoided without detriment to performance, then it is best to do so. Otherwise the number of cavities should be minimised. Hollow profiles often require overlapping reinforcement, and hence thickness changes may be necessary to accommodate the excess material.

Structural design with pultrudates: Typical pultruded structural elements are shown in Figure 3.14. Performance characteristics and data are available on such profiles, allowing structural analysis to be carried out on a 'strength of materials' basis. As when designing with metals, wood etc, pultruded materials and elements are assumed to be homogeneous. This allows the conventional engineering formulae to be used to interpolate test results to the particular design problem.

Glass fibre pultrusions have a significantly lower tensile modulus than steel or aluminium. This results in the design procedure being concerned initially with the deflection limits and only subsequently with the strength requirements, which are generally found to be satisfactory.

The relatively low modulus of glass fibre pultrusions also requires that buckling characteristics be assessed a little more critically than would be the case with a steel component, which has an abundance of stiffness. Carbon fibre pultrusions, on the other hand, have a high modulus, of the same order as steel, and thus buckling is a less significant problem.

The shear stiffness of composite materials is relatively low. This can give rise to deflections in beams due to shear, which are appreciable, possibly of similar magnitude to those due to bending. Therefore it is essential to calculate these deflections to ensure they are insignificant or taken into account. This is particularly important with deep sections, short spans

Figure 3.14 *Pultruded composites – structural elements*

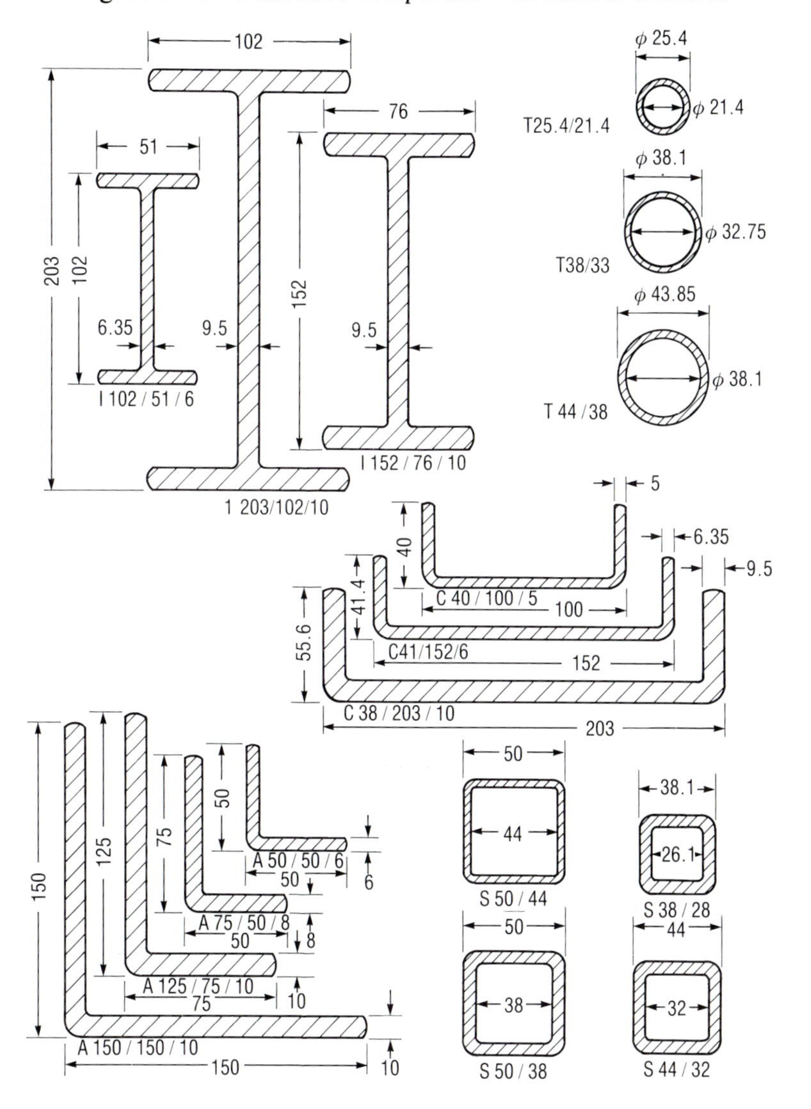

and hollow sections. The general physical characteristics of both glass and carbon-reinforced pultrusions are given in Table 3.6.

Composites do not exhibit a yield zone in their stress/strain response. Therefore 'local yielding' cannot be relied upon to solve certain design problems, as is the case with steel design. Stress concentrations, perhaps due to misalignment of elements, cannot be left to yield, as might be the practice with steelwork. The stress will remain in place and may result in premature failure.

Table 3.6 Physical characteristics of pultrudates reinforced with glass and carbon

	Glass fibre	Carbon fibre
weight	low	very low
strength	high	very high
strength to weight ratio	high	very high
stiffness	low	very high
impact properties	very good	good
corrosion resistance	good	good
dimensional stability	good	excellent
high-temperature resistance	good	good
flame resistance	good	good
pigmentability	good	no
paintability	good	possible
moisture and rot resistance	good	good
maintenance	low	low
electrical insulation	yes	no
heat insulation	yes	no
antistatic or electrical conduction	no	yes
electromagnetic transparence	yes	no
design flexibility	yes	yes
parts consolidation	yes	yes

A further result of the lack of a yield zone in composites is a requirement for the use of large washers with bolted connections to allow local stresses to be spread more evenly.

If the matrix resin in a composite does not have sufficient 'strain to failure' available, then microcracking may occur at a level substantially below the ultimate strength of the component. This allows the ingress of the environment which, if it is particularly aggressive to the fibre, can result in premature failure. Hence, the use of composite materials in aggressive environments such as chemical plant has prompted the assessment of strain-limited design methods. These recognise that, although a structure may be designed adequately within stress limitations, it may be subject to strains which are not acceptable. A strain limitation which is applied in BS 4994 is a maximum of 0.2%. This has been demonstrated to be a reasonable level for chemical plant.

NOTE

1 *High performance fibres: R-glass.* Vetrotex data sheet produced by Dept Nantes Technologies, BP, Chambéry, France (1987)

4 Resin Matrices

George Green MA PhD CChem FRSC

Technical Manager, Ciba-Geigy Plastics
Duxford, Cambridge

4.1 INTRODUCTION

Organic resin matrices for use in advanced structural composites have been developed for a wide variety of applications, and their use today is increasing at 15–20% per annum in some sectors such as the aerospace industry. Both the processing conditions and the properties of the final composite are influenced by the chemistry of the materials used.

The combination of resin matrices with a fibrous reinforcement can be effected in a number of ways such as filament winding, wet lay-up, pultrusion etc. One of the most convenient ways of presenting such materials is in the form of prepreg. This consists of fibres in either unidirectional or woven form that have been pre-impregnated with a resin matrix. The manufacture of composite components from fibre-reinforced matrices takes place by the application of heat and/or pressure with or without the use of a vacuum. The mechanical and physical properties of the component are determined largely by the combination of fibre and matrix resin. The matrix itself has a greater influence upon the less fibre-dominated properties of laminates such as the interlaminar shear strength, transverse tensile strength and fracture toughness. It also influences the temperature at which the composite component can be used, either in the dry or wet condition, and the resistance to fluids such as fuel or paint-stripping solvents.

As explained already in Chapter 1 (pp. 33), there are basically two types of organic matrix systems that can be used in the manufacture of advanced structural composites, namely, thermosetting or thermoplastic materials. Thermosetting systems are composed of reactive components which undergo an irreversible chemical cross-linking reaction upon application of heat, whereby solidification occurs and an infusible material is formed. Thermoplastic systems, on the other hand, are high-molecular-weight materials which do not undergo a chemical reaction but merely melt upon application of heat and pressure to form the desired component when cooled.

Before thermal cross-linking, the composition or type of resin matrix used will influence the manufacture and properties of pre-impregnated materials such as prepreg or single tow, and the processibility of the fibre reinforced system. This can be illustrated by the manufacture and use of unidirectional or woven prepreg which can be made by either a solution or film process. In the solution technique the parallel array of fibres or the fabric is impregnated with a solution of resin matrix and the solvent is removed by heating. With the film technique the unidirectional fibres or woven fabric are contacted on one or both sides by a film of the resin matrix which has been prepared beforehand on a backing paper. This is then gently heated and passed through consolidation rollers in order to impregnate the fibres and form the prepreg. Resin film can be manufactured either from a solution of the resin matrix or by a hot-melt method which involves heating the resin in order to soften or lower its viscosity for the coating process. Not all resin matrices are suitable for this solvent-free process. During the manufacture of prepreg, therefore, the chemistry of the resin matrix is important, especially with respect to its solubility in solvents and its thermal stability. As far as the former is concerned it is an obvious advantage if non-flammable, non-toxic solvents can be used that are capable of easy removal. Any residual solvent in a composite component will tend to lower the mechanical properties. It is also important to design the matrix using constituents that can be made with confidence and can be readily verified by analytical techniques. In this way it is possible to manufacture prepreg that is reproducible, reliable and traceable.

Some properties of the prepreg itself can be influenced by the resin matrix. Prepreg requires good tack if possible so that the individual layers will hold together during the lay-up process, and good drape in the case of woven fabrics so that contoured surfaces can be constructed easily. The resin matrix must also provide adequate cohesion between the fibres for them to hold together prior to the curing process. In the case of thermosetting matrices the usable life or out-time of the prepreg once it has been removed from cold storage will depend upon the stability or latency of the resin used. Ideally it should remain processible after several weeks' storage at room temperature because of the time taken to lay up complicated or thick components. Finally, the use of prepreg demands that the storage life is reasonable, and this will depend on the chemistry of the resin system used in the case of thermosetting matrices.

The type of resin matrix will influence the processibility of the resin-fibre system. It is necessary to control the viscosity during heat-up and cure if one is to avoid excessive flow of resin during the manufacture of the composite, and voids in the final component. With thermosetting resin matrices the viscosity drops on heating, reaches a minimum and then rapidly increases because of the cross-linking reactions which lead to the final cured matrix. This minimum viscosity will depend upon the process conditions and the chemistry of the resin matrix employed. Usually the viscosity profile is measured with neat resin so that the effects of the fibre are not considered. This has to be borne in mind when attempting to correlate rheological data. Not only can the fibrous reinforcement affect the flow characteristics, it can also alter the rate of the chemical reactions in some cases. The cure cycle required for manufacture of the composite component, ie temperature, heat-up rate, pressure and total times, will be determined by the type of resin system used. Another important aspect of the matrix system during processing is the evolution of any volatile materials. These need to be kept to the absolute minimum because they can lead to faults such as voids in the final part.

In composites the resin matrix in its cured or final state serves a number of important functions which can be influenced by its chemistry and physical properties. For example, it provides environmental protection for the reinforcement, helps to avoid crack growth through the fibres by providing an alternate failure path along the fibre-matrix interface, and keeps the fibres aligned in their predetermined directions, which enables them to resist load. An important property of the cured resin matrix is its glass transition temperature. This is the temperature at which the matrix begins to soften, and the mechanical properties of the composite usually begin to decrease rapidly. It is not only an important parameter which decides the dimensional stability of the composite under the influence of heat, but it also has a critical effect upon most of the physical properties of the matrix at ambient temperature (Fischer, Lohse and Schmid 1980). Another important property of the final composite component is its resistance to moisture. In general the matrix-influenced physical properties of composites manufactured from organic resin matrices will decrease with increasing temperature and humidity because of the absorption of moisture from their surroundings. In the case of a carbon-fibre/epoxy composite, for example, a decrease of approximately 20°C (36°F) in the glass transition temperature is experienced for an absorption of 1% moisture.

4.2 TYPES OF RESIN MATRIX

4.2.1 Epoxy resin systems

These thermosetting systems are extensively used in advanced structural composites. For example, more than two thirds of the current aerospace market for composites utilises epoxy matrices reinforced with either carbon, glass or aramid fibres. There is a wide choice of systems from which to choose. They consist of epoxy resins and a curing agent or hardener system. Epoxy resins range from low-viscosity liquids to high-melting-point solids and they can be readily formulated to give suitable products for the manufacture of prepreg from both the solution or hot-melt techniques. They can be modified with a variety of different materials, eg by the addition of high-molecular-weight polymers.

The majority of epoxy resins used in composites are manufactured by the reaction of epichlorhydrin with materials such as phenols or aromatic amines. Addition of the epichlorhydrin yields an intermediate chlorhydrin which ring-closes under alkaline conditions to give the epoxy resin. Epoxy resins all contain the epoxy or glycidyl group, illustrated in Figure 4.1,

Figure 4.1 *Epoxy resin*

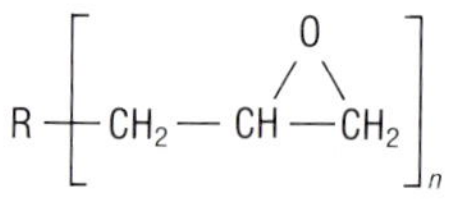

$$R \left[CH_2 - \overset{\overset{O}{/ \quad \backslash}}{CH - CH_2} \right]_n$$

which is capable of reaction with a number of compounds that contain an active hydrogen atom, such as phenols, amines and carboxylic acids. The reaction is an addition which does not produce any side products, and epoxy resins cure to give three-dimensional, stable networks with relatively low shrinkage – typically 2–3%. These resins are also capable of being polymerised in a catalytic manner by certain materials such as Lewis acids, whereby a polyether structure is formed. Different hardener systems give rise to chemically different structures which can possess a wide range of properties and morphologies. Their chemistry is well established (Lee and Neville 1967, Lohse 1987, and May and Tanaka 1973), and typical cross-linking reactions can be illustrated by describing some of the curing agents used. For example, cross-linking of epoxy resins by amines proceeds as illustrated in Figure 4.2, where the epoxy ring is opened by the primary amine group to give a secondary hydroxyl group and a secondary amine which further reacts with another epoxy group. Thus when amines and epoxy resins possessing functionality greater than two react together they cross-link to give a three-dimensional polymer

Figure 4.2 *Reaction of epoxy resin with amine*

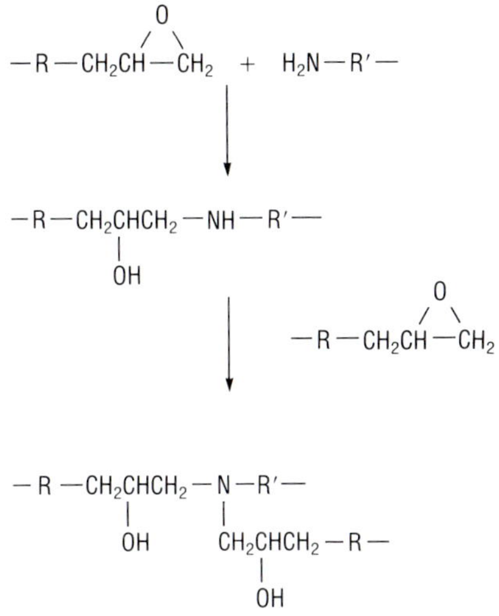

$$-R-CH_2CH\overset{O}{-}CH_2 \quad + \quad H_2N-R'-$$

$$\downarrow$$

$$-R-CH_2CH(OH)CH_2-NH-R'-$$

$$\downarrow \quad -R-CH_2CH\overset{O}{-}CH_2$$

$$-R-CH_2CH(OH)CH_2-N(CH_2CH(OH)CH_2-R-)-R'-$$

network containing ether bridges, secondary hydroxyl groups and tertiary amine groups.

A more latent hardener for the cross-linking of epoxy resins is dicyandiamide, whose mechanism of cross-linking has been shown to be quite complex (Zahir 1982). Not only do the active hydrogens of this hardener react with the glycidyl group by an addition reaction similar to amines, but dicyandiamide is believed to be cleaved during the curing reaction to give 2-amino-oxazoline units. There are also indications that ether bridges are formed by reaction of the hydroxyl groups with glycidyl groups, and the reaction proceeds with less than the expected stoichiometry of one molecule of dicyandiamide to four glycidyl groups.

The curing of epoxy resins with dicarboxylic acids or anhydrides having a cyclic structure leads to the formation of polyetheresters. The reaction with anhydrides is usually catalysed by tertiary amines or other catalysts. Although these cross-linked matrices are more resistant to oxidation than amine cured systems, they are less resistant to hydrolysis, especially under hot conditions.

Unlike the reactions described above, which involve addition reactions between the hardener and the epoxy resin in near-stoichiometric quantities epoxy resins can be catalytically polymerised by such hardeners as boron trifluoride amine complexes or imidazoles to give polyethers.

Because the formulation of epoxy resin matrices is often complex, however, one is rarely concerned with a single type of cross-linking reaction. For example, polymers are often added to the system in order to improve such properties as toughness and damage tolerance of the final cured composite, or to control the viscosity of the system during the manufacturing process. Examples of polymers are thermoplastics such as acrylonitrile-butadiene-styrene copolymers and polyethersulphones or rubbers such as carboxy-terminated butadiene-acrylonitrile copolymers (CTBN). These materials often possess reactive end groups that are capable of addition reactions with the epoxy resin, eg the carboxyl group in the case of CTBN.

Prepregs manufactured from epoxy resin matrices possess good drapability and excellent tack. Typical cure cycles are one hour at 120°C (248°F) to two hours at 180°C (356°F). With 120°C curing systems glass transition temperatures of 100–150°C (212–302°F) are obtained whereas with 180°C curing systems they can range up to around 220°C (428°F).

Composite components of fibre reinforced epoxy resins are capable of operating up to around 150°C (302°F), depending upon the environmental conditions.

Some of the more important epoxy resins used in advanced composites today are N-glycidyl derivatives of 4,4′-diaminodiphenylmethane and 4-aminophenol, and aromatic di- and polyglycidyl derivatives of bisphenol A, bisphenol F, phenol novolacs and tris (4-hydroxyphenyl) methane. These, together with a wide variety of available hardener systems and modifying agents offer a choice of cross-linking reactions that influence the cured matrix properties and the mechanical properties of the composite component (Fischer, Lohse and Schmid 1980).

One of the most important epoxy resin monomers used today in the formulation of advanced composites based on epoxy systems is the tetraglycidyl derivative of 4,4′-diaminodiphenylmethane, whose formula is given in Figure 4.3. This material has been well characterised (Chaudhari, Cobuzzi and King 1984), and a typical system used widely in the aerospace industry consists of this resin, the aromatic amine hardener 4,4′-diaminodiphenyl sulphone shown in Figure 4.4, and a boron trifluoride amine complex as accelerator. By alternative modification of this epoxy monomer using more latent curing agents and polymeric modifiers it is possible to obtain matrices that have improved tack, shelf-life, control over flow during processing and toughness.

Figure 4.3 *Tetraglycidyl derivative of 4,4′-diaminodiphenylmethane*

```
   O                                         O
  / \                                       / \
CH2—CHCH2                               CH2CH—CH2
        \                               /
         N—⟨benzene⟩—CH2—⟨benzene⟩—N
        /                               \
CH2—CHCH2                               CH2CH—CH2
  \ /                                       \ /
   O                                         O
```

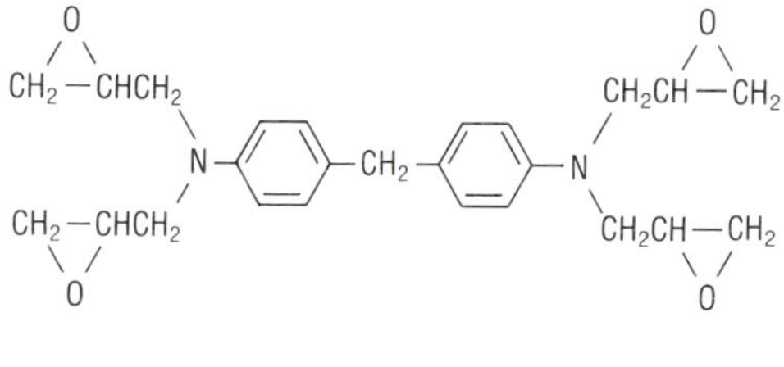

Figure 4.4 *4,4′-diaminodiphenyl sulphone*

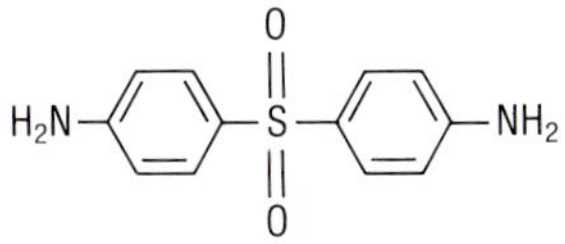

As described earlier the glass transition temperature of epoxy resin matrices can range up to about 220°C (428°F) and is dependent upon a number of factors such as the formulation and cure cycle used. For example, if the diglycidyl ether of bisphenol A is cured with dicyandiamide, 4,4′-diaminodiphenylmethane or 4,4′-diaminodiphenyl sulphone one obtains glass transition temperatures of 120°C, 160°C and 205°C (248, 320 and 401°F), respectively. Thus the change from a methylene group to a sulphone group in the aromatic hardener increases the value quite substantially. Although the use temperature of composites manufactured with epoxy resin matrices satisfies many applications, attempts are being made to improve this property.

4.2.2 Maleimide resin systems

These thermosetting matrices can be regarded as a subdivision of polyimides, but because of their importance in advanced structural composites they are considered separately here. The vast majority of these resins are difunctional, ie bismaleimides whose generic formula is represented in Figure 4.5, where R is predominantly an aromatic group. They are prepared by reaction of maleic anhydride with the corresponding primary amine via the intermediate formation of a maleamic acid and then cyclisation (Hummel, Heinen, Stenzenberger and Seisler 1974). One of the most widely used bismaleimides is that of 4,4′-diaminodiphenylmethane.

Bismaleimides react through the double bond of the maleimide group, which is capable of a variety of reactions. Radical initiated homopolymerisation of the double bond, leading to cross-linking with multifunctional maleimides, can be effected at temperatures of around 200°C (392°F) (Grundschober and Sambeth 1968). These polymerisation temperatures can be decreased with the use of catalysts capable of generating free radicals. It is also possible to copolymerise bismaleimide resins with other unsaturated monomers (Haug 1980 and Yoshikawa *et al.* 1984) such as triallyl isocyanurate. Furthermore, bismaleimides are capable of addi-

Figure 4.5 *Bismaleimide resin*

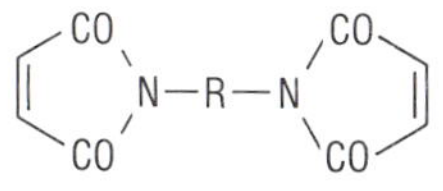

Figure 4.6 *Reaction of bismaleimide with amine*

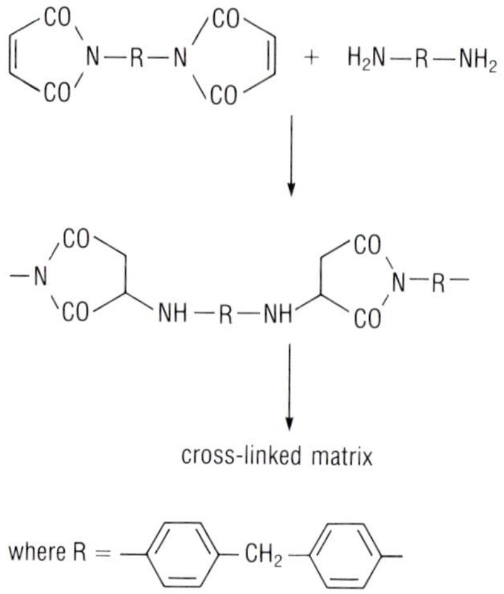

$$\text{bismaleimide} + H_2N-R-NH_2 \rightarrow \text{adduct} \rightarrow \text{cross-linked matrix}$$

cross-linked matrix

where R = $-C_6H_4-CH_2-C_6H_4-$

tion or cross-linking reactions with compounds such as triazine resins (Gaku, Suzuki and Nakamichi 1978), diallylphenols such as the diallyl derivative of bisphenol A (Renner and Zahir 1978) and compounds with reactive hydrogen atoms such as amines, phenols, thiols and carboxylic acids. Once again, cross-linking reactions are often complex and involve more than one mechanism. The Michael addition of active hydrogen compounds can be exemplified by aromatic primary amines which are often used in the formulation of bismaleimide matrices (Grundschober 1970). For example, Figure 4.6 illustrates the reaction between the bismaleimide of 4,4′-diaminodiphenylmethane and the amine itself. Initial addition of the primary amine groups to the double bond of the bismaleimide occurs at temperatures between 100 and 150°C (212–302°F) and this is followed by addition of the secondary amine groups, and thus cross-linking, at temperatures up to around 250°C (482°F) (Di Guilio, Gautier and Jasse 1984). For best results the reaction of bismaleimides with aromatic amines is carried out with less than stoichiometric amounts of the amine. When stoichiometric amounts are used the resulting polymers have poor thermal stability.

Although the range of bismaleimide matrices available today is limited compared to other systems such as epoxy, their number is increasing. In general they are not as easy to process as epoxy-resin-based materials. They have lower solubility in desirable solvents and higher viscosities in the melt form, although it is possible to reduce the latter by the use of eutectic mixtures. The processing conditions required to manufacture composite components from bismaleimide matrices are more severe than those used for epoxy systems; for example, cure temperatures as high as 250°C (482°F) are typically employed for several hours. The resulting composites are more brittle than those of epoxy matrices, although recent developments have produced systems whose toughness is comparable to the first-generation epoxy matrices without sacrificing the advantage of higher-temperature performance. All too often, attempts to improve the damage tolerance of composites results in a decrease of their temperature resistance or other properties. Efforts to improve the processibility and toughness characteristics of bismaleimide systems include the use of diallyl bisphenol A as a co-reactant (Chaudhari, Galvin and King 1985). In general the glass transition temperatures of cured bismaleimides are some 100°C (212°F) higher than cured epoxy matrices, which is reflected in their better retention of mechanical properties under hot/wet conditions. They can be used to manufacture components that are capable of operating up to around 200°C (392°F).

4.2.3 Phenolic resin systems

Phenolic resins are the oldest thermosetting resins and their manufacture from phenol and formaldehyde is well documented (Knop and Scheib 1979). Reaction of phenol with less than equimolar proportions of formaldehyde under acidic conditions gives novolac resins which range in molecular weight from around 400 to 2,000, and contain aromatic phenol units linked predominantly by methylene bridges. Novolac resins are thermally stable and can be cured by cross-linking with formaldehyde donors such as hexamethylenetetramine.

The more significant phenolic resins from the point of view of composites, however, are resoles. These are manufactured by reacting phenol with a greater than equimolar amount of formaldehyde under alkaline conditions. The ratio of phenol to formaldehyde is usually between 1 : 1.0 and 1 : 3.0 and is typically 1 : 1.5 to 1 : 2.0. Besides phenol itself, phenolic resins can be prepared from other phenols such as cresols or bisphenol A. Resole resins are essentially hydroxymethyl functional phenols or polynuclear phenols. Unlike novolac resins they are low-viscosity materials and are easier to process with respect to composite components. On the application of heat the hydroxymethyl groups react, with the elimination of formaldehyde and water, and cross-linked matrices are formed that contain mainly methylene bridges as illustrated in Figure 4.7. The reaction involves the intermediate formation of ethers as shown in Figure 4.7, which is a simplified reaction scheme.

Besides the disadvantage of evolution of volatile materials during cure,

Figure 4.7 *Cross-linking of resoles*

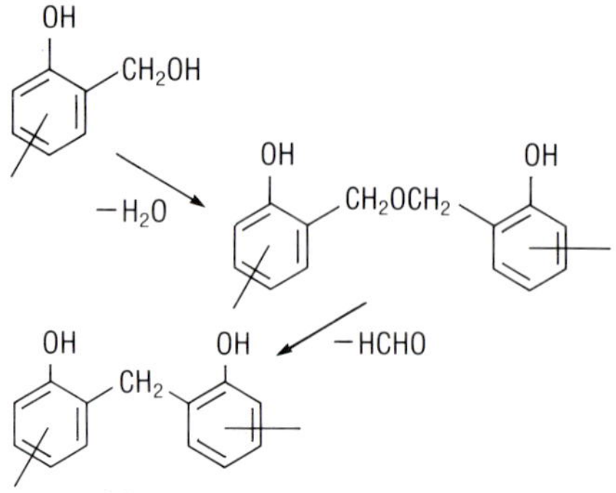

resoles are less storage-stable at ambient temperatures than many thermosetting systems. During the manufacture of prepreg from such systems the thermal input is critical and requires careful control in order to obtain consistent products. The major advantage of composite components made from fibre-reinforced phenolic resin matrices is their inherent stability to thermal oxidation. Most other thermosetting materials require additives, such as brominated resins and antimony trioxide in the case of epoxy systems, in order to achieve good flame retardancy. Coupled with this good fire resistance, phenolic resins emit low levels of smoke on burning and their emissions possess low toxicity. The combination of these properties makes them obvious contenders for the manufacture of composite parts for mass transportation such as the interiors of aeroplanes and rail carriages.

One of the main factors that has limited the use of phenolic resins as a matrix for advanced structural composites is the brittleness of the final components. Attempts to improve this property, either by chemical modification or the addition of various toughening polymers, usually leads to a reduction of their fire, smoke and toxicity advantages. Fibre-reinforced phenolic composites are capable of operating up to 120°C (248°F) with good strength retention. Cure cycles of uncatalysed resole matrices are around one to two hours at 125°C (257°F).

4.2.4 Polyimide resin systems

In addition to the bismaleimide resin systems mentioned earlier, there are a growing number of other polyimide matrices. Some give rise to a thermoplastic matrix through chemical reaction of two difunctional monomers. These condensation polyimides (Cassidy 1980) are typically formulated from an aromatic tetracarboxylic acid dianhydride such as

Figure 4.8 *Formation of condensation polyimide*

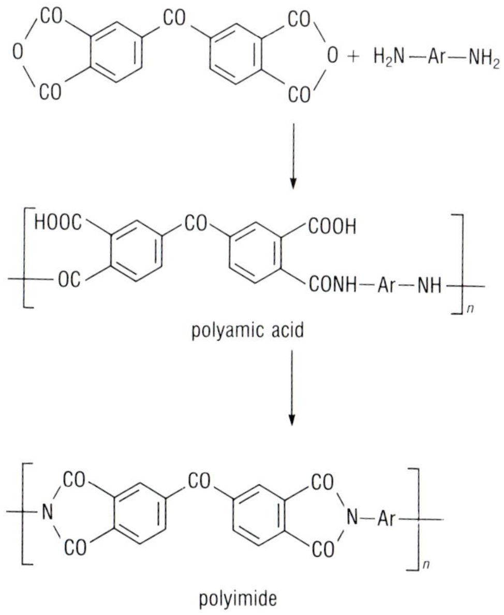

$$\text{BTDA} + H_2N{-}Ar{-}NH_2$$

polyamic acid

polyimide

benzophenone tetracarboxylic acid dianhydride (BTDA) and an aromatic amine as illustrated in Figure 4.8.

Two reaction steps are involved. Initially a high-molecular-weight polyamic acid is formed by reaction of the amine and anhydride groups. The cure reaction which then takes place involves the loss of water and cyclisation to give the polyimide matrix. Prepreg can be made from the intermediate polyamic acid, but this involves the use of high boiling, polar solvents such as N-methylpyrrolidone or diglyme, which is a disadvantage during processing. There are a number of alternative dianhydrides such as pyromellitic dianhydride, and diamines such as diaminodiphenyl sulphone, diaminobenzophenone or diaminodiphenyl ether that can be employed in the formulation of these systems (Hergenrother 1986). An alternative approach to the formation of these types of polyimide matrices is the reaction of the diamine with the ethyl ester of an aromatic dianhydride where ethanol is evolved in addition to water (Gibbs 1985).

Other types of polyimide resin systems are thermosetting matrices that cross-link on cure. One of the best known systems is the PMR 15 matrix which exists only as a few discrete formulations. These are formulated from the monomethyl ester of nadic acid (NE), the dimethyl ester of benzophenone tetracarboxylic acid (BTDE) and 4,4′-diaminodiphenylmethane (DDM), as illustrated in Figure 4.9. A typical stoichiometry is 2:2:3 of NE:BTDE:DDM. The esters are prepared from the

Figure 4.9 *PMR 15 system*

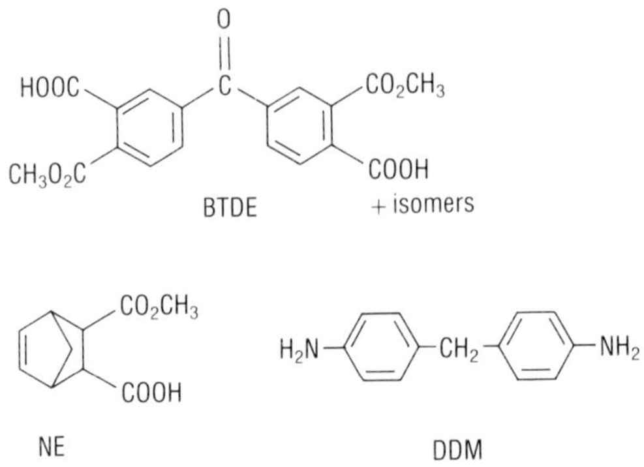

BTDE + isomers

NE

DDM

reaction of methanol with the corresponding anhydrides. Cure of these systems again proceeds in stages. Initially, methanol is released at around 75°C (167°F) with the formation of a polyamic acid which is end-capped with nadic groups. Then cyclisation occurs at around 200°C (392°F) to give a linear imidised structure with the elimination of water. Finally, at higher temperatures, up to 330°C (626°F), cross-linking reactions take place whereby the resin undergoes a reverse Diels-Alder reaction of the nadic end groups involving the evolution of cyclopentadiene which then takes part in an addition polymerisation with the resulting maleimide. Because the curing reaction of the PMR 15 matrix involves the elimination of volatile materials, namely, methanol, water and some cyclopentadiene, problems can be experienced with the processing and manufacture of composite parts from the system unless precautions are taken.

The cure cycles necessary to process polyimides are long and involve a programmed heat-up to high temperatures followed by a long cooling cycle and a post-cure cycle. These processing features have given rise to other disadvantages. For example, laminates manufactured from fibre-reinforced PMR 15 matrices show a tendency to develop microcracks. This microcracking occurs during thermal cycling or mechanical loading and can cause a significant reduction in properties. It is possible by curing at lower temperatures, ie 290°C (554°F) instead of 330°C (626°F), to reduce this tendency whilst still affording acceptable mechanical properties (Hay *et al.* 1987). The reason is thought to be due to decreased thermal stresses because of the lower processing temperature and a modified polymer network.

Another matrix system that is similar to the PMR 15 resins is LARC 160, which exhibits improved flow, tack and drapability of the prepreg, but at the expense of some thermal oxidative stability in the composite.

LARC 160 systems are based upon the diethyl ester of BTDA, the ethyl ester of nadic acid and a polyaromatic amine (Jones 1983). The mechanism of cure is similar to that of the PMR 15 system, except that ethanol is produced instead of methanol.

The major advantage of polyimide resin matrices is their ability to operate at higher temperatures than any other commercially available organic system. Composite components that have been manufactured from carbon fibre reinforced PMR 15 matrices can operate up to around 300°C (572°F) with good retention of properties. However, they tend to be brittle, and attempts have been made to improve their toughness. For example, PMR 15-type matrices have been formulated with flexible aromatic diamines (Bowles and Vannucci 1986), but a reduction in the temperature performance is experienced.

4.2.5 Unsaturated polyester resin systems

These thermosetting resins contain the unsaturated diester group illustrated in Figure 4.10. They are manufactured by the reaction of maleic anhydride with low molecular weight diols such as ethylene or propylene glycol. A variety of other dibasic acids or anhydrides, eg phthalic anhydride, and diols can be co-condensed with the maleic anhydride, and the properties of the cured resin matrix can be tailored to requirements by changing the ratio and nature of the components (Boenig 1964 and Bruin 1976). For example, the use of isophthalic acid gives rise to products with improved chemical resistance, whereas halogenated monomers impart flame retardancy to the final composite. In order to complete the formulation of these systems a comonomer is required such as styrene, diallyl phthalate or triallyl cyanurate, and a free-radical initiator system such as a peroxide with or without an accelerator; fillers and inhibitors can also be added. By dissolving the unsaturated polyester in monomers like styrene a low-viscosity system is obtained that has good processibility. Typical peroxides used are benzoyl and methyl ethyl ketone peroxide, and accelerators for these can be cobalt salts such as cobalt naphthenate.

Unsaturated polyester matrix systems can be cross-linked at ambient or elevated temperatures. This involves radical copolymerisation of the maleic anhydride derived unsaturated polyester and the reactive diluent such as styrene. The reactivity and shelf life can be further controlled by

Figure 4.10 *Unsaturated diester*

```
—R—O—C—CH=CH—C—O—
     ||        ||
     O         O
```

the use of inhibitors, such as hydroquinone, which are free-radical scavengers. These systems can also be modified with fillers, such as china clays, or thermoplastics in order to improve such characteristics as shrinkage on cure or damage tolerance. Depending upon the ratio of diols to anhydrides used in the manufacture of the unsaturated polyester one can obtain either hydroxy- or carboxy-terminated polymers. The latter in general possess higher viscosities and improved physical properties of the final matrix, although the difference diminishes with increasing molecular weight. During the manufacture of these resins the maleate group within the polyester is capable of isomerisation to the fumarate ester, and very high conversions can be achieved, depending upon the diol and co-reactants used. The ratio of fumarate to maleate will have an effect upon the reactivity of the matrix system because maleates copolymerise more slowly than fumarates do with styrene.

Like epoxy resin systems, unsaturated polyester matrices cure with no evolution of volatile materials, and they can be formulated to satisfy a wide range of requirements. In general their cross-linked matrices possess lower glass transition temperatures than epoxy matrices and they shrink by 7–8% on cure compared to about 2–3% for epoxy systems. Attempts to overcome the limitation of unsaturated polyester matrices caused by their brittleness have been made by modifying them with telechelic liquid polymers such as vinyl-, carboxy- and amine-terminated butadiene-acrylonitrile copolymers, and high-molecular-weight rubbers in a similar way to that carried out with epoxy systems, but with more limited success (Siebert 1985).

4.2.6 Other thermosetting resin systems

Vinyl ester resins are another class of thermosetting resins that cure by a radical initiated polymerisation in a similar way to unsaturated polyesters. They are mainly derived from reaction of an epoxy resin, eg bisphenol A diglycidyl ether, with acrylic or methacrylic acid. These hydroxy(meth)acrylates have the general formula shown in Figure 4.11. Like unsaturated polyesters they are copolymerised with diluents such as

Figure 4.11 *Vinyl ester resin*

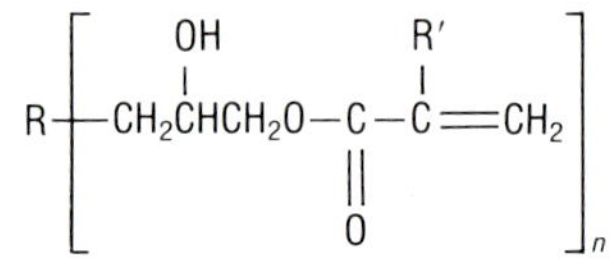

$$R\left[CH_2\overset{\overset{\Large OH}{|}}{C}HCH_2O-\underset{\underset{\Large O}{\|}}{C}-\overset{\overset{\Large R'}{|}}{C}=CH_2\right]_n$$

$R' = H$ or CH_3

Figure 4.12 *Cyanate resin*

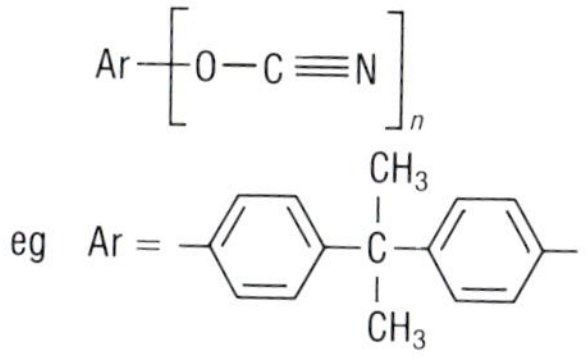

$$Ar\left[O-C\equiv N\right]_n$$

eg $Ar = -C_6H_4-C(CH_3)_2-C_6H_4-$

styrene using similar free-radical initiators. They differ from the polyesters in that the unsaturation is at the ends of the molecule and not along the polymer chain. In general they offer improved chemical resistance (when methacrylates are used) compared to composites manufactured with unsaturated polyester matrices. Their properties can be varied by the choice of epoxy resin used such as the glycidyl ethers of bisphenol A, bisphenol F, tetrabromobisphenol A and novolacs.

It has already been noted that thermosetting resin systems can often be modified by additives in order to improve certain characteristics or to optimise properties. In a similar way optimisation of certain features can be obtained by mixing systems, eg epoxy and bismaleide matrices. Another example is the use of cyanate resin whose general formula is given in Figure 4.12. These resins can polymerise on their own, when heated, to give cross-linked polymers containing the triazine ring, formed by trimerisation of the cyanate group (Kubens *et al.* 1968). The cured matrices obtained have glass transition temperatures of 240–290°C (464–554°F). The resin systems offer advantages such as hot-melt processibility and low shrinkage, and have been compared to toughened bismaleimide systems (Shimp 1987). Triazine resins are also capable of cross-linking reactions with epoxy or bismaleimide resins.

4.2.7 Thermoplastic resin systems

Thermoplastic resins are essentially high-molecular-weight linear molecules. Unlike thermosetting matrices they do not cross-link to three-dimensional networks on heating but remain chemically unreacted. They merely melt and flow when heat and pressure are applied, and resolidify on cooling. Physical changes such as partial crystallisation can occur during their processing.

There are many thermoplastic resins available today but only a few offer the range of mechanical and thermal properties required in order to be truly considered for advanced composite materials. They have been developed by overcoming many of the disadvantages of the earlier thermoplastics such as poor thermal oxidative stability, low glass transition temperature, and hence retention of mechanical properties, and

Figure 4.13 *Thermoplastic resins*

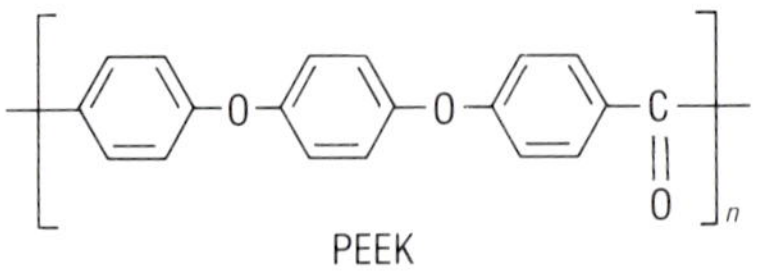

PEEK

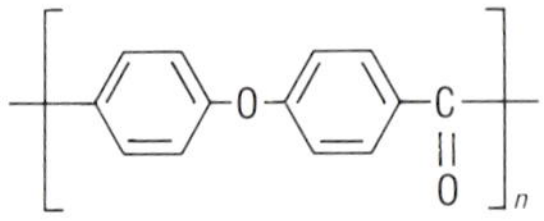

PEK

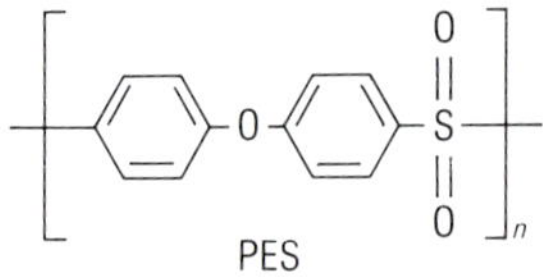

PES

brittleness. By the synthesis of polymers based upon phenylene or arylene backbones linked by such groups as sulphone, ketone and imide, and atoms such as sulphur and oxygen, a variety of suitable matrix polymers have been made. For example, a number of thermoplastics for advanced composites have been manufactured by the use of starting materials such as 4,4′-dichlorodiphenyl sulphone, 4,4′-dichlorobenzophenone, 4,4′-dihydroxydiphenyl sulphone, hydroquinone and 4-chloro-4′-hydroxy-diphenyl sulphone. Amongst these are polyetheretherketone (PEEK), polyethersulphone (PES) and polyetherketone (PEK), illustrated in Figure 4.13, whose glass transition temperatures are 143, 230 and 165°C (289, 446 and 329°F), respectively. Another useful group of thermoplastics used in advanced composites are based upon aromatic units linked by sulphur atoms. Amongst these matrices are polyphenylenesulphide (PPS) and polyarylenesulphides (PAS). The glass transition temperature of PAS resins is 125°C (225°F) higher than that of PPS itself, ie 215°C (419°F), which makes them a contender for applications where good retention of properties is required at higher temperatures. Yet another class of thermoplastic resins used today in advanced composites are polyimides which include polyetherimides. The interest in thermoplastic matrices is increasing, and includes such materials as liquid crystal polymers (Tai-Shung Chung and McMahan 1986) and polyquinoxalines (Hergenrother 1986).

The manufacture of prepreg from thermoplastic resins can also be difficult and sometimes involves the use of high-boiling, polar solvents. They give dry, boardy prepregs which make them difficult to handle during composite manufacture. Because their melting points range from about 250°C to 350°C (482 to 662°F) they require high temperatures and pressure during processing, and efficient wetting of the fibre reinforcement is often difficult compared to thermosetting systems where the low viscosity achieved prior to curing readily allows wetting and impregnation of the fibres. Composite components manufactured from thermoplastics often lack good solvent resistance. Where this is an exception, for example in PEEK, there can be difficulties to overcome in their processing because of fluctuations in the degree of crystallinity that can arise (Seferis 1986, and Ostberg and Seferis 1987).

Despite some of the problems associated with thermoplastics, they do offer some advantages over thermosetting systems. For example, they display improved damage tolerance, infinite storage life and the potential for fast processing. Considerable efforts are being made to improve their processibility, eg by the use of techniques such as spot-welding or hot-melting of the prepreg tape prior to lay-up. An interesting development is the use of hybrid yarns or co-mingled fibres (Clemans, Handermann and Western 1987). In this technique the thermoplastics fibre and reinforcement fibre are interwoven to give a drapable fabric which can be converted under heat and pressure to the composite part, thus obviating the need to manufacture prepreg in a conventional sense.

4.3 RESIN MATRIX/FIBRE INTERFACE

The mechanical properties of advanced composite materials depend not only upon the properties of both the fibrous reinforcement and the resin matrix but also upon the interaction between them. In the case of glass fibres a variety of silane coupling agents are available which are capable of bonding between the glass surface and the matrix. For example, an aminosilane will bond to the glass surface via silyloxy linkages and to an epoxy matrix system by chemical reaction between the amine and epoxy groups as described earlier.

Carbon fibre itself is both surface-treated in order to improve the mechanical properties of the composite, and coated with a sizing agent in order to aid processing of the fibre. Surface treatment, amongst other things, creates potentially reactive groups such as hydroxyl and carboxyl groups upon the surface of the fibre which are capable of reaction with the matrix. Very often the fibre is used in its received form, which is not necessarily optimised for the particular matrix system employed. Furthermore, the sizing agent may not be the most suitable for the resin matrix. Epoxy-based sizing agents are quite common and carbon fibres treated with them are used for a variety of different matrices. Some composite mechanical properties are improved by adhesion between the fibre and matrix, especially transverse- or matrix-dominated properties, but other properties such as open-hole tensile benefit from little or no adhesion. Thus any insight into the nature of the matrix/fibre interaction, which involves a variety of bonding mechanisms other than chemical, could lead to improved systems. To this end a great deal of work is currently being carried out.

5 Joining of composites

F L Matthews BSc (Eng) ACGI CEng FRAeS FPRI

Director, Centre for Composite Materials
and
Senior Lecturer, Department of Aeronautics
Imperial College of Science, Technology and Medicine, London

5.1 INTRODUCTION

Joints in components or structures incur a weight penalty, are a source of failure and cause manufacturing problems; whenever possible, therefore, a designer will avoid using them. Unfortunately it is rarely possible to produce a construction without joints due to limitations on material size, convenience in manufacture or transportation and the need for access. Such considerations apply equally to joints between metallic components or between composite components.

Fortunately the main methods used for joining metallic parts, mechanical fastening and adhesive bonding are also applicable to composites, provided care is taken to allow for the characteristics of composites. Welding is also a possibility for thermoplastics composites, but this technique is not well developed for load-carrying joints.

When choosing between the two basic methods their various advantages and disadvantages must be kept in mind. Mechanically fastened joints are easily disassembled without damage, do not need special surface preparation, are easy to inspect but do have high stress concentrations (at the holes) and are heavy. Whilst bonded joints have lower stress concentrations and weight penalty, they cannot easily be disassembled, adequate surface preparation is essential, inspection is difficult and, in addition, they are sensitive to environmental effects.

An advantage of composites, compared to metals, is the freedom to tailor mechanical properties such as stiffness and strength by judicious selection of fibre type, content and orientation. This can be a major benefit, of course, but can cause problems with joints, particularly with very non-isotropic lay-ups, resulting in extremely complicated and heavy configurations and, also, laminates that are difficult to repair. It is clear that joints must be considered as an integral part of the design process if a cost-effective solution is required.

The objective of this chapter is to describe the basic characteristics of load-carrying joints in composites. The typical data to be quoted should be adequate for preliminary design work.

5.2 MECHANISM OF LOAD TRANSFER

The purpose of a joint is to transfer load between the two items being joined. As a result of this load transfer there will be a stress variation in the components in the joint region, as well as stresses in the joining medium (fasteners or adhesive).

To illustrate the mechanism let us consider the single cover butt joint shown in Figure 5.1 *a*. The load is to be transferred from plate A to plate B via the cover plate C. The stresses in the direction of the load will be zero at the points marked x, the free ends, and maximum at the points marked y. The consequent variation of strain in the plates along the joint means that the adhesive or the fasteners will be loaded in shear, as shown in Figures 5.1 *b* and *c*.

Figure 5.1 (*a*) *Layout of single cover butt joint*, (*b*) *adhesive in shear*, (*c*) *fastener in shear*, (*d*) *bending of plates causing peeling at the joint's ends*

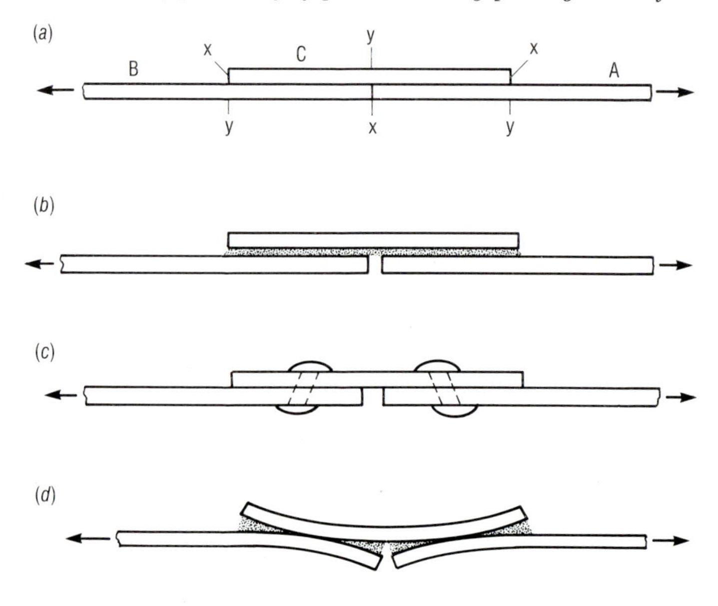

Because the lines of action of the resultant load in the plates and cover plate are not coincident the joint will bend under load, as shown in Figure 5.1 *d*. It is seen that the joint will tend to peel apart at points x, putting the fastening medium there into tension.

The mechanisms described above will occur, to a greater or lesser extent, in all joints. Fairly clearly, then, the ends of the joint are likely to be the critical areas since it is here that the joining medium is most highly loaded, in a combination of shear and tension (peel). Of course not only must the adhesive or fasteners be able to withstand these stresses, but also the plates being joined. A particular problem with composite plates, which are relatively weak through the thickness, is that failure is often caused by the peel stresses; one of the main objectives in design, therefore, is to minimise peeling.

5.3 STRESS DISTRIBUTIONS

A knowledge of the stresses in joints is vital if we are to understand the failures that occur in practice and hence improve designs and predict strength. Even relatively simple theories can be useful if they allow the important parameters to be identified.

5.3.1 Bonded joints

There are many publications concerned with the stress analysis of bonded joints (Matthews, Kilty and Godwin 1982). The early works relating to joints between metal adherends were extended to composite adherends by modifying the theory to include such factors as the low through-thickness shear stiffness. More recently, finite-element analysis has been used as an alternative to classical continuum mechanics methods (Adams 1986).

The simplest approach, known as 'shear lag', considers only the extension of the (uniform thickness) adherends and the shear of the adhesive (Volkersen 1938). Although the method gives a physically infeasible (ie non-zero) result for the shear stress at the ends of the joint, as shown in Figure 5.2, it is nevertheless useful in allowing rapid assessment of the effect of changing some of the major parameters (Matthews, Kilty and Godwin 1982). The correct result for the adhesive shear stress distribution is only obtained, with considerable increase in mathematical complexity, by including through-thickness direct strains in the adhesive.

Volkersen's (1938) analysis is a reasonable approximation to the behaviour of (one half of) a double lap joint. However, as indicated by Adams

Figure 5.2 *Shear stress distribution in adhesive arising from simple ('shear lag') theory*

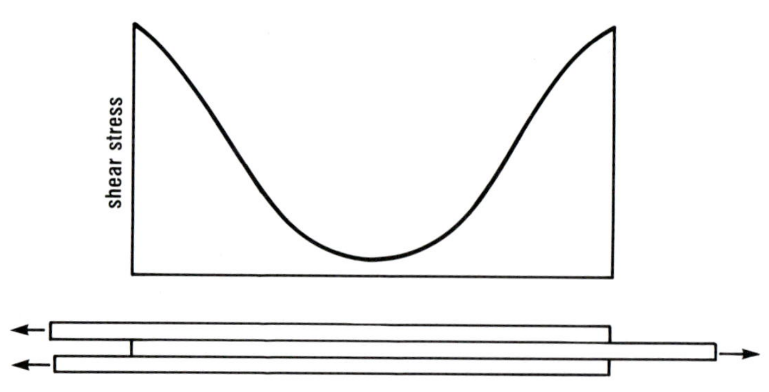

(1986) it ignores the inherent eccentricity of the adherends which causes through-thickness tensile (peel) stresses in the adhesive (and the adherends). It can be shown that the peel stresses are reduced by tapering the adherends.

If joint strength is to be predicted it is important, as recognised by Hart-Smith (1973a), to include in the analysis the fact that adhesives are not purely elastic (ie a linear shear stress-strain curve). Hart-Smith predicts that joint strength increases with the area under the stress-strain curve of the adhesive. Stepped lap joints have stress distributions closely related to those in double lap joints, and scarf joints can be regarded as a special case of stepped joints (Hart-Smith 1973b).

Single-lap joints, although the simplest to make, are complicated to analyse due to the eccentricity of the adherends. On application of the load, the joint will rotate, and the adherends bend, causing high peel stresses at the ends of the joint. This non-linear situation was first solved by Goland and Reissner (1944). Hart-Smith (1973c), who extended the approach to composites, showed that the peel stresses are reduced significantly by increasing the length of the overlap.

5.3.2 Mechanically fastened joints

In contrast to bonded joints, and certainly for joints between metallic parts, little attention has been devoted to the stress analysis of mechanically fastened joints (Godwin and Matthews 1980). The advent of composites, with their essentially linear stress-strain behaviour up to failure has, however, led to the need for accurate stress distributions.

The basic load transfer mechanisms described above still apply, and hence the fasteners at the ends of a uniform thickness joint will be highly loaded in shear and tension (peel). In contrast to a bonded joint there are, additionally, significant stresses that vary in the plane of the laminate, ie around each fastener hole, caused by the contact pressure of the fastener on the edge of the hole.

The simplest approach, the equivalent of the shear lag analysis, is obtained by assuming the contact pressure to vary in a cosinusoidal fashion on the loaded half-circumference. A rigorous analysis, which does not make any assumptions about the contact pressure but establishes this as part of the solution, shows that the pressure distribution is strongly influenced by the proximity of the sides and end of the joint, and by the lay-up of the composite plate (Matthews 1986). The simple cosine distribution is rarely obtained in practice.

In terms of joint performance there are three significant in-plane stresses in the laminate: the compressive (bearing) stress on the loaded side of the pin, the tensile stress across the net section and the shear stress on the shear out planes. Typical distributions of these stresses, which can be obtained from continuum analyses, are shown in Figure 5.3.

To obtain information on through-thickness stresses it is necessary to use finite-element analysis. It can then be seen that, in the case of pin-loading, the bearing stresses give rise to through-thickness tensile strains (no through-thickness restraint). In view of the relatively low strength in this direction one might expect failure at low loads. Any mechanism which restrains this expansion will, therefore, be beneficial. Such restraint will be afforded by rivets and bolts, although the latter, by virtue of the

Figure 5.3 *Typical in-plane stress distributions around a pin-loaded hole: (a) radial compressive stress, σ_{rr}, (b) tensile stress on net section, σ_{yy}, (c) shear stress on shear out planes σ_{xy}*

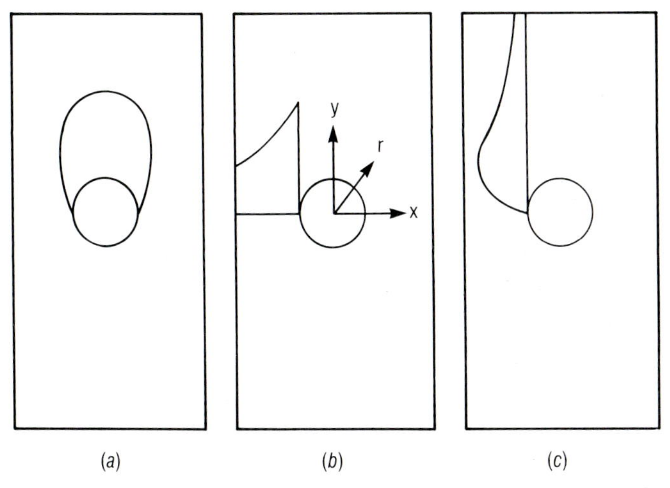

Source Matthews 1987

washers under the nut and head, will give most enhancement. The analysis shows that even a bolt with a finger-tight nut will have through-thickness compressive stresses. With a fully tightened bolt, high through-thickness shear stresses occur between the washers (Matthews, Wong and Chryssafitis 1982).

A further complication arises when there are several fasteners aligned in the direction of the applied load. Apart from the fasteners at the ends of the joint the stress distribution around each hole will arise from the combined effect of the bearing pressure, described above, and the 'by-pass' stresses, arising from the load introduced at adjacent fasteners. The by-pass stresses can be obtained by analysing a plate, containing an unloaded hole, subjected to a tensile force. This situation will clearly give rise to very high tensile stresses on the net section.

5.4 BONDED JOINTS

5.4.1 Surface preparation and adhesive selection

Whether bonding composite-to-composite or composite-to-metal, correct preparation of the surfaces prior to bonding is essential if acceptable strength and durability are to be obtained.

Procedures for the pre-bonding treatment of aluminium alloy are well established and should be followed when bonding to polymer matrix composites. Titanium alloys, with their higher stiffness and strength and lower coefficient of expansion are, perhaps, better matched to carbon-fibre reinforced plastics (CFRP). However, titanium also needs correct pretreatment, particularly if operation is expected in conditions of high humidity (Cotter 1978). The pre-treatment required for the composite's surface is less complex than for the metal's, the essential requirement being to remove contaminants transferred by the chosen release system during moulding of the laminate. Three release methods are in common use: release cloths, coated metal sheets or silicone rubber. The influence of contaminants has been studied extensively by Parker and Waghorne (1982), and it is apparent that simple hand abrasion is generally inadequate, especially after moulding against coated metal. Best results are obtained by blasting with dry alumina grit. There is no evidence that surface contamination significantly affects joint strength reduction arising from exposure to humidity, although the choice of adhesive does have an influence. Also, the strength reduction associated with different levels of pretreatment will be affected by the adhesive used.

Composites can be bonded by most commonly used epoxy adhesives. For adequate durability it appears that hot-setting adhesives give better results than the cold-setting types. However, for both variants the quality of the bond will be dependent on the amount of absorbed water in the composite.

5.4.2 Configurations

The design of bonded joints has been investigated in great detail by Hart-Smith (1986a), who shows clearly the importance of adherend thickness on the relative strength of the different joint configurations, as illustrated in Figure 5.4. The weakest joints are those where failure is limited by interlaminar failure of the adherend or peel of the adhesive. The next strongest joints are those in which the load is limited by the shear strength of the adhesive. The strongest joints will fail outside the joint area at a load equivalent to the strength of the adherend. Such failures can be related to the stress distributions described earlier.

Now, adhesive layers are at their most efficient in the thickness range 0.1–0.25 mm (0.04–0.01 inch). Although attractive from a theoretical viewpoint, thicker bonds are not practicable because of the impossibility of making them without intolerable levels of flaws or porosity. The inability

Figure 5.4 *Effect of adherend thickness on bonded joint strength*

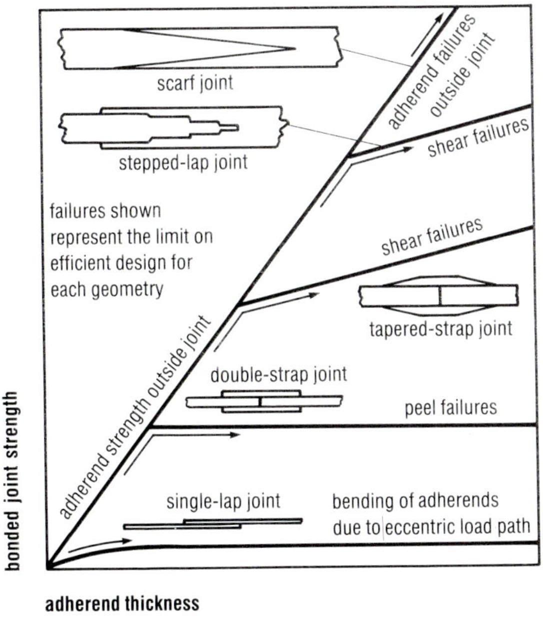

Source Hart-Smith 1986a

to vary adhesive thickness means that the above three failure modes arise mainly because of the influence of adherend thickness, adherend strength being directly proportional to thickness, shear strength proportional to the square root of thickness and peel strength proportional to the fourth root of thickness (Hart-Smith 1986a).

An unsupported single-lap joint is the weakest configuration and, for practical lay-ups, will never be as strong as the adherends being joined. However, acceptable efficiencies can be achieved provided the overlap-to-thickness ratio is sufficiently large. Thicker adherends would need to be tapered at the ends of the overlap.

For thicknesses of about 1.5–1.75 mm (0.06–0.07 inch) for quasi-isotropic lay-ups (less for unidirectional laminates) a double-lap configuration is needed to transfer the strength of the adherends. The optimum overlap-to-thickness ratio is about 30:1. If the adherends are uniform, thickness is limited to about 4.5 mm (0.18 inch), whilst tapering can increase the limit to 6.35 mm (0.25 inch).

For thicknesses greater than 6.35 mm stepped or scarf joints should be considered. Theoretically, merely by making the angle small enough, it should be possible to make a scarf joint stronger than the adherend being joined. In practice, particularly for wide joints, it will be impossible to make the required knife-edge.

5.4.3 Single-lap joints

The strength of single-lap joints is limited by their inherent eccentricity. The resulting bending of the adherends causes peel stresses that lead to failure at low loads. Although tapering the adherends will effect some improvement, the main method of reducing the peel stresses is to use large overlap-to-thickness ratios; at least 50:1. External support of the joint, to prevent overall rotation, will also increase the strength.

As a general rule a more efficient joint will be made with a ductile, rather than a brittle, adhesive. For large overlaps it is possible to more than double the thickness of quasi-isotropic adherends by using a ductile, as opposed to a brittle, adhesive.

It is worth noting that, because the adhesive will fail in peel, the standard 12.5 mm (0.5 inch) overlap specimen is not appropriate for determining shear properties, whatever its merits for quality control. To characterise the adhesive it is preferable to use a thick adherend or napkin-ring specimen.

The severe bending of the adherends makes single-lap joints particularly sensitive to stacking sequence. For a given lay-up, and therefore axial stiffness, large changes in bending stiffness can be obtained by altering the stacking sequence. The objective should be to maximise the bending stiffness, this implying that 0° plies should be placed on the outside of the laminate (Matthews and Tester 1985).

5.4.4 Double-lap joints

As noted above, Hart-Smith (1973a) showed that the strength of double-lap joints is determined by the area under the shear stress-shear strain curve of the adhesive. Consequently a stronger joint will be obtained with a weaker, but ductile, adhesive if the area under its stress-strain curve is larger than that of a stronger, but brittle adhesive. A further consequence is that although failure stress and strain will be altered by environmental effects, such as temperature change, the joint strength will not be affected if the area under the stress-strain curve stays the same, and provided the integrity of adherend/adhesive interface is maintained.

Ductile adhesives will exhibit very high shear strains at failure, and for the purposes of strength prediction can be idealised as elastic – perfectly plastic. As shown previously (Figure 5.2), high strains occur at the ends of the joint, and using the idealised stress-strain behaviour will give the adhesive shear stress distributions shown in Figure 5.5.

It is important to note that the length of the overlap has a significant effect on the shape of the shear stress distribution, as also illustrated in

Figure 5.5 *Effect of lap length on strength and adhesive shear stress distribution (*) of bonded double lap joints: (a) short overlap, (b) intermediate overlap, (c) long overlap*

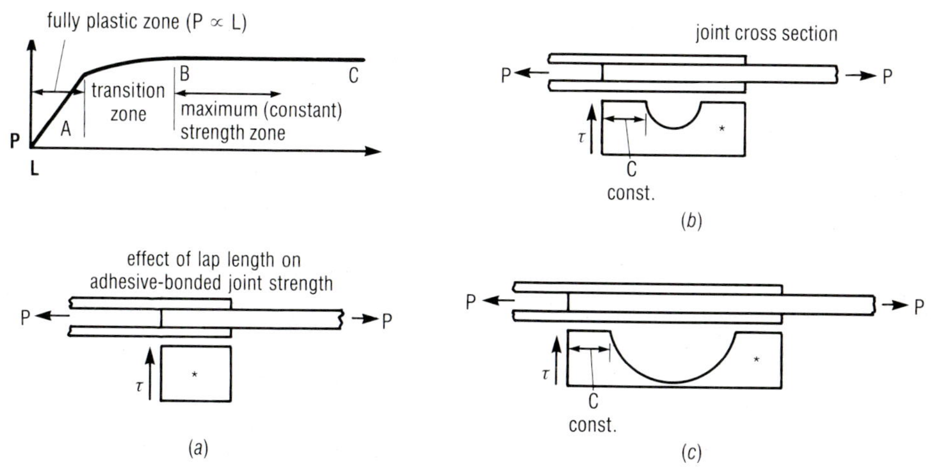

Source Hart-Smith 1986a

Figure 5.5. Because the load transfer zones occur at the ends and eventually reach a constant length, no increase in joint strength will be achieved once these zones are fully developed, at an overlap-to-thickness ratio of about 30:1. There is no advantage to be gained by using a longer joint. It is vital that the joint is proportioned to give a non-uniform stress distribution with a very low stress in the middle of the joint. Without this the joint is likely to fail from creep rupture at low lifetimes (Hart-Smith 1986a).

As already seen, peel stresses are also present in double-lap joints and, although not as severe as in single laps, can lead to failure. Usually the adherend rather than the adhesive will fail, ductile adhesives in particular having a greater tensile strength than current laminates in the through-thickness direction. The peel stresses can be alleviated by tapering the outer laps to a tip thickness of about 0.5 mm (0.02 inch), with a taper of 1 in 10.

5.4.5 Stepped-lap joints

Stepped-lap joints have been extensively used for joining CFRP (up to 25 mm (1 inch) thick) and titanium, subjected to high load intensities (up to 5.25 MN/m (0.3×10^6 lbf/inch) of width). They have found application in many modern American military aircraft, such as the F-14, F-15, F-16 and F-18.

These joints are very sensitive to stiffness imbalance or thermal mismatch between the two adherends. Insufficient attention to these factors can lead to strength reductions of 50% from the ideal, balanced case (Hart-Smith 1986a). The other factors influencing joint strength are the number of steps and the length and thickness of each step. The step length is relatively unimportant since, as for double laps, there is little increase in strength once a certain minimum length is reached. However, the length of the end steps may need to be limited to restrict the amount of the load transferred at these points. Their thickness should also be kept small to minimise peel stresses (Hart-Stress 1986a).

The strength would be maximised by having one step for each ply (0.125 mm (0.005 inch) thick). Such an approach would be viable for the most highly loaded joints. In other cases steps of thickness equal to four plies ($0/90/\pm 45°$ say) and about 12.7 mm (0.5 inch) long, would be more appropriate. One step must be sufficiently long for the deep elastic trough of shear stress to develop thus, as in double laps, preventing creep failure. At each step the ply next to the metal should, if possible, be oriented along the joint (ie at 0°).

5.4.6 Scarf joints

Scarf joints are often regarded as ideal because they are thought to exhibit constant adhesive shear stress along the joint. Such uniformity, however, is not obtained if there is stiffness or thermal expansion imbalance between the adherends. Also the adherends must have knife edges if the ideal state is to be attained. In practice, especially for wide joints, this will not be achievable and the joint must be made with a finite tip thickness, causing strength reductions of up to 25%.

Owing to the non-uniform adhesive shear stress, scarf joints between dissimilar adherends will fail in fatigue at the tip of the stiffer (and stronger) adherend. The same situation can occur with stepped joints. To restrict creep deformation the scarf angle must be kept small, resulting in long joints (Hart-Smith 1986a).

5.5 MECHANICALLY FASTENED JOINTS

5.5.1 Selection of fasteners

Four types of fastener are available: pins, screws, rivets and bolts. Pins are not viable for standard joints, although they may be used for linking (thick) lug joints. Screws, because of the ease with which they can be pulled out of the laminate, offer very low efficiencies and are not recommended for load-carrying joints.

(*a*) *Rivets*: In general, rivets are suitable for joining laminates up to 3 mm (0.12 inch) thick. The basic choice is between solid and hollow types, the latter being needed when only single-sided access is available. The installation process, which involves using a closing force that may not be well controlled, can result in wide variation of joint strength. Care must be taken not to cause excessive damage since this could weaken the joint.

On external surfaces countersunk rivets may be required. The countersink angle should be as large as possible, say 120° or 130°. As a general rule non-countersunk are to be preferred to countersunk rivets, and solid rivets produce stronger joints than hollow rivets.

(*b*) *Bolts*: With bolts the clamp-up force is better controlled than with rivets, resulting in less scatter of test results. The high through-thickness force in the laminate, caused by tightening the bolt, means that such fasteners produce the strongest joints (Godwin and Matthews 1980). There is no limitation to the thickness that can be joined, provided the diameter-to-thickness ratio is kept within certain limits (see below). Bolts with countersunk heads will give weaker joints than will conventional bolts.

5.5.2 Hole preparation

Carbon and glass fibre/epoxy (GFRP) composites can be drilled using adaptations of standard equipment and methods (Collings 1986). The abrasive nature of composites means that high-speed steel tools are not recommended because cutting edges are rapidly blunted, causing increased heating, tearing of the hole surface and delamination of surface plies. Improved hole quality is obtained with tungsten-carbide- or diamond-coated drills. Vacuum extraction should be used to remove the dust generated during drilling.

A major difficulty with laminated materials is drill break-through, causing delamination of the rear surface plies. The problem can be avoided by using a sacrificial backing plate to support the rear surface, or by the use of 'controlled-feed' drills. Also the use of a pilot hole will reduce the force needed to push the point of the drill into the material. Conventional twist drills can be employed if the usual included angle of 120° used for metals is reduced to about 60° for thin laminates and about 100° for thicker parts.

Because of the nature of the fibre, it is difficult to achieve a clean-cut edge when machining Kevlar fibre-reinforced epoxy (KFRP), particularly unidirectional laminates. However, better results are obtained with multi-directional and woven fabric laminates. Also, incorporating one layer of glass fabric in the laminate surface improves the quality of the hole, or other cut edges, as do changes in drill design (Collings 1986).

5.5.3 Lay-up and stacking sequence

Consideration of the stresses around a loaded hole lead us to expect that a joint in a unidirectional laminate will fail at a very low load, when loaded parallel to the fibres, by shear-out. When the fibres are normal to the load we would expect failure to be in tension at the net section.

More practical, multi-directional, laminates behave in a more complex fashion, depending on the proportions of fibres in the various directions. The general trends can be illustrated by considering a $0/\pm 45°$ lay-up. As the proportion of $\pm 45°$ material is increased the shear strength increases until bearing becomes the predominant failure mode. Further increase of this constituent will eventually change the mode of tension, since $\pm 45°$ laminates have a low tensile strength. Typical results are shown in Figure 5.6. Examination of all main failure modes shows that a near-quasi-iso-

Figure 5.6 *Influence of fibre orientation on failure mode of bolted joint in 0/±45° CFRP*

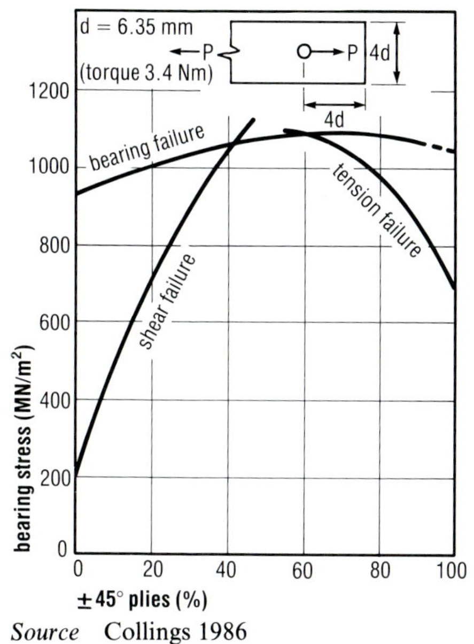

Source Collings 1986

tropic lay-up will give the best performance for a tightened bolt (Collings 1986).

If the plies of a laminate are homogeneously mixed, ie the fibre direction changes from layer to layer, the stacking sequence has little effect on the bearing strength of bolted joints. Pinned joints, and to a lesser extent riveted joints, are affected, however, the former showing a 30% difference between the strongest and weakest for a quasi-isotropic lay-up in GFRP (Quinn and Matthews 1977), but rather less difference in CFRP (Smith 1985).

Bolted joints are, however, significantly weakened if plies of the same orientation are grouped together (a 'blocked' laminate). Reductions in bearing strength of up to 50%, compared to the 'homogeneous' case, have been reported (Collings 1986).

5.5.4 Joint dimensions

Very wide joints will fail in bearing. As the width is reduced the mode will eventually change to tension. The width at which the mode change occurs will depend on the lay-up, ie there is a hole size effect as well as a material effect.

Figure 5.7 *Effect of width on bearing failure stress of single-hole joints in CFRP*

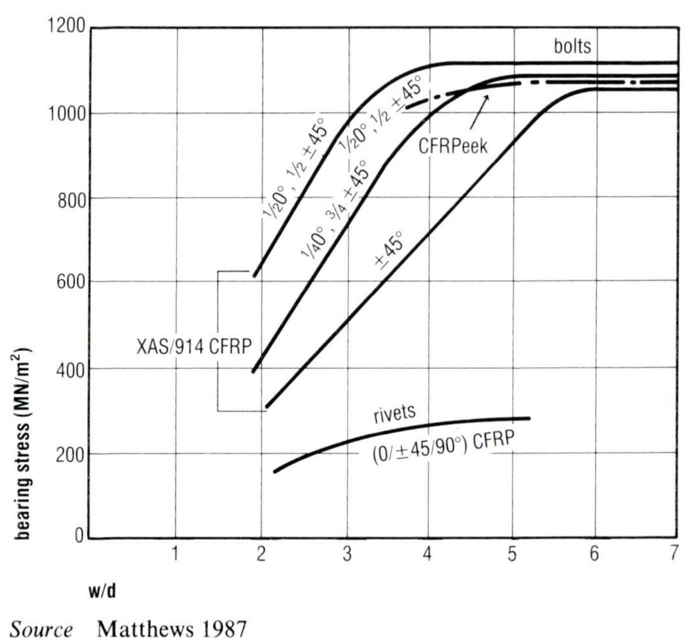

Source Matthews 1987

Figure 5.8 *Effect of width on bearing failure stress of single-hole joints in GFRP and KFRP*

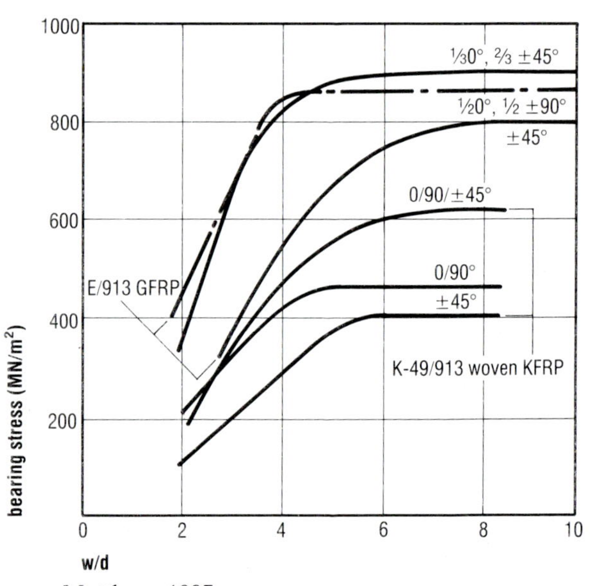

Source Matthews 1987

Changing end distance has the same effect on shear-out failure as width has on tensile failure. Clearly a joint must have adequate end distance if it is to achieve its full bearing strength. As an approximate guide the end distance should be equal to the width at which the failure mode changes from tension to bearing.

Typical data for bearing failure strength (failure load/(diameter × thickness)) are given, for single-hole joints loaded in double shear, in Figures 5.7 and 5.8, where the influence of lay-up is clearly seen. The 'plateau' regions of the curves correspond to bearing failure and the 'knee' to the change in mode.

Although gross tensile strength is reduced as hole diameter increases, for bolted joints in all types of FRP there is a minimal effect of hole size on net tensile and shear strengths. The bearing strength of CFRP is similarly unaffected (Collings 1986). Low-modulus materials such as GFRP and KFRP, however, show a reduction in bearing strength for $d/t > 3$. Nevertheless, if fastener failure is to be avoided d/t should not be too low, values in the range 1–1.5 being a recommended minimum. Even if it does not fail, a small-diameter fastener will undergo severe bending, thus giving rise to non-uniform bearing stress through the thickness and reduced clamping due to head and nut rotation, both effects causing a reduction in strength. For rivets (and pins) hole size effects will be more significant.

5.5.5 Design considerations

Unfortunately there is no universally accepted method for designing a joint to achieve a specified strength. Of the several approaches suggested, the semi-empirical method of Hart-Smith (1986b) of McDonnell-Douglas appears to be the most useful:

(*a*) *Validity of single-hole tests*: For reasons of simplicity and cost, most testing will be carried out on single-hole double-shear joints loaded in tension; much original work in this field was undertaken by Collings, of the Royal Aircraft Establishment, whose findings are described in greater detail elsewhere (Collings 1986). The data obtained, examples of which are given in Figures 5.7 and 5.8, must then be applied in practical, ie multi-hole joints. For a row of holes, ie at right angles to the load, the data from single-hole tests is applicable provided the fasteners are not too close, say greater than $4d$ apart, so width and pitch are equivalent. For a line of holes, ie parallel to the load, the situation is complicated by the presence of the by-pass load, mentioned above.

Available data are also limiting in that they rarely apply to joints loaded in compression or edgewise shear (as at a web/flange junction). Limited information suggests that the compression strength is perhaps 10% higher than that in tension. Shear-loaded joints appear to have similar bearing strengths to those in tension provided pitch and edge distance are greater than $3d$.

(*b*) *Allowance for bolt-clamping:* A further consideration is how to account for bolt-clamping. In service, bolt tension may be reduced due to creep in the laminate and, possibly, vibration of the component. It may, therefore, be prudent to use strength values for finger-tight rather than fully tightened bolts. This would imply dividing the typical data of Figures 5.7 and 5.8 by a factor of 2. Such a reduction seems reasonable since, in addition, damage around the hole is unlikely and the associated bearing strain (about 0.004) is similar to that currently used for general design purposes.

As noted above, if the fasteners have too small a diameter, say $d/t < 1$, the associated bending will reduce the through-thickness constraint. To compensate for this effect in configurations such as double cover butt joints it is recommended that the thickness of the cover plates be increased by 20% over that for a balanced design (Hart-Smith 1986b); the central plate is not affected because it is sandwiched between the covers.

The eccentric nature of single-lap joints will also give rise to fastener rotation, and strength reductions of some 20% can be expected from the double-shear case; the bending of the laminate will cause further weakening. For tensile loading of single-lap joints it is recommended that two, or more, rows of fasteners are used to react the eccentric loading (Hart-Smith 1986b). Note that the joint must be stabilised if the loading is compressive.

(*c*) *Choice of lay-up*: It was seen above that a near-quasi-isotropic lay-up will give optimum joint performance. As a general rule there should never be less than 1/8 nor more than 3/8 of the fibres in any one of the basic directions 0°, +45°, −45° and 90°. The implication of this is that highly orthotropic lay-ups, chosen to satisfy a stiffness requirement, say, will need significant reinforcement at the joints, to achieve the required load transfer, and hence be extremely heavy. Equally, it will be difficult to carry out a cost-effective repair of such a laminate (Hart-Smith 1986b).

(*d*) *Efficiency and multi-bolt joints*: An important factor in Hart-Smith's approach is to present data, such as that in Figure 5.7, in terms of joint efficiency, ie (failure load of joint)/(failure load of laminate). It is apparent

Figure 5.9 *Influence of design on structural efficiency of bolted joints in CFRP*

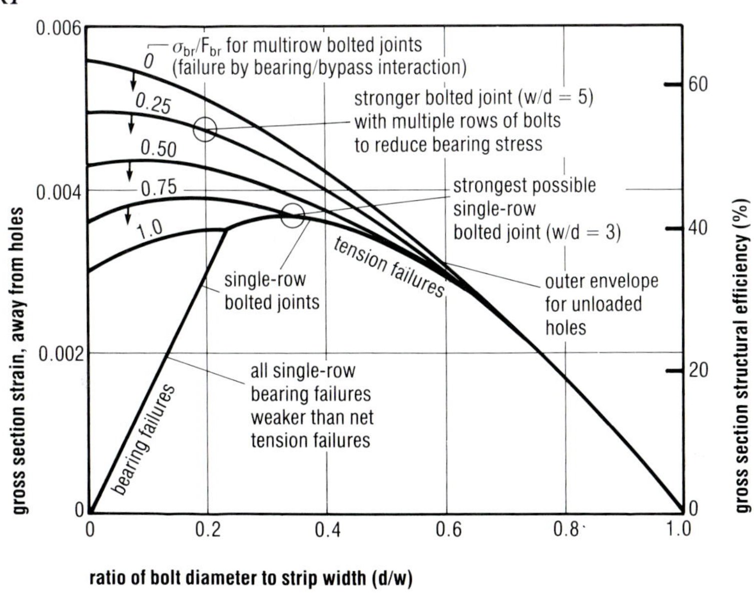

Source Hart-Smith 1986b

that if the results for one of the lay-ups of Figure 5.7 are replotted, a curve similar to the lowest in Figure 5.9 will be obtained. The peak in this curve will correspond either to the point of failure mode change from tension to bearing, the 'knee' of Figure 5.7, or to tensile failure. Clearly, increasing width, or pitch, to ensure bearing failure will reduce the efficiency since the laminate strength will increase whilst the joint strength is unchanged.

When multi-row joints are accounted for the upper curves in Figure 5.9 are obtained. It is obvious that it is difficult to improve upon the optimum single-hole (row) joint. Indeed, a line of two bolts will only be about 10% stronger than the single-bolt case because, although the bearing stress is halved, the stress concentration arising from the by-pass load is nearly as high as that caused by the bearing load (Hart-Smith 1986b).

Significant improvements in efficiency can only be achieved with complicated configurations employing several rows of fasteners, tapering of the butt plates and using different-sized fasteners. The objective of such designs is to reduce the bearing stress, to perhaps 1/4 of its capability, on the fastener where the by-pass load is greatest. It is necessary to employ a stress analysis that will accurately predict the load distribution between the various fasteners (Hart-Smith 1986).

5.6 REPAIR

5.6.1 Introduction

Repair is an obvious application of joining methods. The approach to repair of composites to be described in this section is based on experience in the aircraft sphere (airlines, manufacturers, military), where most work has been done. The philosophy is quite general, however, and could obviously be applied in other fields.

Many types of damage are possible in composite structures. Defects may be introduced during composite manufacture due to errors in manufacturing procedure. Damage may be initiated during the life of the component as a result of exposure to the service environment, by accumulation of minor damage sustained in the normal use of the composite, from occasional exposure to an abnormal environment, or from a local mechanical overload due to misuse. The damage may affect the component in two simple ways. First, the defect may degrade the strength or stiffness of the component to a level below its design limits. A defect of this nature must be detected and the component repaired or replaced. Of more concern are smaller defects which do not immediately degrade mechanical performance but which will grow under service conditions to a point where strength or stiffness is unacceptable.

Most of the service damage experienced with composite structures is due to impact with a foreign object or mishandling. The damage resulting from an impact can range from small surface dents to internal delaminations and through-penetration. Because of the tendency for delaminations to form, it is necessary to inspect the damaged area for such occurrences. One of the most important aspects of a repair action is to establish the area which is defective. Ultrasonics and X-radiography are the most commonly utilised inspection techniques.

The objective of a repair action is to restore structural integrity to a damaged component (Armstrong 1983). In this context the use of composite patches, particularly when using hybrids (eg glass and carbon), to repair metal structures is a powerful technique (Rogers, Kingston-Lee and Phillips 1980). The particular procedure used will depend on the type of component, amount of joint efficiency required and considerations of surface smoothness.

The repair environment will also affect the technique used in repair operations. At a depot facility, major repairs are possible, but at a field position the limits on repairability are much more restricted. The depot is equipped with freezer storage for materials, autoclave capability for processing, and tools designed for repair applications. Also, the level of expertise available in a depot is higher than that available in the field, Thus, parts which are easily detachable from the aircraft can be repaired under ideal autoclave conditions. Manufacturing materials, both prepregs and adhesives, can be used in the repair procedure. The depot is capable of performing all required repairs.

The field repair environment is more severe. No freezer storage may be available, and the materials used should be capable of ambient temperature storage. There is no access to autoclave facilities, and thus repairs must be performed with heating blankets. Compaction and conformational pressure is usually limited to vacuum bag pressure of 1 bar (14.50 lbf/in^2). The personnel and tooling available restrict the repairs to simple bolted patches and external bonded patches, probably using cold-setting adhesives. The procedures of Rogers, Kingston-Lee and Phillips (1980) have proved particularly effective for the emergency repair of metal components.

5.6.2 Bonded-repair techniques

Bonding techniques can be used in situations which range from cosmetic to primary structural repairs. Cosmetic repairs refer to damage which is not structurally significant (eg dents or missing surface plies). The repair is made to restore surface smoothness. In these repairs, a potting compound or a liquid adhesive is spread into the damaged area and formed to the component's contour.

Injection repairs are another type of repair procedure which is used for minor disbonds or delaminations. In this procedure a number of holes are drilled to the depth of the damage. Filler resin is heated to decrease its viscosity and injected under pressure until the excess flows out of adjacent holes. Pressure can be applied to the repaired area to ensure mating of adjacent regions. If serious damage is encountered, then more rigorous repair procedures must be employed. In this case there are two types of bonded patch which can be used to repair structural damage, namely, flush scarf patches and external patches.

After assessing the damage and before deciding upon a repair, the question of the parent laminate moisture condition becomes important. Moisture absorbed in the laminate and/or entrapped moisture in honeycomb can be very detrimental to the integrity of bonded repairs. Examination of cured carbon/epoxy patches bonded to substrates containing moisture, similar to long-term service experience in a high-moisture environment, show a porous bond line if elevated cure ($>100°C$ (212°F)) is used.

This absorbed moisture has had detrimental effects on repairs in the following four ways:

(i) local delamination or blistering in parent laminates
(ii) reduced strength of the repair and repair bond line, resulting from porosity
(iii) expanding moisture in honeycomb cells, creating sufficient pressure to separate the skin from the core
(iv) reduced effectiveness of ultrasonic inspection due to strong signal attenuation, making it difficult to verify bond line integrity.

Pre-bond drying (a minimum of 48 hours at 76–93°C (169–199°F)), slow heat-up rates, reduced cure temperatures, and selection of adhesives less sensitive to moisture can minimise or eliminate the above problems. The 120°C curing adhesives, as a group, are more sensitive to prebond moisture at higher temperatures (above 70°C (158°F)) than 175°C curing adhesives. Drying the parent laminate to an average moisture content of less than 0.5% is recommended. This can be very time-consuming, taking over 24 hours for a 16-ply laminate, as shown in Figure 5.10.

(*a*) *External patches*: The external patch technique is simpler to apply and less critical in nature than a scarf approach. External patch repairs require less preparation than scarf repairs. Limited back-side access or substructure interference in a damaged area would favour the use of an external patch repair.

Figure 5.10 *Influence of temperature on drying time for CFRP laminates*

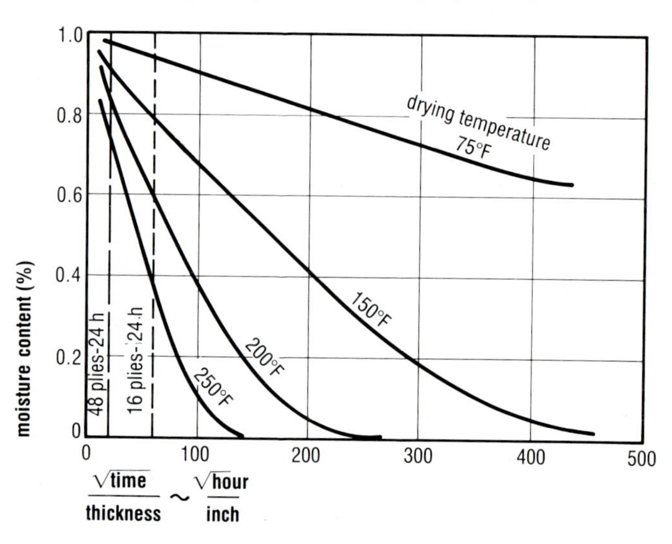

Source AGARD 1984a and b

Figure 5.11 *Effect of detailed design in reducing peel stresses at patch edge*

FM-400 bonded joints (width = 1.00 inch)	failure mode	joint eff
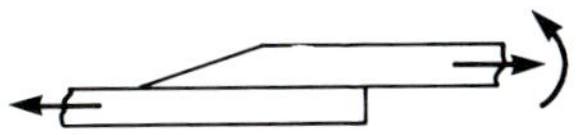 control	peel and shear failure	0.52
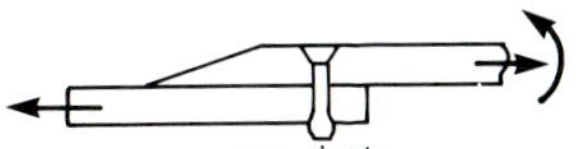 one rivet	shear and patch net tension failure	0.73
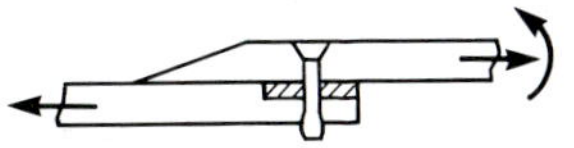 undercut and one rivet	shear and rivet head pull-through failure	0.78

Source AGARD 1984a and b

In this approach the load is taken over and around the damaged area. The bending strains due to the eccentric load path must be considered in the patch design. The patch must also be capable of withstanding the high peel and shear stresses which develop at the edge of the damage. In order to minimise this effect and 'fair in' loads to the substrate, the patch plies are stepped in diameter. The low through-thickness tensile strengths of laminates impose a thickness limitation which means that the technique is restricted to thinner laminates. Peeling can be minimised by small fasteners pitched at 25 mm spacing around the hole and a router cut about 0.4 mm (0.016 inch) deep filled with adhesive and a ply of fibreglass prepreg. This concept, shown in Figure 5.11, on a 16-ply $[(\pm 45/0/90°)_2]_s$ laminate using 3.18 mm (0.125 inch) blind rivets, raised the joint efficiency from 52% to 78% (AGARD 1984a and b).

Details of a typical 22-ply pre-cured patch design are shown in Figure 5.12 (AGARD 1984a and b). An alternative approach is to place the largest diameter patch ply next to the parent laminate. Loads of between 1 and 1.25 kN/mm (5,710–7,137 lbf/in) can be carried by such repairs.

(*b*) *Scarf repairs*: Flush scarf-type bonded repairs are used where surface smoothness is essential. This approach provides the highest joint efficiency of any repair technique. Scarf repairs are used on critical components where load concentration and eccentricities, especially for compressive loading, must be avoided. Thick monolithic structures lend themselves to such repair since an external patch would cause excessive

Figure 5.12 *Details of pre-cured patch design*

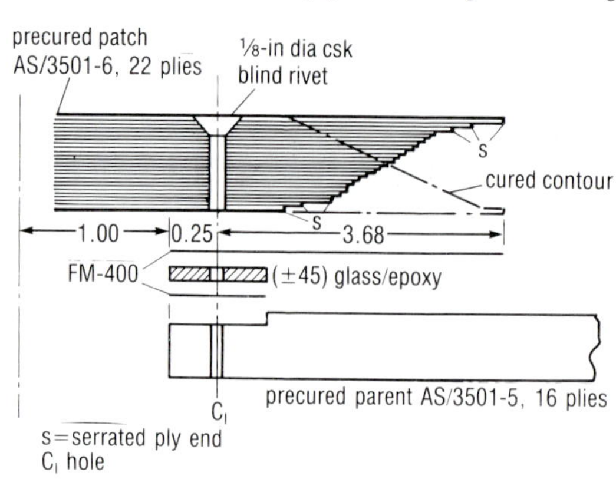

Source AGARD 1984a and b

out-of-mouldline thickness and unacceptably high bond line peel and shear stresses.

The flush repair procedure requires careful preparation of the damaged area to obtain the correct scarf angle and dimensional tolerances. The laminate orientation of the patch must match that of the damaged section which has been removed. Scarf patches can be co-cured (cured on the damaged laminate) or pre-cured (cured and then secondarily bonded on to the damage) and they can either be single or double scarfs as shown in Figure 5.13. The double scarf permits a shorter joint length and requires less material to be removed from the substrate. Taper ratios of 20:1 are typical. Figure 5.14 illustrates a typical repair for a 16-ply laminate (AGARD 1984a and b). This basic scarf joint employs an 18:1 taper ratio since a 2.5 mm (0.10 inch) step is 18 times the nominal ply thickness. The scarfing is accomplished quite readily with the use of a portable router to cut 2.5 mm concentric steps, each successively larger, followed by a port-

Figure 5.13 *Single and double scarf patch design*

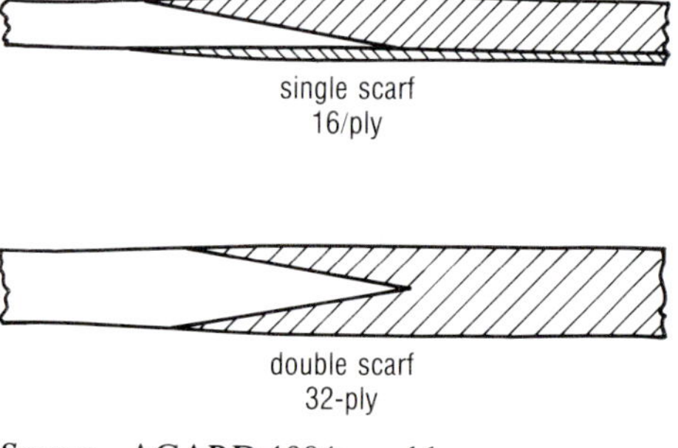

Source AGARD 1984a and b

Figure 5.14 *Typical scarf repair for a 16-ply laminate*

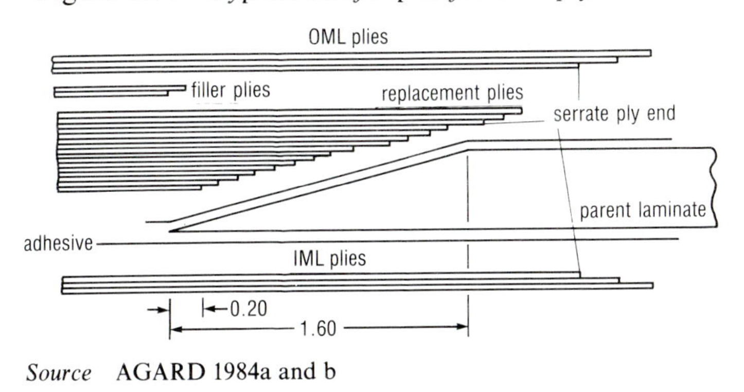

Source AGARD 1984a and b

able power driven sander to provide a finished scarf. A good scarf patch design practice is to extend the outermost plies over the ends of all the other plies and to serrate these plies to minimise ply end peeling. This can easily be done to the edge normal to the fibre direction with a pair of standard 3 mm (0.12 inch) V-notch pinking shears. It is also good practice to avoid placing unidirectional material as the outermost ply. An outside cover of woven material, or for balanced laminate lay-ups consisting of 0°, 90° and ±45° layers, the high-strain low-modulus (±45°) layers, should go on the outside. This makes surface defects such as cuts, scratches and abrasions less strength-critical. One other point to remember is that laminates cured only with vacuum pressure tend towards void contents of about 5%, as compared with less than 1% voids for laminates cured at 100 psi in an autoclave. The higher void content reduces strength properties by approximately 15% for the vacuum-cured material, and this strength reduction should be considered when developing the repair.

Repairs up to 250 × 250 mm (9.84 inch square) in area in 16- and 24-ply laminates have restored 80–100% of the parent laminate strength utilising the techniques described above. This has been verified through a series of repair joint coupons, sandwich beams, flat panels and box beam tests (AGARD 1984a and b).

5.6.3 Bolted-repair techniques

Bolted-patch composite repair is an alternative approach to the bonded-repair concept. Bolted repairs can be used in cases where bonded patch repair of a thick laminate may result in shear stresses beyond the limit of the adhesive strength. This method is also appropriate when a bonded scarf approach would be too complex in terms of preparation and material removal.

Bolted patches can be applied from one side, or from two sides with a backing plate. These backing plates can be inserted through the damage to gain access to the back-side. If the plates are thick, and bolt tolerances are tight, they can also carry load. The patch must be thick enough in some cases to accept flush head fasteners. The patch may have bevelled edges to improve surface conformability.

Another approach to bolted-on repair is a flush-type patch. In this case, the damage is 'cleaned up' and a section is inserted which is now flush with the surrounding undamaged area. A doubler is mandatory with this approach. Fasteners are applied through the patch to the doubler, as well as from the undamaged area into the doubler. One of the problems associated with the flush-type patch is potential difficulty with limited back-side access to install large doublers. The large doublers arise from the high number of fasteners required with such a repair.

As well as composites, aluminium and titanium can be used as patch materials. Aluminium is lighter than titanium and offers less of a weight penalty. It is also, however, significantly electronegative, or anodic, relative to carbon. This leads to possible corrosion problems. Aluminium patches must be physically separated from the underlying composite by scrim cloth or sealant, or both. Titanium will not corrode in the presence of carbon but presents additional difficulty with regard to machinability and formability.

Bolted repairs are not without limitations. The drilling operation can be time-consuming, and improper drilling may introduce additional damage. Also, bolted repairs cannot be used on honeycomb structures. A further limitation may be the parent laminate; certain lay-ups may give too low a bearing strength to enable an effective repair to be made.

5.6.4 Unresolved problems

1 Environmental effects

Environmental degradation due to moisture has to be considered in the case of permanent repairs. A reduction of about 20% in strength is likely both for repair laminates and adhesives. It has to be assured that the environmental degradation of the repair is not greater than that of the parent laminates. The environmental problem will be of great concern when using 120°C (248°F) cure prepreg systems and adhesives, or room-temperature curing systems. For permanent repairs no reduction in serviceability with regard to the maximum design temperature and design life can be allowed. Therefore the repair methods have to be evaluated with regard to the design conditions.

2 NDT techniques for repaired structures

As mentioned above the quality of bondlines and of the repair laminates may be inferior to the materials in the unrepaired structures. Therefore, it is very important to have good control of the quality of the repair patches and the repair bondlines. Unfortunately, the possibilities for non-destructive testing (NDT) inspection on repaired structures are more limited than those after manufacture of structural parts.

In most cases NDT of repaired parts will have to be done *in situ.* Therefore methods like through-transmission and reflector-plate technique are not applicable. The design of structural repairs (tapers, patches etc) is leading to configurations where parent laminate, bondline and repair laminate have to be tested together. This means that the detection of critical defects in the bondlines of repair patches is very complicated and the probability for proper detection is low. Most of the approved NDT techniques which can be used in the field are done by hand. This means that, in practice, the resolution is poorer than that of automated NDT methods. Further detrimental conditions for inspection of repairs can be accessibility and limitations due to safety problems (X-ray).

It is not sufficient to improve repair techniques alone; NDT methods also have to be developed and improved.

3 Repair of joints

The repair of joints can be divided into two areas:

- bonded joints of metal parts to composite structures
- bolted joints.

At present, there is little literature available about these problems. The repair of bonded joints of metallic to composite parts is mainly a problem of the surface treatment of the metallic surface prior to bonding. The surface treatment of metal parts is complicated enough without the repair situation (Allen, Chan and Armstrong 1982).

It has not yet been shown whether the approved surface treatments for bonding of metallic structures are applicable in the repair situation. Some work has been done with respect to 'field' surface treatment of clean aluminium surfaces. For titanium a similar process has not been demonstrated.

The specific problem of bolted joint repair is that of a worn fastener hole. Simply refilling worn holes with resin and redrilling is not adequate for a permanent repair. Worn holes are always an indication that the design of the part is inadequate. The repair problem becomes more difficult as greater strength is required for the bolted joint.

A similar problem occurs for impact damage close to fasteners. It is not desirable that the scarf, which is necessary to remove the impacted materials, extends into the bolted area. For both situations a development and validation of acceptable fastener hole repair is currently needed.

5.7 CONCLUSIONS

It is possible, using currently available data and design methodology, to construct joints with acceptable efficiency using high-performance composites.

Bolts give the strongest mechanically fastened joints. A near quasi-isotropic lay-up gives the highest strength, provided width (or pitch) and end distance are sufficiently large. Grouping of similarly oriented plies in the stacking sequence should be avoided.

Stepped lap or scarf joint configurations give the strongest bonded joints. Double-lap and single-lap joints have similar strength provided the latter is restrained against rotation. Tapering the adherends reduces the peel stresses. In general a ductile adhesive is to be preferred to brittle adhesive, even if the latter is stronger.

Repairs can be effected using knowledge of the behaviour of bonded or bolted joints, although the efficiency of the repair depends critically on the circumstances under which it is carried out. It is clear that requirements for repair should be considered at the outset of design; it is quite possible to select an unrepairable laminate!

NOTES

1 AGARD. Report no. 716. February 1984
2 AGARD. Addendum to Report no 716. August 1984

6 Specifications, Test Methods and Quality Control of Advanced Composites

James M Methven BSc PhD

Lecturer in Polymer Engineering
Department of Mechanical Engineering
UMIST, Manchester

6.1 DEFINITIONS AND NOMENCLATURE

The general class of polymer composites contains those materials with a composition which consists of a polymeric matrix (resin), fibrous reinforcement and particulate filler(s). Both thermoplastics and thermosetting polymeric materials are used as matrices. The general perception of advanced composites is that these materials are a subset of this class wherein particulate fillers are absent and the volume fraction of reinforcement in the form of continuous fibres exceeds about 50%. For the purposes of this book the definition is relaxed to include composites which contain both particulate fillers and reinforcing fibres which may be discontinuous.

A composite artefact can be manufactured by some moulding or conversion operation carried out either directly on precursors (resins, hardeners, fillers, reinforcements) or on some intermediate moulding material (eg preform, prepreg or moulding compound). In the former case the artefact is manufactured in a single step which combines both the mixing of the constituents of the composite and the required shaping of the mixture. In the latter case the mixing operation is first carried out in the manufacture of the intermediate moulding material, and a second, separate, shaping step is employed to produce the finished artefact. This clear disjunction of classes of manufacturing processes is reflected in the observation that the properties of a composite depend in some manner on its mode of production. This is illustrated in the use of different factors of safety ascribed to different processes in BS 4994.

For completeness, the relationships between the various raw materials,

Figure 6.1 *Process and materials classification*

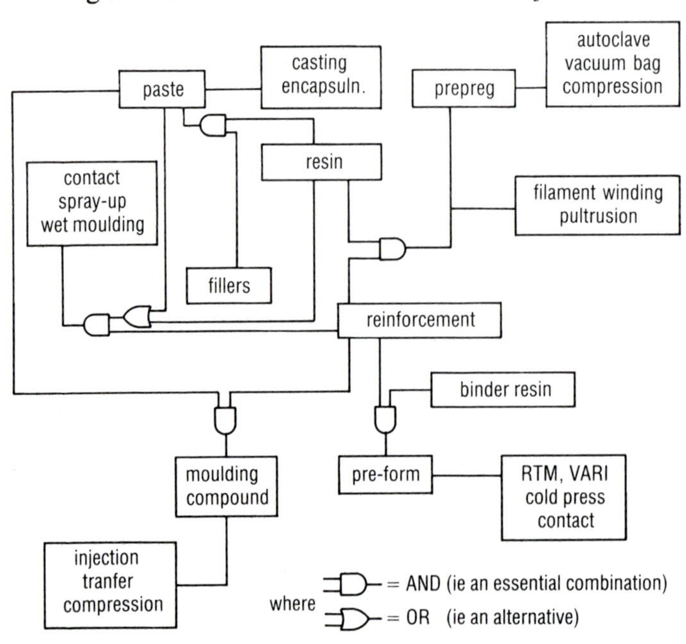

intermediates and processing routes commonly encountered in the manufacture of composites artefacts and intermediates are illustrated in Figure 6.1.

6.2 THE NEED FOR SPECIFICATIONS AND TESTING PROCEDURES

Over the past three decades the growth of polymer composites has been consistently higher than that of plastics in general (Frost and Sullivan 1986). Composites based on carbon/epoxy, aramid/epoxy and carbon/vinyl ester are commonly used in the aerospace industry while composites based on glass/polyester, glass/vinyl ester and glass/thermoplastics find widespread use in both the automotive and engineering industries.

This rapid growth is attributable to at least two factors. The first factor is the ease with which individual composites can be tailored to provide unique properties or property combinations such as high specific strength and stiffness, retention of mechanical properties at elevated temperature, corrosion resistance etc. The second factor is the freedom with which polymeric materials in general can be designed. With this comes the facility for parts consolidation and the reduced cost of materials, tooling and manufacture.

The objectives of any design exercise are to define the dimensions of an artefact and the materials from which it is made, so that it can perform some task acceptably. They may also include proposals on methods of manufacture. The design process generally starts by dividing the overall task into a set of simpler functions. It proceeds by a process of iteration, perhaps focusing on one or two functions during each cycle until converging to an acceptable solution.

During this process it is desirable to express each function in terms of constraints such as stress, temperature and time, and it is this set of constraints which forms the basis of a formal specification. The functions themselves subsume some measure of reliability, and this is expressed in terms of the confidence with which limits of the constraints may be expressed. It follows that, in order to comply with any specification, there must be available some knowledge of not only the properties of materials but also the degree of precision of the values which these are ascribed. This requires procedures for both measurement of such properties and the statistical interpretation of the results. Such procedures are embodied in formal standards. They are also fundamental to the routine monitoring of product quality, which is described in Section 6.5.

Since standard procedures are focused on the measurement of properties it is constructive to recall that the properties of a composite are given by a more or less complex function of the properties and concentrations of its constituents (Ashton, Halpin and Petit 1969, Nicolais 1975, Chamis 1980, and Kardos *et al.* 1983). Ideally this relationship must take account of the nature of the interface between the matrix and the other constituents, the anisotropy of the structure, the presence of any adventitious materials such as voids, moisture etc, and the magnitude and direction of any residual stresses. All of these features may be influenced by the process of manufacture of the composite, and it is useful to consider this in greater detail before describing formal test procedures.

6.3 COMPOSITE MANUFACTURE

6.3.1 General considerations

It is reasonable to expect that all the processing routes of Figure 6.1 involve a sequence of operations including heating, melting, mixing, metering, pumping, cooling, crystallisation and cross-linking. It is convenient to distinguish between operations which apply to thermoplastics and to thermosets.

Melt processing is primarily concerned with the manipulation of a (thermoplastic) melt and, as such, involves the application of some of these operations to non-Newtonian fluids at relatively high (>200°C (392°F)) temperatures. Melt processing is essentially a reshaping operation which involves the reversible transformation of starting materials (granules, sheet etc) to a finished product by the application of heat and pressure.

Thermoset processing, or, more generally, reactive processing, by contrast, is concerned primarily with the chemical transformation of essentially linear precursor(s), which contain a greater or lesser number of functional groups, into a cross-linked three-dimensional network. The transformation may be described in the simple form

$$\underset{(P)}{\text{precursors}} \Rightarrow \underset{(N)}{\text{network}} \tag{6.1}$$

Typically (P) is a combination of more or less viscous liquids at the appropriate processing temperature while (N) is an infusible solid confined to a particular shape by some sort of mould or die. For example, (P) might be a polyurethane system and (N) a shoe-sole or a car bumper, or (P) might be a carbon fibre prepreg and (N) a helicopter main rotor blade. (P) might even be paint and (N) a painted wall. Irrespective of the nature of the materials, the transformation $P \Rightarrow N$ is irreversible.

There is a perception that the reversibility of melt processing makes it more tolerant of errors than reactive processing. An appropriate analogy is that of rearranging hydrogen bonds in a shirt by ironing – a bent crease does not imply a new shirt. Reactive processing, on the other hand, is more akin to frying an egg – any problems need another egg. With advanced thermoplastics composites, particularly those containing continuous fibres, this tolerance is considerably reduced. While these materials can be reprocessed this is rarely possible by the simple expedient of repeating the original manufacturing operations. It most commonly involves granulation of the reject artefact to a moulding granule or powder which can then be converted into a (different) artefact by injection-moulding. This is an extremely expensive way in which to manufacture a moulding compound.

In general, errors in manufacturing are reduced only as an understanding of the different processes involved is developed.

6.3.2 Chemorheology

Chemorheology (Aronhime and Gillham 1986, Han and Han 1983, and Apicella 1986) is concerned with the interplay of transient heat transfer,

Figure 6.2 *Features of chemorheology*

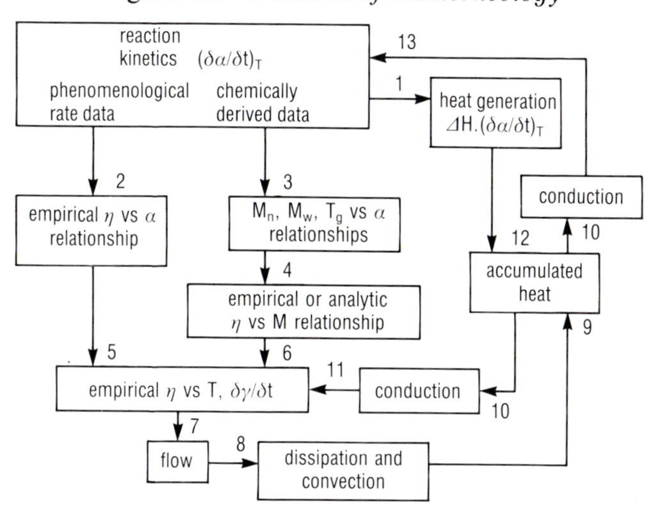

chemical kinetics and chemical energetics on the rheological behaviour of a changing system. The relationships between these parameters is shown schematically in Figure 6.2 for a reacting system. An analogous scheme may be described for a melt system where the reaction kinetics of Figure 6.2 are replaced by a function relating to the rate of crystallisation of the polymer or some rate of change of specific heat capacity.

Figure 6.2 also illustrates manufacturing sources of property variability. Matrix flow (links 2–11) can affect fibre wetting, distribution, alignment and orientation. The cross-linking reaction and rate of heat transfer (links 1, 9–11, 13) dictate the through-thickness development of network modulus and temperature, and through these the development of residual stresses.

Explicit relationships for each of the 13 links shown in Figure 6.2 allow any process to be modelled by solving the appropriate conservation equations for mass, momentum and energy. Since these equations are invariably coupled through time, reaction extent and temperature the solution must be derived numerically. Various illustrations of this approach have been published for both thermoplastics (Pittman 1986) and thermosetting systems (Han, Lee and Chin 1985, Ma *et al.* 1987, Broyer and Macosko 1978, Silva-Nieto, Fisher and Birley 1985, Loos and Springer 1983, and Batch and Macosko 1987).

These analyses are extremely complex and time-consuming exercises, and can be fraught, not only with problems of numerical stability but also with problems of quantifying physical parameters which are difficult to measure accurately and are certainly difficult to measure with an acceptable degree of precision (Loos and Springer 1983, and Batch and Macosko 1987). It is in these areas that the empirical test procedure can make a valuable contribution. Providing these are designed with due respect to the meaning of the links of Figure 6.2, it is possible to combine the substance of various links into a reduced number of lumped parameters. This approach has been applied to thermoplastics processing with some success (Parnaby *et al.* 1981).

Many standard test methods have been developed in this fashion; for example, information on reaction kinetics and energetics (links 1, 9, 12 and 13) may be obtained by measuring the gel time and peak temperature rise (exotherm) of a prescribed sample under prescribed conditions. By the same token the flow behaviour of a material (links 2, 5, 7, 8 and 10) may be determined by measuring directly the flow length and/or flow time at a prescribed temperature and pressure.

However convenient, lumped parameters should be recognised for what they are and not be too closely identified by name with only one particular contributing factor.

6.4 STANDARD TEST METHODS

6.4.1 Perspectives

It is clear from the foregoing that in order to assess the quality of a composite it is necessary to characterise both its constituents and its composition before and after moulding. Accordingly, standard test methods have been developed to measure

- the quality of precursors and/or intermediates
- the homogeneity of the composition
- the shaping processes
- the consolidation processes
- the cross-linking and/or cooling processes

These are described in the remainder of this section. No attempt has been made to provide a complete concordance of standards from different countries, although derivative procedures and those which differ significantly are noted.

A comprehensive guide to quality testing, which incorporates routine monitoring and assessment and procedures for tracing materials, is described in BS 5750.

In general terms, all quality control measurements, particularly those based on lumped parameters, should not be used out of context for the purposes of design. Deviations from formal standards should be reserved for internal use where information can be correlated empirically with a pass/fail classification or used to 'fingerprint' a material, perhaps at various stages of manufacture.

6.4.2 Reinforcements

1 Glass fibres

The commonest reinforcement used in composite construction is glass. Various grades of glass which differ in both the nature and content of metal oxides are commercially available (Bosshard and Schlumph 1987). The most common grades are designated by the letters A, C, E, R and S, and the latter three form the basis of composites reinforcement. The properties of these materials are shown in Table 6.1 together with the properties of other high-performance fibres used in the manufacture of composites.

Table 6.1 Physical and mechanical properties of reinforcing fibres

Type	Density (kg m^{-3})	Failure elongation (%)	Tensile modulus (GPa)	Tensile strength (MPa)
E-glass	2,540	2.5	70	2,000–3,500
S-glass	2,490	2.8	86	4,570
R-glass	2,758		85	4137
[1]Aramid	1,440	3.3–3.7	63–67	3,000–3,150
[2]Aramid (HM)	1,450	2.0–3.0	120–130	2,700–4,000
[3]Carbon (XA)	1,800	1.5–2.0	230–240	3,400–4,500
[3]Carbon (IM)	1,750	1.1–1.5	290–300	3,400
[4]Carbon (IM)	1,770	1.8	305	5,500
[3]Carbon (HM)	1,860	0.7–0.8	360–405	2,700–3,100
[4]Carbon (HS)	1,800	2.0	270	5,500
[5]Carbon (HM)	2,180	0.27	827	2,200
[6]Silicon carbide	3,190		400–700	3,000–14,000
[7]Alumina	3,300		300	1,000
Boron	2,630	0.9	400	2,100

The various forms of textile glass used in the construction of composites are described and classified in ISO 6555:1980, ISO 2078:1985 and ISO 2559:1974. US nomenclature is described in ASTM D3878. The basic reinforcing element is a monofilament which is commonly between 9 μm and 23 μm (0.35 and 0.91 $\times 10^{-3}$ inch) in diameter. Filaments are combined without twisting to form strands or can be twisted into yarns.

Strands, in turn, may be chopped or combined to form rovings, while yarns are used to manufacture woven, knitted or braided fabrics.

Strands, rovings and yarns are measured by their linear density (count), which in Europe is expressed in units of grams per kilometre (tex). An alternative measure is the fibre yield or length per unit mass. In the USA, yield is also known as yardage (yards per pound), while in Europe the yield is expressed in metres per kilogram. With count expressed in tex (g/km), yardage = 2.01 × yield (m/kg), yield (m/kg) = 10^6/count (tex) and hence yardage × count (tex) ≏ 50,000.

Individual forms of glass reinforcement are specified in ISO 2113:1980 (woven fabrics), ISO 2797:1986 (rovings), ISO 2559:1980 (mats) and ISO 3598:1986 (yarns). Glass reinforcement is also subject to MIL specification R-60346. Some aerospace specifications are given in DIN 65060 (rovings) and DIN 65066 (E-glass fabrics).

Sampling procedures for batch testing of glass reinforcements are described in ISO 1886:1980 and procedures for measuring the mechanical properties of individual reinforcing elements are described in ASTM D2256 (unimpregnated yarn) and ASTM D2343 (impregnated strand and yarn).

Test methods for the form and style of glass reinforcements are described in ISO 4602:1978, and ISO 1888:1974, and methods of measuring the physical properties are described in ISO 4603:1978, ISO 4605:1978, ISO 5025:1978, ISO 3374:1980, ISO 1889:1975, ISO 3616:1977 and ISO 1889:1975. The determination of mechanical properties is described in ISO 4604:1978, ISO 5025:1978, ISO 3375:1975, ISO 3616:1977, ISO 3342:1974 and ISO 3341:1984. The form of reporting the properties of glass reinforcements is described in ASTM D3544.

All glass reinforcements contain a greater or lesser quantity of organic material in the form of a binder (size) and/or surface treatment (coupling agent). The size protects the glass during manufacture and subsequent textile processing, while the coupling agent provides some chemical attachment to the matrix. The total level of this organic (combustible) matter may be determined using the methods of ISO 1887:1980.

The nature, level and distribution of the binder is also of considerable importance in determining the ease of processing and the mechanical properties of composite derived from the reinforcement. For example, binders which are readily soluble in a matrix facilitate exposure of individual reinforcing elements (strands, filaments or yarns), providing good wet-out. At the same time the integrity of the reinforcement is considerably reduced and it becomes difficult to handle. Relatively insoluble binders allow rapid penetration of the matrix (wet-through) and may even preserve the original fibre or fabric structure beyond gelation. Only ISO 2558:1974 describes a procedure for determining the rate of binder dissolution (in styrene).

Specifications for surface-treated woven fabrics for use with epoxy resins are described in ASTM D2408 (amino silane) and ASTM D3098 (epoxy silane). Analogous specifications for woven fabrics for unsaturated polyesters are described in ASTM D2409 (vinyl silane) and D2410 (chrome complexes).

Reinforcement for use with unsaturated polyesters is specified in BS 3496 (chopped strand mats), BS 3691 (rovings), BS 3749 (woven roving fabrics) and BS 3779 (woven tapes). The specifications in BS 3691 are also applicable to epoxy resins.

Manufacture of composites is influenced by the drape of the reinforcing fabric, its compressibility (Gutowski, Morigaki and Cai 1987) and permeability (Springer 1986). Drape may be deduced from procedures described in DIN 54306, ISO 4604:1978 and ISO 3375:1975, while the compressibility expressed in terms of thickness under load and recovery is described in ISO 3616:1977. A more direct method for measuring compressibility has been described for pultrusion (Quinn 1987). No formal procedures for measuring permeability have been published.

It is becoming more common for manufacturers to provide standard labels on packages of glass reinforcement so that the contents can be readily identified. For example, a roving package may be described as

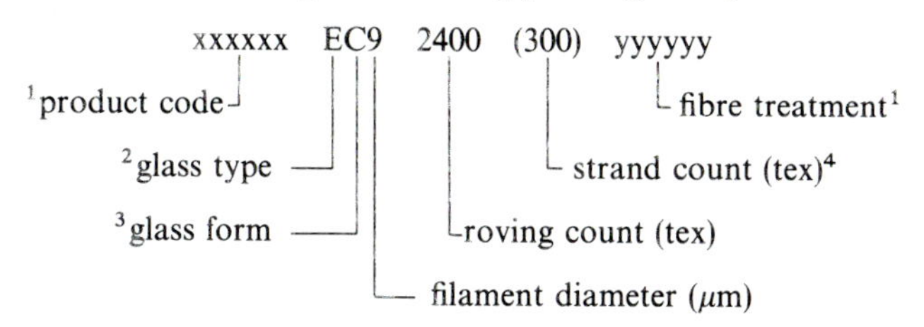

1 Unique to each manufacturer
2 Typically E, S, R or D
3 Continuous filaments (no twist)
4 It is not essential for this information to be provided. The ratio of the roving count to the strand count is sometimes known as the split

2 *Carbon fibres*

Carbon fibres are manufactured by controlled pyrolysis and cyclisation of certain organic fibre precursors (see, for example, Bacon and Moses 1986). The most common precursors are polyacrylonitrile (PAN) and mesophase pitch (MPP). Carbon fibres are classifed according to their mechanical properties (Table 6.1) as type I (high-modulus), type II (high-strength) or type III (intermediate-modulus). These grades are also known as HM, XA and IM. XA fibres are further classified as HP (high-performance) and HS (high-strain). All the latter designations are invariably followed by the letter S, signifying that the fibre has been surface-treated (eg HM-S).

Carbon filaments are typically between 5 μm and 8 μm (0.20 and 0.31×10^{-3} inch) in diameter and are combined into tows containing between 3,000 and 12,000 filaments. A typical tow count is thus between 200 tex and 900 tex. These tows can be twisted into yarns and woven into fabrics analogous to those described for glass. Combined fabrics containing glass and carbon (hybrid fabrics) are also commercially available.

A specification for carbon filaments is given in DIN 29965, and test methods based on those described in ISO 1883:1977 are given in DIN 61584 and LN 29964 for determining the dimensions and surface densities of carbon yarns and fabrics. A method for determining the thermo-oxidative resistance of carbon fibres is described in ASTM D3552.

Procedures for measuring the mechanical properties of carbon fibres and yarns are described in ASTM D4018. More extensive information is available from manufacturers whose testing procedures invariably have the approbation of interested parties; for example, the Courtauld's (Grafil) test method for measurement of tensile strength (103.25) is in conformance with EN 3075 described by the Association Européene de Constructeurs de Material Aerospatiale (AECMA).

Various documents described as 'Werkstoffehandbuch der Deutschen Luftfahrt' and listed with DIN standards apply to carbon reinforcements. These include WL8.3610-WL8.3614, WL8.3621–WL8.3623, WL8.3632, WL8.3642, WL8.3651–WL8.3653, WL8.3659, WL8.3661–WL8.3663, WL8.3672 and WL8.3673.

Carbon fibre packages are commonly described by a standard code such as

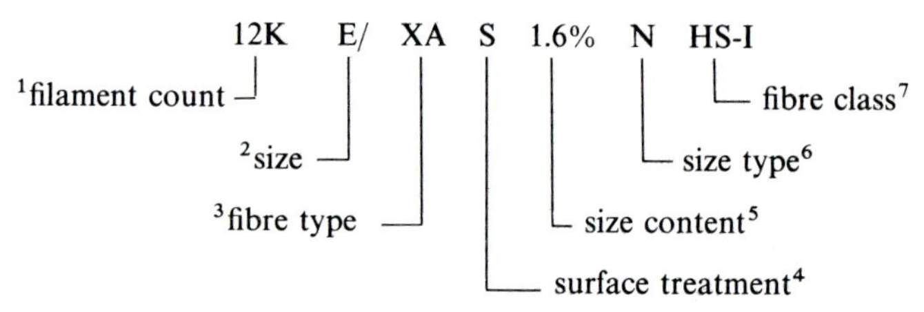

1 Thousands of filaments
2 Blank indicates no size
3 See text and Table 6.1
4 S = surface treated, U = untreated
5 Percent by weight
6 N = DGEBA resin type (Section 6.4.3 3), A = flexible resin type 823 = Shell Epikote
7 See text

3 Aramid fibres

Fibres which are based on aromatic polyamides, where at least 85% of the amide groups are connected directly to an aromatic ring, are known generically as aramid fibres. Aramids may also include partially imidised structures. World-wide there are perhaps ten companies manufacturing aramid fibres, but the most common materials in the USA and Europe are manufactured by Du Pont as Nomex and Kevlar. The latter material is used as a reinforcement both for tyres (Kevlar 29) and composites (Kevlar 49), and the properties of both are shown in Tables 6.1 and 6.2. The discussion which follows applies exclusively to Kevlar 49.

Table 6.2 Thermal properties of reinforcing fibres

Material	[8]Expansivity (μ/K)	Specific heat capacity ($Jkg^{-1} K^{-1}$)	[8]Thermal conductivity ($Wm^{-2} K^{-1}$)
E-glass	5	840	1.3
S-glass	2.9		
R-glass			
[3]Carbon (XA)	−0.26 (26)	710	24
[3]Carbon (HM)	−0.13 (25)	710	105
[2]Aramid	−2 (59)	1,420	0.53

1 Kevlar 29 (Du Pont)
2 Kevlar 49 (Du Pont)
3 PAN precursor (Grafil – Courtaulds)
4 PAN precursor (Apollo – Courtaulds)
5 Meophase pitch precursor (MPP)
6 TOKAMAX (Tokai carbon)
7 Mitsubishi carbon
8 Figures in parentheses are measured perpendicular to the fibre axis

Kevlar 49 is made by condensation of terephthaloyl chloride and *p*-phenylene diamine in a mixture of solvents such as hexamethyl phosphoric triamide and *N*-methyl pyrrolidone. If the molecular weight of the condensation product (poly *p*-phenyleneterephthalamide) is sufficiently high (Elias and Vohwinkel 1986), the polymer solution becomes anisotropic, with characteristics of a lyotropic liquid crystal. The anisotropy not only orients the polymer molecules prior to spinning into fibres but also reduces the viscosity of the melt to facilitate the actual spinning process. Subsequent hot-drawing of the spun fibres in the absence of air causes extensive alignment of the polymer chains, resulting in a polymer which is almost completely crystalline.

Kevlar reinforcement is supplied as rovings or as yarn for weaving into fabric. Yarn is available from about 20 tex to 300 tex, while rovings are around 500 tex to 1,000 tex.

Specifications for aramid reinforcements are described in ASTM D3317 (yarns and rovings), DIN V65427/2 (yarn-based fabrics) and DIN 65356/2 (yarns). Note that ASTM D3317 and ASTM D3318 have recently been withdrawn. Test procedures for measuring the physical properties and surface density of aramid reinforcements are described in DIN 654271/1 and DIN 65356/1. Aramid fibres are also covered by the 'Werkstoffehandbuch der Deutschen Luftfahrt' under WL8.2600, WL8.2610–WL8.2615 and WL8.2620–WL8.2627.

4 Boron filaments
Boron filaments are specified in LN 29974 and test methods provided in LN 29966.

6.4.3 Resin matrices

1 General features of thermosetting resins
Typical thermoset precursors (resin systems) consist of mixtures of species containing differing functionalities. A suitable general case involves the intermolecular reaction of x moles of an f-functional species A_f with y moles of a g-functional species B_g. The reaction may be written

$$x[A_f] + y[B_g] \Rightarrow \text{products} \tag{6.2}$$

The stoichiometric ratio of the reactants r is given by $fx[A_f]/yg[B_g]$, where the symbols in square brackets are the reactant concentrations. It can be shown (Miller and Macosko 1976) that the average molecular weight M_w of the system diverges when

$$\alpha_A \alpha_B = 1/(f-1)(g-1) \tag{6.3}$$

where α_A and α_B are the (fractional) conversions of the functional groups on A and B and $\alpha_B = r\alpha_A$. For exact stoichiometry $r = 1$ and $\alpha_A = \alpha_B = \alpha$; hence

$$\alpha^2 = 1/(f-1)(g-1) \tag{6.4}$$

The point at which M_w diverges is called the gel point. Inspection of equations 6.3 and 6.4 shows that for gelation to occur at least one of the species involved must have a functionality greater than 2.

If the kinetics of the reaction of equation 6.1 are known, then the time taken to reach gelation (gel time) can be found explicitly. Since flow of material ceases at the gel point, the gel time may be taken as the maximum time allowable for mould filling (shaping) and consolidation. Further analysis of the development of the cross-linked network after gelation results in an albeit idealised relationship between conversion and the glass transition temperature T_g of the system (Aronhime and Gillham 1984). It is convenient to define a vitrification conversion as that conversion at which the network T_g reaches the mould temperature. This coincides approximately with the network having sufficient mechanical integrity to facilitate demoulding and handling. The corresponding vitrification time is exactly analogous to the time taken for a thermoplastics melt to reach its (constant) T_g, at which point it, too, can be demoulded.

Gelation and vitrification are characteristic features of any moulding cycle, and as such provide valuable information about the efficiency of the process. While these parameters are extremely difficult to isolate from the other factors present in a real moulding operation (Figure 6.2), suitable lumped parameters are the subject of various empirical test methods, and in essence these form the core of the process-related standards for thermosetting systems.

Thermosetting resins release between 200 J/g and 450 J/g (1,354 and 3,047 cal/oz) and shrink by between 1% and 15% on cross-linking. Both features are composition- and structure-dependent. The rate of energy release and the rate of change of volume are dependent on the kinetics of the cross-linking reactions. In order to minimise residual stresses in a composite it is essential to maintain uniformity of temperature and matrix flow before and during cross-linking. Since the thermal diffusivity of composites is small (deduced from Table 6.2) this is possible only when both the imposed rate of change of material temperature and the rate of heat generated by reaction are low. These are characteristic features of, for example, the autoclave moulding of epoxy prepregs.

The rate of cross-linking of thermosetting resins may be measured by differential scanning calorimetry (ASTM E698), from which results it is possible to obtain at least phenomonological rate constants. The analysis procedure of E698 represents a gross oversimplification of the cross-linking reactions which take place in both unsaturated polyesters (Stevenson 1986) and epoxies (Lee, Loos and Springer 1982), and is best used solely as a complement to the DSC fingerprint for a specific material. Rate data based on specific reaction mechanisms can be obtained from DSC determinations, although the procedures involved are quite complex (Nixon and Hutchinson 1985, and Ryan 1984) and are, as yet, not standardised. It is also possible to use adiabatic measurements to obtain data on both the reaction rate and the reaction energy (Richter and Macosko 1978).

Thermosetting resins are classified in DIN 16946/2, and general methods of test are described in ISO R75:1974 and DIN 16946/1. Standard procedures for measuring the mechanical properties of cross-linked matrices specified in the latter are in essence the same as those which apply to composites. These are described in Section 6.4.5 rather than in this Section.

2 Unsaturated polyesters

These materials are condensation products of a stoichiometric mixture of glycols and a mixture of saturated and unsaturated dibasic acids. The product of this condensation reaction (alkyd) is dissolved in an unsaturated solvent (monomer) such as styrene or methyl methacrylate and the resulting resin solution is cross-linked by free-radical copolymerisation in the presence of organic peroxides. Closely related materials are vinyl esters (Anderson and Messick 1980) and oligourethane acrylates (Sayers, Howard and Holland 1986).

General testing and analysis procedures for alkyd and polyester resins are described in ASTM D2689. The identification of specific materials is covered in ASTM D1306 (phthalic acid – gravimetric determination), D1307 (phthalic acid – spectroscopic determination), D2455 (carboxylic acids), D2456 and D2998 (both for polyhydric alcohols), and D2690 (isophthalic acid).

The molecular weight of an unsaturated polyester is measured either by viscometry (ASTM D1824, ISO 2555:1974 – Brookfield, ISO 2884:1974 – cone and plate) or by titration with alcoholic potassium hydroxide. An idealised alkyd has one carboxylic acid group and one hydroxyl group at each end of the molecule, and determination of either gives the average molecular weight directly. The hydroxy content (value) is found by acetylation of the –OH groups and titration with alcoholic KOH (ISO 2554:1974).

The acid value (ISO 2114:1974) is found by direct titration and gives the number of milligrams of KOH required to react with the COOH groups per gram of resin. Assuming one acid group per molecule, the molecular weight of the polyester is numerically 56,100/A, where A is the acid value. An alkyd based on one mole of fumaric, phthalic acids and two moles of propylene glycol has a repeat unit molecular weight of 362, giving for an acid value of 20 mg (0.31 grain) KOH a molecular weight of 2,800 and about 7 double bonds per molecule. This is equivalent to a functionality of 14 (Equations 6.3 and 6.4).

Characteristics of the cross-linking reaction such as the time to reach gelation can be determined directly under prescribed conditions of temperature and catalyst level (ISO 2535:1974). A measure of both the energy of the cross-linking reaction as well as its rate can also be measured directly from the temperature rise of a prescribed sample (ISO R584:1982, also known as the 'SPI gel test').

The volume shrinkage of an unsaturated polyester depends on the (homopolymerisation) shrinkage of the reactive solvent (monomer), the solvent concentration and the level of unsaturation in the main chain. The volume shrinkage of a styrene-based resin is between 8% and 10%, while that of a resin based on methyl methacrylate is between 12% and 15%. These reflect the shrinkage of the respective monomers (15% and 21% – see Richter and Macosko (1978)). Volume shrinkage can be determined directly by dilatometry or more commonly by geometry (ISO 3521:1976).

Cross-linked castings from unsaturated polyesters are classified (ISO 3672:1979) by tensile strength (three classes from less than 40 MPa (5,800 lbf/in^2) to more than 70 MPa (10,150 lbf/in^2)) and temperature of deflection under load (nine classes from less than 40°C (104°F) to more than 180°C (356°F)). Other classification schemes based on the same and differing criteria are described in DIN 16946 part II and BS 3532.

3 Epoxy resins

Epoxy resins contain at least two oxirane (epoxy) or glycidyl groups within their main chain. Oxirane groups may be incorporated directly into alicyclic, fused aromatic and linear structures, while glycidyl groups are commonly pendant or terminal to a linear backbone, which may itself contain aromatic, aliphatic or alicyclic moieties:

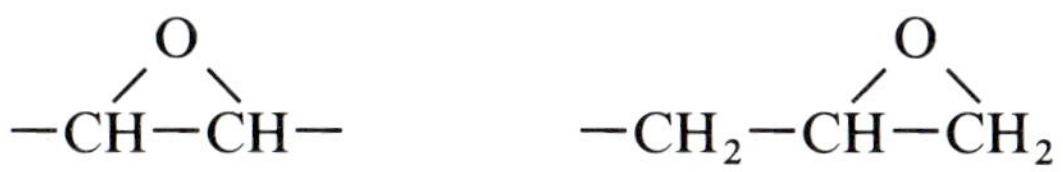

$$\underset{\text{oxirane group}}{-\overset{\diagup^{\text{O}}\diagdown}{\text{CH}-\text{CH}}-} \qquad \underset{\text{glycidyl group}}{-\text{CH}_2-\overset{\diagup^{\text{O}}\diagdown}{\text{CH}-\text{CH}_2}}$$

Epoxy resins may be classified according to the procedures of ASTM D1763.

In this scheme the resins are grouped into six types (I–VI) according to their structure:

Type I	diglycidyl ether (dian resins)
Type II	cresol novolac derivative
Type III	aliphatic diglycidyl ether
Type IV	diglycidyl amine derivative
Type V	cycloaliphatic diglycidyl ester
Type VI	cycloaliphatic oxirane derivatives

The types are subdivided according to the presence (grade 1) or absence (grade 2) of reactive diluents (such as a monofunctional epoxy), and each is then further subdivided into classes (A, B, C ...) according to their epoxy-equivalent weight (EEW – see below).

ISO 3673 Part 1:1980 describes a classification scheme using up to ten criteria which, in addition to those above, include the resin viscosity, density and the presence of additives. Resins may be formally described by a sequence of digits corresponding to each class.

The epoxy (oxirane) group in an un-cross-linked resin can be determined quantitatively by titration in dioxane or for a cross-linked material by acetylation using perchloric acid. Procedures are described in ISO 3001:1978, ASTM D1652 and DIN 16945. The epoxy equivalent weight (EEW) of a resin is defined as the weight of resin (in grams) containing one mole of epoxy groups. The epoxy value (EV) of a resin is the number of moles of epoxy groups contained in 100 g resin.

For an epoxy resin of molecular weight M and functionality (number of epoxy groups per molecule of resin – see equation 6.4) f, it follows that $EEW = M/f$ and $EV = 100/EEW$.

Epoxy resins are cross-linked by the addition of materials which contain active hydrogens. In the context of epoxy resins, materials in this class are generally known as hardeners. These include polyfunctional primary and secondary amines, amides and anhydrides. Up to 10 PHR (parts per hundred of resin) of tertiary amines and up to 5 PHR of certain amine complexes of boron trifluoride accelerate the (cross-linking) reactions between epoxy resins and anhydrides and epoxy resins and primary and secondary amines by a mechanism involving proton transfer.

A classification scheme for hardeners and accelerators analogous to that described for epoxy resins is given in ISO 4579 Part 1:1983. Amine hardeners are characterised by their H-active equivalent weight (AHE) or amine value (AV). AHE is the quotient of the molecular weight of the hardener and the number of active hydrogens per mole. AV is measured by reaction with hydrochloric acid or perchloric acid followed by back tritration with KOH. AV is expressed as the number of milligrams KOH per gram of hardener. For an amine of molecular weight M, having a functionality (number of replaceable hydrogens per molecule) f it follows that $AHE = M/f$. Anhydrides are classified according to the anhydride equivalent weight (AEW), which is the quantity (grams) containing 1 mole of anhydride. For exact stoichiometry (Equations 6.3 and 6.4) the weight of hardener required for 100 g (1,543.24 grain) of resin is given by $EV \times AHE$ (for an amine) and $EV \times AEW$ (for an anhydride).

Epoxy resins (and composites) are widely used as electrical insulators and dielectrics where the presence of impurities, particularly ionisable materials, is undesirable. These contaminants can be estimated using the procedures of ASTM D1762 (hydrolysable chlorine), ASTM D1847 (total chlorine) and ISO 4573:1978 (inorganic chloride).

The determination of the viscosity of epoxy resins is described in ASTM D2293. The procedures of ISO 2884:1974 and ISO 2555:1975 may also be used. The cross-linking shrinkage is determined by the procedures of ISO 3521:1976.

Epoxy resin systems other than those based on accelerated primary amine hardeners cross-link relatively slowly at room temperature, and for optimum mechanical and thermal properties they are generally cross-linked for fixed periods of time at fixed increasing temperatures. These cure-cycles are system-specific and are described in manufacturers' data sheets. It follows that a universal procedure for measuring the gelation time of an epoxy system is not only difficult to establish, but also may give misleading information. Such data are invariably established on an *ad hoc* basis.

6.4.4 Intermediate materials

1 Moulding compounds

Combinations of resins, fillers and reinforcement in a form suitable for processing by injection, transfer and compression are called moulding compounds. The materials may be granular, powders, sheets or doughs, according to both the constraints of manufacture and the intended application. Compounds based on thermoplastics resins are generally either granules or powders, while the form of those based on thermosetting resins is resin-specific. Most moulding compounds contain less than 25% by volume of fibrous reinforcement in the form of relatively short fibres (<1 mm–25 mm (0.04–0.98 inch)) which are randomly distributed.

Procedures for determining moulding shrinkage are described in ASTM D955, BS 2782 (method 640A) and ISO 2577:1975.

The mechanical properties of moulding compounds are described in BS 5734 and DIN 16849/2. These standards refer to formal methods of test for mechanical and thermal properties, all of which are described in detail in Section 6.4.5.

2 Prepregs

A prepreg is a combination of reinforcing fabric and resin. Historically only thermosetting resins have been used in prepregs, but thermoplastics materials have been developed recently as precursors for stamping

(Margolis 1977). Thermosetting prepregs are considered exclusively in the remainder of this Section.

Carbon-fibre epoxy prepregs are described in DIN 65146/2. Standard test procedures include DIN-V-EN 2331 and ASTM C613 (resin content), DIN-V-EN 2330 and ASTM D3530 (determination of volatiles), ASTM D3259 (solids content), DIN-V-EN 2332 and ASTM D3531 (resin flow), and ASTM D3532 (gel time). LN 29656 provides methods for measuring the dimensions of unidirectional carbon epoxy prepregs. The bond strength or tack (cohesion) of carbon epoxy prepregs can be measured according to DIN E65142.

Glass epoxy prepregs are specified in DIN 65090 and DIN 65142 and described in LN 29530. Methods for measuring the dimensions and linear density are provided in DIN-V-EN 2329 and LN 29549.

Aramid epoxy prepregs are described in DIN E65426/2, and methods for measuring dimensions and linear density are described in DIN 65426/1.

Prepregs are commercially available in a range of widths from 20 mm (tape) to 300 mm (0.8 to 12 inches). They are supplied on rolls in lengths of up to 250 m (273.4 yards) between moisture barrier films (typically aluminium foil). A typical prepreg description is (Ciba Geigy)

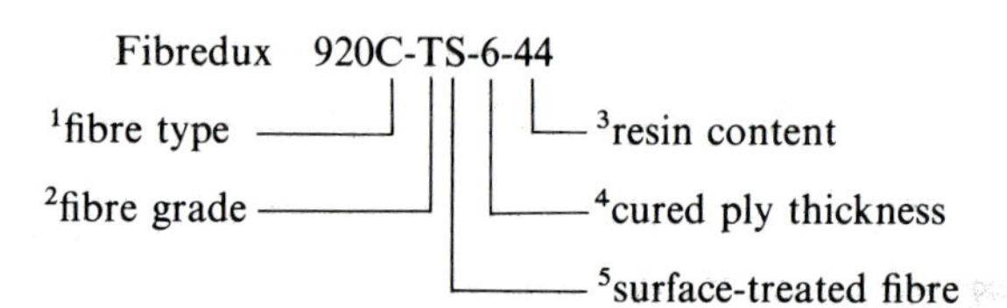

1 ⟨C⟩arbon, ⟨G⟩lass, ⟨K⟩evlar (aramid)
2 ⟨E⟩-glass, ⟨R⟩-glass, ⟨49⟩ Kevlar (aramid) grade high ⟨T⟩ensile carbon, high ⟨M⟩odulus carbon
3 Nominal resin content (% by weight)
4 Theoretical cured thickness (in 1/000 inch; 0.025 mm)
5 Surface treatment for carbon fibres only

6.4.5 Composite test methods

1 Categories of test methods

It is useful to classify tests or groups of tests into two broad groups. One group contains test procedures which are used to characterise a particular class of composite structures such as laminates, tubes, pipes etc. This group, by definition, or at least by implication, also includes procedures which are applications-specific. The other group contains tests which are designed to measure the mechanical and thermal properties of a composite material and in particular its elastic constants. Many of the tests in the second group, perhaps with some minor modifications, are prescribed as test procedures by the standards which apply to the first group.

2 Methods relating to structures and applications

This group contains methods which describe laminates in general, eg ISO 1268:1974 (preparation of specimens by low-pressure lamination), ISO 4899:1982 (properties and test methods for unsaturated polyester-based GRP), ISO 4901:1985 (residual styrene in GRP) and DIN 16944 (moulded GRP). In addition, there are standard methods of design and test for pipes and pressure vessels (BS 4994, ASTM D2585, ASTM D2586) and moulding compounds (Section 6.4.4 1).

Specifications for industrial laminates are described in ISO 1642 and DIN 7735. Both standards describe the National Electrical Manufacturers Association (NEMA) classifications consistent with both the International Electrotechnical Commission (IEC) document 249 and the Verband Deutscher Elektrotechniker (VDE) document 318.

Applications oriented standards include methods of test for fire described in ISO 4589, ASTM D2863 (LOI), ASTM D163, ASTM D1433/VDE 1340, ISO 1210 (horizontal flame spread), ASTM D568 (vertical flame spread), ASTM E162, ASTM E648 (radiant panel), UL 94, DIN IEC 707/VDE 0304 (Bunsen flammability). Specific fire tests for buildings are described in ASTM E84 (tunnel test), BS 476, DIN 4102 (Brandschacht), ISO 1182, NF P92-501 (Epiradiateur) and ASTM E119. Smoke measurements are described in ASTM D2483 and ISO DP8887. Toxicity testing is proposed in BS 6503 Parts 1 and 2 (ISO/TC 9122).

Other pertinent procedures include

	ASTM	*DIN	ISO
dielectric strength	D149	53981	IEC 243
tracking	D3638	53480	IEC 112
resistivity	D257	53482	IEC 167
permittivity	D150	53483	IEC 250
arc resistance	D495	53484	
weathering (daylight)	D4459	53388, 53389	877, 4582
(natural)	D1435	53386	4607
(artificial)	D2565	53387	4892
chemical resistance	D543	533476	175
stress cracking	D2552	53449/1/2/3	4600, DP4599

* All methods are consistent with VDE 0303.

The thermal properties of composites are of considerable importance both during processing (Loos and Springer 1983), and in the calculation of thermal stresses in the fully cross-linked composite (Kabelka 1984). The thermal conductivity (ASTM C177, DIN 52612 – both use the guarded hot-plate procedure) and specific heat capacity (ASTM C351), together with a knowledge of the density of the material, provide the thermal diffusivity and thence the rate of heat transfer. A measure of the reduction in mechanical properties with temperature is given by the temperature of deflection under load (3-point bending), measured according to ASTM D648, ISO 75:1974 and DIN 53461. This gives some indication of a maximum use temperature for the material, and is used, for example, in BS 4994 for this purpose.

3 Measurement of mechanical properties

Procedures for the measurement of the mechanical properties of solid polymers are described in ASTM D638, DIN 53455, ISO R527:1966, BS 2782 Part 320E (tension), ASTM D790, DIN 53452, ISO 178:1975, BS 2782 Part 335/A (flexure), ASTM D695, ISO 604:1973, DIN 53454 (compression) and ASTM D4255 (in-plane shear). Methods derived from these general procedures which are specific to composite materials, and in particular composites containing continuous reinforcement, include ASTM D3410 (compression), ASTM D3039, BS 2782 Part 10 Method 1003 [EN 61] (tension), BS 2782 Part 10 Method 1005 [EN 63] (flexure), BS 2782 Part 3 Method 341, ASTM D2344 (short-beam ILSS), ASTM D3518 (in-plane shear) and ASTM D3479 (tensile fatigue). The procedures described in these standards are directly suitable for routine quality testing, but most require some modification (*vide infra*) to provide elastic constants.

The preparation of test specimens by compression moulding is described in ASTM D796, ASTM D956, ISO 295:1974, DIN 53451 and DIN 53457. Methods for preparing laminates for test purposes have been identified in Section 6.3. Test specimens derived from prepregs should be moulded using the procedures and curing cycle given by the prepreg supplier. Ideally such samples should be prepared during the manufacture of a component by accommodating an appropriate blank in the moulding tool.

Specimens containing continuous reinforcement can be made by impregnating the dry fibres with resin and curing the mixture in a 'leaky mould'. In its simplest form this consists of a tee-section punch which fits fairly tightly into an open-ended U-channel die (Grafil method 201.15).[1] The depth of the channel is at least the sum of the depth of the tee flange and the specimen thickness. The quantity of dry fibres to give the specified fibre volume fraction at the required specimen thickness is weighed and impregnated with resin. Alternate layers of impregnated fibres are then laid along the die length. The punch and die are then squeezed together slowly so that excess resin exudes from the open ends of the die, before it is clamped at the appropriate separation. The assembly is then subjected to the required cure cycle.

Machining of specimens containing glass and carbon should be carried out with clean, diamond-tipped tools at a feed rate which minimises heat generation in the specimen. Machining of aramid-based composites is more difficult.[2] Drilling should be performed with a sacrificial material under the specimen to reduce the risk of delamination (Curtis 1984). Where tests require waisted specimens (eg tension of unidirectional laminates), these can be prepared by grinding. In general, all machined specimens should be free from defects which can act as stress concentrations, and in particular specimen edges should be abraded to eliminate saw marks.

The fibre content of specimens can be determined by the procedures of ASTM D3171 (matrix digestion with sulphuric acid and hydrogen peroxide) or ISO 1172:1975 (matrix combustion). The latter procedure cannot be applied to composites containing either carbon or aramid fibres.

For normal test purposes specimens should be preconditioned to $23 \pm 2°C$ ($73.4 \pm 3.6°F$) and $50\% \pm 5\%$RH (eg ISO 219, DIN 50014, ASTM E171 and BS 2782 Part 10 Method 1004 [EN62]). Cabinets which can maintain alternative environments before and during testing are specified in ASTM E197.

Methods for the measurement of elastic constants are based on specific modifications to the appropriate standard procedures identified earlier for the determination of mechanical properties. Common to all such tests is the use of strain gauge combinations rather than extensometers or cross-head movement to determine the strain(s) in the specimen and hence Poisson's ratio(s). Specimen dimensions and parallelism must be appropriately controlled (± 0.05 mm (1.97×10^{-3} inch)). The time over which the test is carried out should be kept short enough to exclude any effects of creep of the specimen – this may be difficult to accomplish with tests at elevated temperatures and high humidity (Curtis 1984).[1] In all these tests the specimen failure mode must be consistent with the analysis which relates the test result to the parameter which is being measured. If this is not the case then the result is rejected and the test repeated with another specimen.

Specific test procedures include:

(*a*) *Tension*: To minimise slip in wedge-action grips it is common to attach 50 mm (2 inch) end tabs on each side and at both ends of the specimen. Aluminium tabs suitably cleaned in chromic acid are commonly used for standard tests, while ±45° GRP tabs are used for tests at high temperatures to minimise interference from differences in thermal expansivity. The tabs are attached with a prescribed adhesive.

Specimens of unidirectional composites which are used to measure ultimate strength should be waisted across their width to minimise stress concentrations. These same specimens can also be used to obtain a secant modulus at 0.25% strain. The minimum thickness at the waist should be 1 mm and the radius of waist is specified (Curtis 1984) as 125 ± 5 mm (4.92 ± 0.20 inches) for specimens having a shear strength less than 5% of the ultimate tensile strength and 1,000 ± 10 mm (39.37 ± 0.39 inches) otherwise. Parallel-sided specimens should be used for measurement of the tangent modulus E_{11}.[1]

Parallel-sided specimens are used for tensile measurements on multidirectional (axially orthotropic) and angle-ply composites. In the latter case the length between end tabs is selected so that no individual fibre goes under both end tabs.

(*b*) *Shear*: The in-plane shear modulus is measured by loading a ±45° laminate in uniaxial tension (ASTM D3518). Alternative procedures are under continuing review and have recently been discussed in some detail (Summerscales 1987a).

The interlaminar shear strength of a composite may be measured by three-point bending over a short span, providing its flexural strength is at least ten times its shear strength. A suitable span:depth ratio is 5:1 (ASTM D2344). Careful assessment of the fracture surfaces of the specimens is essential for the measurements to be valid.

(*c*) *Flexure*: The flexural properties of a composite are generally used only for purposes of quality assessment rather than for the purposes of design. The commonest test procedure involves three-point bending (EN 63) of a thin parallel-sided strip. This in itself is attractive for the purposes of routine analysis since the specimens require little machining. Generous loading noses are preferred (25 mm (1 inch)), and the span:depth ratio should be adjusted according to both the composition and fibre orientation of the specimen. The ratio ranges from 40:1 for unidirectional carbon tested at 0° to 25:1 (same material at 90°) to 16:1 for woven 0/90 aramid.

(*d*) *Compression*: Compression loading is notoriously difficult to apply to a (thin) laminated material without introducing out-of-plane deformation by buckling. Various loading arrangements have been developed to minimise this problem, and the two most common arrangements are those described in the Celanese test[1] (method 403.12), ASTM D3419 and the anti-buckling guided method (Curtis 1984). ASTM D685 uses a shallow dog-bone specimen which is also restrained in a fixture.

The Celanese method employs split tapered collet grips which house the specimen ends in a machined slot. The specimen is typically 10 mm (0.4 inch) wide, 2 mm (0.08 inch) thick and 110 mm (4.3 inches) long fitted with 2 mm (0.08 inch)-thick alloy tabs 50 mm (2 inches) long attached to each face and on each end (see (*a*) *Tension*, above). The grips, in turn, are inserted into tapered sleeves and the whole assembly fitted into a guide cylinder. One end of this rests on the load cell of the testing machine while the other is attached to the cross-head. As the cross-head moves, the specimen ends are wedged by the tapered sleeves so that the free length of the specimen (between the collets) is placed in axial compression. This free length (10 mm (0.4 inch)) is sufficient to accommodate strain gauges while still eliminating Euler buckling. The Celanese method is suitable for unidirectional laminates.

For multiaxial laminates the antibuckling guided arrangement (Curtis 1984) is necessary. This method uses a wider sample (25 mm (1 inch)) with alloy or GRP end tabs. The free length of the specimen and the fitting of the end tabs is the same as that described in paragraph (*a*) for tensile tests. Since the free length of the specimen may now be up to 100 mm (4 inches), the out-of-plane bending has to be restrained rather than eliminated. This is accomplished by housing the specimen in a PTFE-faced fixture which supports most of the free length. The separation between the guide plates on the fixture is such that the strain at the onset of unstable buckling is greater than that of the failure strain in compression.

(*e*) *Fatigue*: Fatigue tests may be carried out using specimens and equipment described for the static tests (*a–d*), provided heating is minimised. This can be accomplished either by using a low test frequency (5 Hz (5 c/s)) or by conducting away the heat through aluminium end tabs attached with an aluminium-powder-filled adhesive.

6.5 INSPECTION AND CONTROL PROCEDURES

6.5.1 Inspection

Inspection and quality control procedures are concerned with the routine characterisation of materials and structures. Any property of a material or structure may be used provided it is both representative and meaningful. While the mechanical property tests of Section 6.4.5 are commonly used for these purposes, they require the preparation of dedicated specimens which are destroyed during the test. The specimens are, at worst, presumed to be representative of the artefact which is being manufactured or, at best, correlated empirically with its performance. Moreover, since these tests tend to be used frequently, the preparation of specimens can become a tedious and expensive exercise. This, of course, applies to the 'destructive' measurement of any parameter.

For these reasons there is a continuing interest in non-destructive inspection (NDI) techniques which do not require sacrificial specimens (Duke 1983). The two most common NDI techniques employ acoustic emissions (Hamstad 1985) and ultrasonics (Serabian 1985), while other methods include procedures based on thermography (McLaughlin, McAssey and Dietrich 1980), radiography (Alberti 1985) and chemical spectroscopy (Liao and Koenig 1984, and Galiotis *et al.* 1984). These procedures have recently been reviewed (Summerscales 1987b), and are detailed in Chapter 8. All of the procedures may be used for the detection of flaws (Pritchard 1982), while Fourier transform infra-red spectroscopy (Sprouse, Halpin and Sacher 1978, and Buckley and Roylance 1982) can be used to follow cross-linking during manufacture.

6.5.2 Control schemes

Quality control schemes are based on the statistical interpretation of results from a series of formal tests. Such procedures are described in the various parts of BS 2846 and in the ISO Standards 2602:1973, 3207:1975, 3494:1976 and 2854:1976. Test results and tolerance limits appropriate to various confidence levels derived from the statistical analysis are most conveniently presented in the form of control charts. Deviation from the prescribed tolerance limits may then be used as a means of a pass/fail/hold classification procedure. Simple pass/fail classifications prevent both embarrassment to a supplier and risk to an end-user, but since a finite quantity of output is rejected, these procedures can be extremely expensive unless the rejected components can be reused in some way.

It is perhaps less clear that routine quality procedures may be employed to provide feedback which can be used to influence or even directly control the process of manufacture. However, if this feedback is based solely on a pass/fail criterion, then by definition it can only begin to have an effect after a material has failed. In other words, this feedback is retrospective (see, for example, ASTM STP797).

Over recent years there has been interest in developing and exploiting control procedures which can identify trends in quality during manufacture so that potential rejection problems are identified before a component is made. Since the process is influenced continuously, these schemes are known as real-time or statistical process control procedures and are fundamental to formal schemes such as just-in-time (JIT) manufacturing.[3] Cusum charts can form the basis of a relatively simple system.

In many respects the JIT approach is the only satisfactory one to adopt with advanced composites since these materials are difficult to recycle or otherwise reprocess (Section 6.3.2).

NOTES

1 Courtaulds Graftil test methods. Courtaulds Grafil Limited, PO Box 16, Coventry CV6 5AE.
2 Machining of Kevlar composites. Du Pont de Nemours SA, Textile Fibres, CH-1211, Geneva 24, Switzerland
3 See, for example, *Automation*, April/May 1988, pp 17ff

7 Applications

Leslie N Phillips OBE FRIC CChem FPRI

formerly Head of Section
Plastics Technology, Materials Department
Royal Aircraft Establishment, Farnborough

and

Professor (ret.) in Department of Civil Engineering
University of Surrey

7.1 THE GENERAL PHILOSOPHY

The introduction of a new material into engineering practice needs a clear strategy on where to concentrate the effort and how to handle the widening spread of applications.

In the case of carbon fibre and similar advanced composites the prime market was in aeronautics. This field demands the best quality and performance, and puts a heavy premium on the saving of weight.

It was also probable – knowing the favourable properties of composites – that other industries would soon take note and explore possibilities. Again, it was entirely probable that fields quite remote from aeronautics would suggest uses – medicine, communications and sports gear come to mind – and where effort and material can be spared, such explorations should be encouraged.

This is enlightened self-interest. Aerospace materials tend to be – at any rate initially – fairly small scale in production and high in price. Anything which tends to broaden the market, increase production and therefore lower the price is for the general good.

This philosophy, which came to be known as 'spin-off', applies to all new invention, but is particularly obvious in the case of new materials.

In the author's case, over a period of several years, 70% of his time was spent in direct aerospace application (primarily on the military/civil side) with some 30% devoted to non-aerospace initiatives.

We may distinguish three successive stages in the introduction of a new material:

7.1.1 Minimal response

This is a simple geometrical replacement of a component by the new material, for example an angle is replaced by an identical angle, a flat plate by an identical flat plate etc. This replica technique has very little to recommend it. There is no attempt to redesign, and any weight savings are the direct result of using a material of lower specific gravity.

7.1.2 Redimensioning

If, by reference to published data on mechanical properties, the 'replica' of the first approach is found to be both unnecessarily heavy and strong, this indicates the need for a redimensioning exercise.

This is a cautious paring away of unwanted material, so that what remains should have the same strength and stiffness as was originally present. In contrast with the old metals, which are isotropic, anisotropy – the 'grain' of fibrous composites – has unexpected weaknesses in shear and in torsion, which put load on to the matrix rather than the fibre; and moreover, since the composites are made differently their shape will also be different.

The characteristic appearance may be summarised as follows. We notice shell-like structures of uniform or gently tapered wall thickness. There are rarely flat planes, no sharp corners and no abrupt changes of cross-section. Simple and double curvatures are common, with very smooth external surfaces and an absence of undercuts or re-entrant angles. Joints are inconspicuous and will be bonded adhesively rather than by using a large number of mechanical fasteners as with metals.

Whereas metal components are often made from a large number of parts and subassemblies, the composite product carries forward the integration of parts to a high degree.

7.1.3 Designing *ab initio*

With the increasing confidence that comes from the education provided by stages 1 and 2, the engineer casts around for something to 'get his teeth into', ie a new structural component that has not previously been made from traditional materials.

Designing *ab initio* brings greater scope for innovation and the greatest risk of failure. It calls for a close consideration of every aspect of behaviour. It includes the distribution of loads on the structure and the pattern of fibre arrays best able to resist them. It checks the allowable strains on different parts of the structure, the reductions in strength which may arise in adverse environments and the stress concentrations which may arise at cut-outs, joints and changes of section.

Of great importance is the susceptibility to fatigue, ie the resistance to repeated cycles of stress. Although there are fatigue laboratories where materials, representative joints and small structural units can be tested separately, the most convincing way to obtain the real performance of a complex structure is in field trials, or 'demonstrators'.

Many such trials take place in various industries. For maximum benefit the performance should be monitored not for a short period but for at least the required lifetime of the structure, and preferably until failure takes place and the failure mode is established.

The sections which follow begin with the primary objective of new materials research and then address other branches of engineering.

7.2 AIRCRAFT ENGINEERING AND SPACE HARDWARE

We distinguish carefully between *secondary structures*, the failure of which might be uncomfortable or embarrassing but which does not jeopardise the mission (eg failure in a luggage rack, galley or toilet), and *primary structures*, where failure at the very least aborts the mission and often leads to loss of the aircraft (eg loss of a wing, helicopter blade or tail unit).

Over the past 45 years we have seen a steady growth in the use of composites in secondary structures (which used glass and asbestos reinforced plastics) with a great upsurge in application to primary structures in the past 25 years, using the advanced composites primarily based on glass, boron, carbon and aramid.

There are many types of aircraft and it is convenient to deal with civil and military types separately.

7.2.1 Civil aircraft

There are operational reasons why demonstrations should be mounted on civil, rather than military aircraft. Large civil airliners run regular long-distance routes, day-in and day-out rather like a bus, logging-up a huge number of flying hours every year. The system of inspection and reporting is a well established routine. Not only is military flying less intensive, but in cases of damage and failure it may be impossible to recover the aircraft and the evidence.

Many unobtrusive programmes have been carried out, using civil aircraft, for the demonstration of components in glass, carbon and aramid.

NASA has sponsored many of these on either an ACEE ticket (aircraft energy efficiency) or a flight service evaluation (FSE) programme.

Inspection is carried out on many items:
- the nose radome and avionics bay
- leading edge and trailing edge panels
- wheel wells and landing-gear doors
- pylon fairings, body fairing and wing tips
- tail cones, elevators and horizontal stabilisers (in whole or in part)
- locations in the power plant for nacelles, blade-containment rings etc.

To give an illustration of the growth in the use of glass fibre components in successive Boeing planes, we note that a total area of 19 m^2 (22.7 sq yds) was typical for a 707 aircraft; 167 m^2 (200 sq yds) for the 727; 279 m^2 (333 sq yds) for the 737; and 929 m^2 (1111 sq yds) for the 747 types.

With the increase in availability of aramid fibres, the makers, Du Pont, hoped at one time to replace carbon in primary structures and also glass in secondary ones. However, it is clear that aramid alone is unlikely to compete with carbon because aramid is intrinsically low in compressive strength. In consequence there is a move to make hybrid constructions by weaving, usually in a 50:50 C/Ar ratio. In the Boeing 767 there are carbon/Kevlar/epoxy landing-gear doors, wing-to-body fairings and the stabiliser fixed trailing edge.

With respect to glass, aramid has the better stiffness and a lower density but the poor compressive strength is still a disadvantage. Substitution of glass by aramid is fairly simple because the latter can be woven in the same styles (or fabric constructions) that have long been standardised for glass.[1]

Thus it is possible, in a glass fibre fairing made with six layers of a satin-weave cloth, to construct six more prototypes, using the same resin matrices and the same moulds, wherein there is a progressive substitution of one layer at a time by an equivalent aramid cloth.

A series of comparative tests, for strength and stiffness, will soon reveal which is the most effective combination. While not matching the scale of the NASA effort, there have been several demonstrators in the UK and Europe. Normally the good news is delayed in publication for several years.

Jones (1985) reported on the satisfactory conclusion of a lengthy trial (of 5 years plus) of a series of carbon epoxy panels installed on British Airways aircraft. (It is interesting that before BA gave their consent, extra insurance premiums were negotiated – Jones, personal correspondence.)

When it comes to light-aircraft construction, the American amateur

Table 7.1 Structural arrangements at the Oshkosh meeting

Construction	Structural form	% present
tube and fabric	triangulated truss	48
all-metal	mainly thin-walled tube	27
all-wood	mixed	14
composite	thin walled tube with foam-stabilised skins	8

builder has always been most knowledgeable and innovative. Table 7.1 was arranged by Stinton (1983) after an analysis by Cox of an Experimental Aircraft Association meeting at Oshkosh, Wisconsin, in 1978.

Of the many hundreds of aircraft which were flown in for the meeting, the distribution of structural arrangements found to have been used was as listed in Table 7.1. It is quite certain that in the decade since this analysis, the proportion of small aircraft made with composite materials has more than doubled.

Pioneers such as Walters and Nelson used a conventional glass/epoxy/foam sandwich for making their Viking 'Dragonfly' which won the 'outstanding new design' award at Oshkosh in 1980. The power plant was a 1,600 cc Volkswagen engine.

Bert Rutan has made bold use of carbon and hybrids in his very elegant canards, and more recently his team have designed and produced 'Voyager', a most astonishing machine in which his brother and a co-pilot (Rutan R. and Yeager) circumnavigated the globe, flying non-stop for nine days on one filling of fuel from 14 to 23 December 1986 (Mordoff 1986, 1987).

This one-off design consisted of a central fuselage, the forward position housing the cramped cockpit and the rear containing the two 110 hp (82 kW) Teledyne liquid-cooled engines driving the Hartzell constant-speed propeller. The fuselage was attached at the forward end to twin streamlined booms containing fuel; at the aft end, a long thin through-wing, also containing fuel, joined both booms and the fuselage together.

Most of the structure is built of carbon fibre and honeycomb core but the outer wing has a foamed plastic core.

During take-off the weight of fuel in the wings caused them to sag and drag along the runway. This removed about a metre of carbon fibre skin and damaged the core; at the same time it damaged the right-hand winglet, which had to be removed by a side-slip manoeuvre. In order to fly evenly, the left-hand winglet was also amputated by a corresponding manoeuvre.

In the event, by re-routing to avoid some of the worst tropical storms, 40,145 km (24,950 miles) were flown in about 216 hours (an average speed of 185.9 kmh (115.5 mph). This was possible because the design enabled a basic structural weight of 425 kg (938 lb) to carry 3,175 kg (7,000 lb) of fuel in 17 tanks.

A rather different concept of the role of new materials is their use in the design, first of all of the 'Gossamer Condor' (Miska 1978), designed by Paul McCready, which won the Kremer Prize for single man-powered flight in August 1977, and then, two years later, of the 'Gossamer Albatross', which won the prize for crossing the English Channel, with the same method of propulsion.

Although both machines were approximately the same size and had the same canard configuration, there were some interesting differences in detailed layout and in the materials of construction.

The 'Condor' weighs 31.7 kg (70 lb) and has a main structure which is made from aluminium tubing with piano wire. The wing uses Sitka spruce for ribs and spars and is covered by Mylar (polyester) film. The leading edge is corrugated cardboard and polystyrene foam. The pilot lies/sits in a covered enclosure and pedals furiously to drive a 2-bladed propeller, the spar of which is again an aluminium tube.

In order to collect the Kremer Prize (value approximately $86,000) the flight had to fulfil the following conditions:

(i) The take-off had to be unassisted, from ground level, in a wind of 16 kmh (10 mph) or less.
(ii) A figure-of-eight pattern had to be made round two pylons 0.8 km (0.5 mile) apart and a 3.0 m (10 ft) hurdle had to be flown over at both start and finish.
(iii) There was to be no assistance from the stored power of buoyant gases.
(iv) There were to be no stops or other assistance on the way.

The conditions for winning the subsequent new competition, worth £100,000, were deceptively simple. 'Fly – man-powered – from the mainland of the United Kingdom to the mainland of France!' There were slight differences in wing span between 'Condor' and 'Albatross'. The former had a wing span of 29.3 m (96 ft) and a total lifting area of 77.6 m^2 (835 ft^2). The latter had a wing-span of 28.6 m (93 ft 10 in) with only two-thirds the lifting area of 'Condor', with much less taper, sweeping back at 7.5 degrees instead of 9 degrees.

The structure was carbon fibre/epoxy tubing made by wrapping uncured material round an aluminium tube as the mandrel, curing in an electrically heated tubular oven at 143°C (289°F) and then dissolving away the metal with hydrochloric acid (Grosser 1981). The wing ribs were made from polystyrene foam sheet to which were cemented individual tows of carbon fibre to make a series of diagonal trusses. The rib was capped in the same simple fashion, fitted with a thin plywood spar ring and bound with a cord of Kevlar aramid at the trailing edge. The wing and canard were covered with stretched Mylar[2] film which could be heat-shrunk and tightened to give a very smooth surface.

The propeller was of dense polystyrene foam with an internal tubular spar of carbon fibre, the contour being skinned with a Kevlar cloth and epoxy resin. Vigorous pedalling drove the aircraft at a speed of 10 mph.

Bryan Allen, the crew, was a professional cyclist whose body weight was somewhat greater than that of the aircraft. Although a still, windless day was chosen for the flight with an early morning take-off, conditions inside the cockpit soon became overbearingly hot because of the 'glass-house' effect. Only determined concentration enabled the crossing to succeed.

The change from a supine position as in 'Condor' to a more normal upright seating had enabled the crew to generate a higher peak horse-power for the 2.5-hour crossing (Figure 7.1).

Very recently (23 April 1988) the Channel crossing has been exceeded by another astonishing feat of man-powered flight. Once again, using carbon fibre as the structural material and a polyester film wing and tail covering, a man-powered flight has taken place, from the island of Crete in the Eastern Mediterranean to the small island of Santorini (Thira, Kalliste) 117 km (73 miles) to the North.

The journey was accomplished in 4 hours of steady flying/pedalling and demonstrates remarkable stamina on the part of the pilot, Kanellos Kannellopoulos, a professional cyclist. The layout is conventional, with a large monoplane wing, a tail at the rear and a rudimentary 'fuselage' between.

However, the wing is designed to have a pronounced upward sweep at both wing tips, so that it looks like a may-fly in flight. It will be interesting to see whether there are convincing technical reasons for this general curvature, as a striking success of this kind can lead to many imitators.

7.2.2 Military aircraft

An enormous effort, financial and technical, has gone into the development of applications for composite materials in military aircraft. The fol-

Figure 7.1 *Gossamer 'Albatross'. The first human-powered flight across the English Channel*

lowing are pioneering examples:

(i) A boron/epoxy construction was used for the skins of the horizontal stabiliser box of the US Navy F-14.

(ii) The outer wing of the A7D was a demountable primary structure. It consisted mainly of carbon/epoxy with hybrid covers of boron and carbon/epoxy. This proved to be outstanding in fatigue, lasting at least twice the normal lifetime of the metal version. It added evidence to the growing belief about the outstanding fatigue properties of carbon fibre reinforced plastics.

(iii) For the S-3A aircraft a carbon fibre spoiler was produced in a limited edition and flight-tested in service. It consisted of a sandwich construction of thin carbon/epoxy skins stabilised over a core of 'Nomex' honeycomb.

Like UK practice on the Fairey Delta wing, the moulding and curing of the skins and their adhesion to the core was carried out in a single operation known as 'co-curing'. It is a great simplification in manufacture.

(iv) With the adoption of the Harrier jump-jet by the US Marines a virtual 'redesign *ab initio*' became possible (Riley 1986). In the event, the wing-box skins, the forward fuselage, the horizontal stabiliser, elevators, rudder, flaps, ailerons and overwing fairing were all redesigned and manufactured in carbon/epoxy material. Carbon/epoxy was also used to make an extra wing-tip for the Hawker Harrier. This is added when the fighter assumes a ferry role, and it increases range for cross-country delivery trips. This optional extra was an ideal demonstrator. After all, in the event of failure, the Harrier would fly well without it (Figure 7.2).

(v) The German Alpha-Jet is intended as a troop-carrier in its military role. The elevator for this machine is an ideal demonstrator for carbon fibre material to replace the original metal construction. The unit is a pendulum fin having a massive, continuous through-spar box. As well as the aerodynamic flight loads there are large bearing loads taken by swivel bearings. The objectives were to save weight, to show that carbon was better in practice and to obtain unlimited flight approval. The structure chosen was a much simplified box-spar supporting a nose end box and a glass-fibre wing tip.

Two interlocking shells in carbon of U-section comprised the spar box, which enclosed 11 ribs. Two of these had integrated bearings and were made of metal; another two, also of metal, acted as connecting ribs for the wing-tip edges; the remaining seven ribs were carbon.

Figure 7.2 *Carbon/epoxy wing-tip of Hawker Harrier. Clever lighting enabled the photographer to display fibre orientation in different parts of the lay-up*

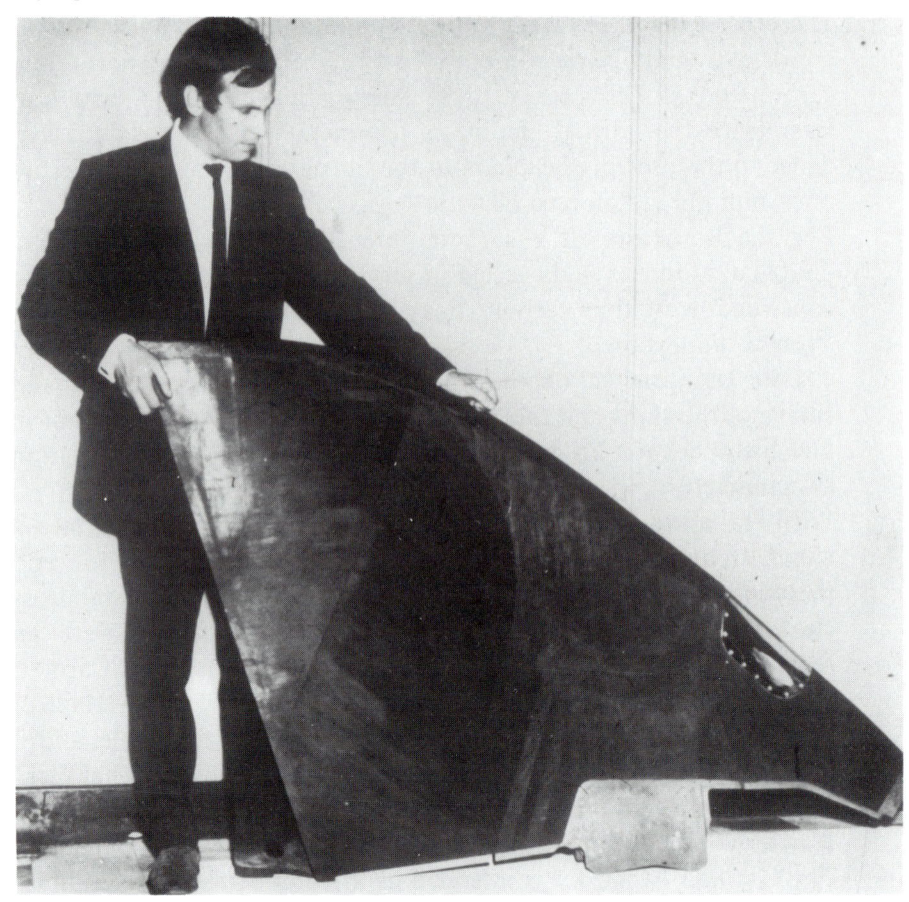

There was a careful cost analysis and a calculation of the weight saved, so that a true comparison could be made of metal versus composite. The first observation is the reduced number of parts needed in the composite (Table 7.2).

Table 7.2 Number of separate parts in metal and composite

Name of part	Metal	CFRP
Spar box	107	52
Nose	36	4
End box	54	6
Wing tip	12	12
Connecting elements (standard units)	4,938	1,201

Figure 7.3 *Breakdown of production costs for typical metal structure and the CRP elevator unit*

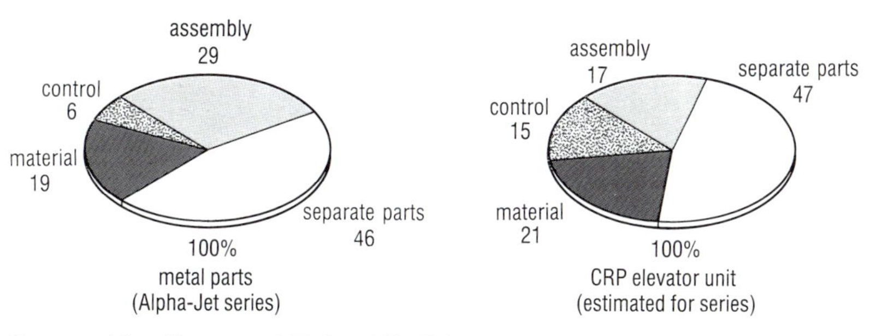

Source After Conen and Kaitatzidis 1981

Figure 7.3 demonstrates production costs: materials and quality control and inspection are slightly costlier in the new material, while the integration of parts has gone further than in the metal (Conen and Kaitatzidis 1981). There are fewer parts, they are larger and this simplifies assembly.

Although such differences were already appreciated and understood by those familiar with composites, the facts came as rather a surprise to aircraft engineers trained to think in terms of metal construction.

(vi) The 173-S has a large horizontal tail unit in carbon fibre/epoxy, which is a good example of the way that European technology is moving towards good design, standardisation of materials and processes, and true interchangeability. Figure 7.4 shows the machine in flight with a UK tailplane on one side and a German-made tailplane on the other.

In recognition of the high-quality work of the European airframe makers, much subcontract work is being placed by US companies such as Boeing and McDonnell Douglas.

(vii) It is interesting to see how the amount of both primary and secondary structures in composite materials has increased steadily in American military aircraft since the early 1970s.

Thus, looking at the fighter series from F-15 through to the F-18, Schwartz (1984) has estimated that the F-15 comprised 1% of the structure weight and the F-16 comprised 2%.

By the time that the F-18 fighter was being designed, both experience and confidence had increased considerably; a massive switch from light alloy to composite construction had begun.

The F-18 uses carbon fibre-reinforced plastics for the wing skins, horizontal and vertical tail boxes, both wing and tail control sur-

Figure 7.4 *The 173-S in flight with UK tail-plane on one side and a German-made tail-plane on the other. Europe has a very good capability in composites technology*

faces, leading-edge extension, speed brake and several doors. Composites now make up more than 50% of surface area and about 10% of the structure weight (Figure 7.5).

(viii) The near-monopoly of such extremely large military transport aircraft as tank- and troop-carriers has meant a great reliance on light-alloy construction and conservative design. However, there are occasional opportunities for the use of composites retrospectively.

Because of heavy duty in Vietnam, there were signs of severe fatigue damage to the centre-wing boxes of the C-130 transport. In the event, a considerable amount of unidirectional boron/epoxy strip material was bonded to the skin and stiffeners of the box, reducing the stress level and correspondingly increasing the fatigue life.

A total weight of 159 kg (350 lb) of boron/epoxy reduced the weight from 2,241 kg (4,941 lb) in light alloy to 2,041 kg (4,500 lb) in the mixed construction.

The large petal-opening cargo doors of the C-141 aircraft suffered frequent damage in use, the original construction being of light-alloy

Figure 7.5 *Composite distribution on an F-18* (*McDonnell Douglas, St Louis*)

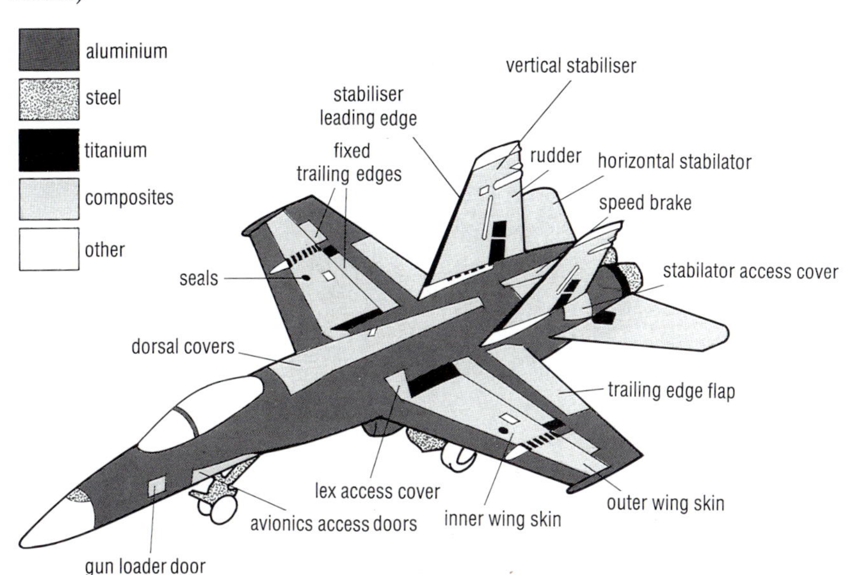

Source Schwartz 1984

skins and a core of light-alloy honeycomb. The modest redesign comprises a sandwich skin of carbon/epoxy over a lightweight core of synthetic foam. A so-called 'syntactic foam' comprises an epoxy liquid resin filled with hollow glass spheres from fly-ash and a small amount of thixotropic silica. The resultant mix has the consistency of butter and can readily be pushed, smeared and injected into irregular cavities. After a few hours it sets to a tough, light foam of good compressive strength which can be cut and drilled like wood and receives fasteners without splitting.

Hat stiffeners and edge members are being produced in the same materials, while chord-wise frames are made in foam-filled glass epoxy to locate with the existing frame structure.

(ix) Following successful development work on the Experimental Aircraft Programme (EAP), the UK Government has announced the decision to go ahead with production of the new Eurofighter (Figure 7.6). Like so many initiatives in the aviation field today, the aircraft is a joint venture (UK 33%, Germany 33%, Spain 13%, Italy 21%).

This advanced fighter brings together several new features: aerodynamic, instrumental, materials and power plant (Turbo-Union RB199). It possesses a co-bonded compound sweep wing (moulded rather than machined) with all-moving foreplanes, and the construc-

Figure 7.6 *The EAP (Experimental Aircraft Programme) Technology Demonstrator*

COURTESY BRITISH AEROSPACE

tion employs carbon-fibre composites throughout. Active controls, with artificially maintained stability of pitch, make it easy to fly and extremely manoeuvrable. The pilot is considerably helped by the provision of coloured 'head-up' displays for presentation of the data.

It is clear that when designers are presented with the opportunity to design *ab initio*, the skills now exist to produce the hardware on a production scale.

7.2.3 Powered aircraft and their engines

In power-assisted flight, there was a constant complaint from amateur builders of lightweight machines – ie those with weight below 455 kg (1,003 lb) – that all the engines available were too heavy. The chunky commercial stock engine, with comparatively large nuts and bolts, alternators and spark plugs was altogether too massive, powerful and heavy for their needs, so that a neat matching of airframe and engine was impossible (Stinton 1983).

However, engine makers have widened the choice over the past few years, with modified snow-mobile, chain-saw and automobile engines being pressed into service. These are often petrol engines, but turbojets which normally run on kerosene are both compact and powerful. When combined with improved propellers, they make the turbo-prop an attractive proposition.

There is a division first of all into: (*a*) lightweight aircraft, where reciprocating piston engines are the norm, at low altitudes; (*b*) high-speed jets for jumbo (large-sized) passenger jets, where the range increases markedly with height, being twice as good at 9,144 km (30,000 feet) as at sea-level; (*c*) intermediate turbo-propeller aircraft, filling the gap between.

1 Propeller blades

The incentive towards turbo-props has been boosted by the development of improved propeller blades using composite materials.

Since the weight-saving potential rises with increasing diameter, one can pinpoint the useful range as between 1 and 6 m (3 ft 3 in and 19 ft 8 in) diameter. In the middle of the range a four-bladed propeller having a diameter of 4.5 m (14 ft 9 in) can reduce the weight from 454 kg (1,000 lb) in metal to 181 kg (400 lb) in composite material. The centrifugal load on hub and control gear is reduced by the absence of metal components of higher specific gravity.

In a very full account of the work of Dowty-Rotol, McCarthy (1985) has pointed out that new moulding techniques have made it much easier to create new shapes in composite materials having better aerodynamic efficiency than was feasible in metal. For example, they use the ARA-D profile instead of the NASA-16 section formerly used; this change leads to improved take-off and climb characteristics.

The Dowty-Rotol blades use unidirectional cloth and tape aligned along the blade axis to act as spars, with glass-cloth laid at $\pm 45°$ orientation to resist torsion, although local variation is used in some areas to alter the torsional frequency. Blade shape is maintained with an internal core of polyurethane foam which is formed *in situ* (Figure 7.7).

The carbon spars undergo a gradual change in shape from flat to round. They are separated, at the blade root, into annular rings by inserts of glass fibre, so that when the whole assembly is impregnated with resin and cured the result is a massive annular wedge which is securely bonded to (and confined within) the inner and outer metal sleeves.

Figure 7.7 *Dowty-Rotol composite blade structure. Design and manufacture at Gloucester in the UK is underway*

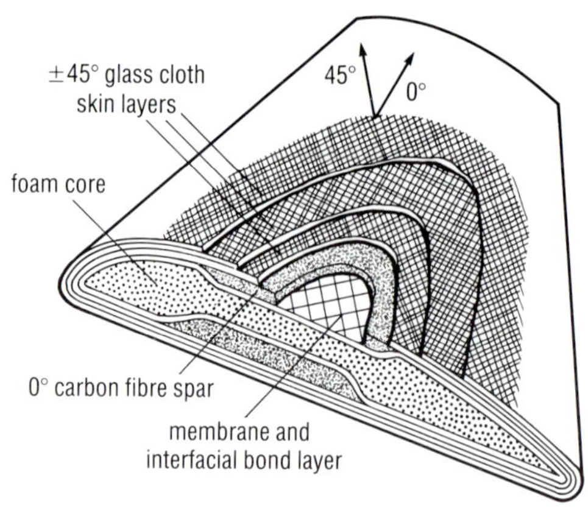

Source McCarthy 1985

Among the advantages of composite construction is a great improvement in fatigue life. It is well known that surface damage sustained by light-alloy blades reduces their limited fatigue life even further, and that the composite blade is tougher in service.[3]

A similar situation exists with the propellers for air-cushion vehicles, popularly known as hovercraft. The environment is very severe as the blades have to cope with small stones thrown up from the beach as well as the salt-water spray and wind-borne sand.

On the US Navy hovercraft, a metal propeller has an average life of about 100 hours, whereas a composite blade, finished like the Dowty-Rotol propellers with a tough polyurethane coating and replaceable leading-edge shoes, is almost unmarked and is fit to continue in service after a similar period.

2 Helicopter blades

There is a general appreciation of the fact that, with proper design, composite helicopter blades have a much better life than their metal equivalent. Metal blades rarely survive 1,500 flying hours, whereas 3,000 hours or more is becoming usual with composite blades.

Specialist firms such as MBB-Munich, Aerospatiale 13 and Westland Helicopters are all producing superb blades. Once it became clear that a loss of strength was always preceded by a loss of stiffness, simple comparative tests could be devised to measure any reduction in stiffness. For example, blade deflections at the tip can be measured under a standard load while the blade is still mounted on the hub. Suspect blades can be removed for more detailed inspection.

While some blades are designed wholly in glass fibre, there are considerable advantages of a hybrid construction. In as much as the tensile modulus (in the 0° direction) of a ±45° carbon fibre laminate is virtually the same as that of a 0° (unidirectional) glass lay-up, this can be exploited to obtain the best combination of flexural and torsional stiffness. At the same time, performance with carbon fibre, used for the trailing edge and rear portion of the blade behind the main spar, is improved enormously with respect to fatigue resistance in flexure.

3 Jet engines

The largest jet engines are turbo-fans rather than turbo-props, and there is apparently much scope for composite materials, particularly at the cooler end.

The use of glass-epoxy and glass/polyimide has already been mentioned in Chapter 2. Carbon/polyimides will be used for both rotors and stators.

Carbon fibre laminates are used for inner and outer cowls and for certain of the ducts where the temperature regime is suitable, ie up to about 250°C (513°F) (Ault and Fresche 1979). Reverse-thrust flaps and sound-absorption panels are also good applications.

Aramid fibres are preferred for the container rings which surround the engine, and are a necessary safety precaution in case of blade failure.

One of the important potential areas for saving weight is in the large front fan. The financial troubles at Rolls–Royce in 1971 were compounded by the bird-impact clauses applying to the RB-211 engine. There is no doubt that technical improvements such as stitching of the laminae and armouring of the leading edge came too late to ensure early adoption; airline operators are so impressed by the good performance record of the titanium blades that there would be a resistance to precipitate change.

However, no reader should imagine that big jet engines will not evolve further. Personnel at the world's great engine makers (Rolls–Royce, General Electric, Pratt and Whitney, Turbo-Union, SNECMA, FIAT and others) are aware of pressure to make their power plants larger, cleaner, quieter and more fuel-efficient (Newton 1980). Figure 7.8 shows the production of by-pass ducts in carbon fibre composite at Rolls–Royce.

There will be improvements in strength and stiffness, in better vibration damping and fatigue resistance. There will be materials with a lower coefficient of expansion leading to lower tip clearances and greater efficiency. Small but significant savings in weight anywhere on the engine reflect in lower shaft, disc (Johansen, Lilholt and Lystrop 1980) and containment, and nacelle weights.

Figure 7.8 *Production of Tay by-pass ducts in carbon fibre composite, using computer control*

COURTESY ROLLS-ROYCE LTD

The official organisations who monitor and encourage engine development (RAE, NASA (Lewis) and others) are well aware of these trends and sustain good design on a long-term basis.

7.3 WIND TURBINES

Composites may be used as materials for wind-driven blades to create electrical power.

Where the diameters of such blades remained small, individual installations were erected for homes, farms and small businesses. Now that the generation of power is a more serious and larger-scale enterprise, we see wind turbines growing both larger and more numerous.

By their very nature, such turbines would tend to be positioned in high, stormy and remote areas. It would be sensible to group them in clusters, known as 'farms', to reduce the costs of installation, to assist in the collection of current and for routine inspection and maintenance.

In such a situation, any consortium would have three types of expertise. There would be an aeronautical partner with knowledge of the design, manufacture and testing of the blades, an electrical partner for providing the dynamos and regulators and providing connections to the grid system, and a civil engineer concerned with providing the tall towers needed for the 'farm', which would be made of reinforced concrete.

While the practice of the Central Electricity Generating Board has been to think in terms of steel or laminated wood for the blades, in countries other than the UK there is a lively interest in composite construction.

The earliest blades were made from glass fibre and polyester or epoxy resins as the matrix. These have become increasingly large over the years, the maximum diameter built so far being 46 m (150 ft) diameter by Hamilton-Standard.

However, there have been a number of structural failures, mainly in fatigue, so that in modern designs there is specification of carbon fibre for the main spar, oriented at both 0° and 90°, because of its outstanding resistance to fatigue. The torsional stiffness is also improved by the substitution of aramid fibre for the skins, instead of the glass previously used. This skin lay-up is at $\pm 45°$.

The aeronautical engineer J. Cools has suggested (Cools 1981) that a life of 10 years should be the minimum target for a viable wind-turbine blade. During this time it would experience about 26.5 million cycles of stress composed of 20 million cycles at lower stresses (3–8 kN/cm^2 (4,350–11,600 lbf/in^2) with wind speeds averaging 6 m/s (19.7 ft/s), and 6.5 million cycles at 4.5–8 kN/cm^2 (6,525–11,600 lbf/in^2) stress from winds peaking to 17 m/s (55.8 ft/s)).

In the Cools design (Fokker Aircraft) the blade has a diameter of 25 m (82 ft), is of conventional propeller shape and spins almost vertically. There are some aerodynamic advantages in running the blade at an angle of 85° to the rotor axis. This 'blade coning' reduces the peak bending loads somewhat. The rotor axis, in turn, is mounted at a slight angle to the horizontal.

The blade design finally chosen comprises a main spar consisting of twin units facing each other, capped by massive carbon/epoxy

Figure 7.9 *Final design of wind-turbine blade designed by J. Cools for Fokker aircraft. A long fatigue life is essential for this application*

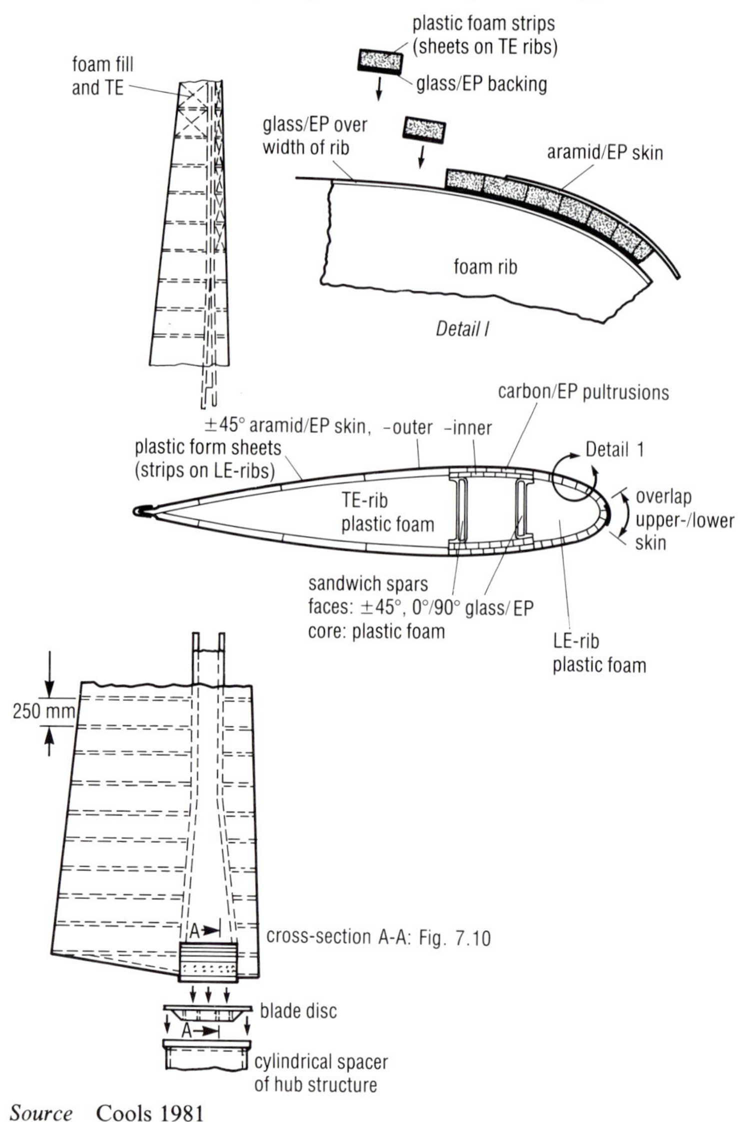

Source Cools 1981

pultrusions, and joined by ±45° aramid/epoxy skins top and bottom (Figure 7.9).

There are numerous leading-edge and trailing-edge ribs made from expanded-foam plastics, reinforced over the width of the ribs with glass/epoxy tape. The outer halves of each blade carry higher centrifugal loads, and the cells between the ribs for both leading and trailing edges are foam-filled.

Figure 7.10 *Root structure of carbon/epoxy panels and sandwich spars. The attachments of blade, root and hub are important areas for detailed stressing and design*

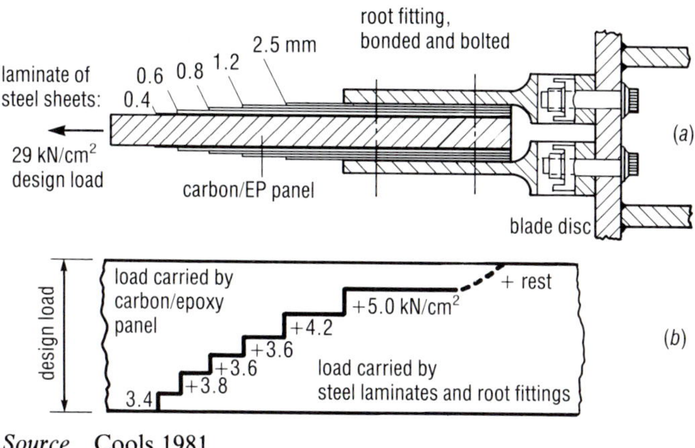

Source Cools 1981

The root structure comprises a thickening of the sandwich spars to a solid carbon/epoxy panel sandwiched between a laminate of steel sheets, with an increasing number of layers towards the root. A bonded and bolted root fitting secures both the carbon/epoxy panel and the laminated steel sheets together, and this in turn is bolted to the blade disc which attaches to the hub structure (Figure 7.10).

A totally different lay-out, which has many supporters, has two blades mounted vertically together on opposite sides of a shaft, which rotates about the vertical axis. Because each blade is supported in two places, the bending moments at each tip are much smaller than in the conventional propeller design.

The Central Electricity Generating Board has recently stated its intention to build a fairly large experimental farm consisting of some 26 towers (there are numerous windy sites to choose from) so that the economics of wind power can be evaluated in detail.

7.4 MARINE CRAFT

The term covers a wide variety of floating vehicles, ranging from the smallest dinghies, through sailing craft and motor boats of every description, to sea-going vessels up to 61 m (200 ft) or more in length.

The choice of composites for marine craft is one of the great post-war successes, and their use has transformed the boat-building industry, from a craft business in numerous small yards, to factory mass-production.

Forty-five years ago (the early 1940s), wood and metal were the traditional materials. Thirty years ago the debate began, not simply as to whether composites were more convenient, but also to consider the remote possibility that composite boats could be built for the same costs.

Today it is freely acknowledged that the price is lower; new materials in many cases reduce the costs of maintenance and increase the possibility of creating new shapes and new uses which would be very difficult in wood and metal.

Although some early dinghies were made from 'no-pressure Durestos', which required warm ovens for cure, the great technical advance was the invention, around 1947, of cold-setting polyester resins together with new forms of rovings cloth and chopped strand mat, that made the creation of moulded hulls both simple and attractive.

The process starts from an original master of wood, plaster of Paris, concrete or similar easily worked material. From this a glass fibre and resin female mould is produced, which is of one-piece construction in the smaller sizes but frequently split at the keel-line for the larger versions.

As well as wide use of the traditional 'hand lay-up', there is nowadays much use of 'spray-up' for rapid, simultaneous, deposition of glass and resin (Figure 7.11). On the larger vessels there is automated lay-up of the reinforcement in a temperature- and humidity-controlled environment.

Because of the relatively low modulus of glass, there is a natural tendency to improve the stiffness of beams, panels and other elements by adding more material and thickening the cross-section. Other ways of doing this are by the extensive use of ribs and stringers and by the use of sandwich construction.

By the late 1960s it became possible to augment these general methods by adding a proportion of carbon fibres to the now traditional glass, either in the form of U/D (unidirectional) tows, U/D fabrics and by weaving carbon and glass together into bidirectional fabrics.[4] Figure 7.12 shows a convenient classification of these hybrid possibilities.

The weight savings made possible by these constructions are very impressive. In one project, in which the writer was involved, the weight of racing kayaks was reduced from 15.9 kg (35 lb) to 10 kg (22 lb) by using only 227 g (0.5 lb) of carbon fibre. In the development of racing sculls (pairs, fours and eights) the author found comparable weight savings.

These racing sculls incorporated large amounts of sandwich construction, as in the best aircraft practice. There were:

(i) Aluminium honeycomb between skins of glass cloth and U/D carbon cloth for the hull

Figure 7.11 *Spray moulding of a boat hull. This method deposits composite considerably faster than by hand lay-up*

COURTESY BINKS-BULLOWS LTD

Figure 7.12 *Classification of hybrids. The designer has great freedom to specify the material and its construction*

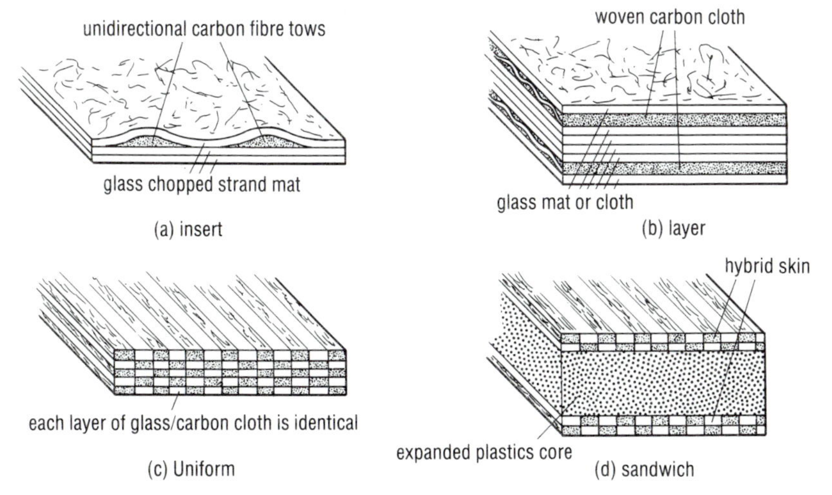

(ii) Polymethacrylimide foam (Rohacell) between skins of carbon fibre cloth for the deck of the stateroom, where the power is generated

(iii) Glass skins over marine plywood cores for the transverse ribs and the main bulkheads, which divide the eights boats into two portable sections

(iv) Marine plywood between extra-thick carbon skins in the area of the sax-boards where the riggers feed load from the oars into the hull.

In the construction of catamarans, a point of potential weakness lies in the cross-beams or bridging units which join both hulls together. Some years ago there were several failures among ocean-racing catamarans, where the racking forces, proving much larger than had been thought possible, had torn the hulls apart. The design of the bridging units incor-

Figure 7.13 *Blom Surveyor. The hull employed for the twin-keel survey vessel is based on a fibre-reinforced plastics sandwich incorporating Du Pont Kevlar high-strength aramid fibre. Major benefits include increased strength and reduction in radiation 'noise' during surveying*

COURTESY DU PONT DE NEMOURS INTERNATIONAL SA

Figure 7.14 *Glass fibre composites have proved very suitable materials for the construction of minesweeper hulls. This vessel, of the Hoddesden class, is 61 metres (200 feet) in length and here the bridge-section is being joined to the hull-section. Hulls of this type are the largest GRP mouldings yet developed*

porates a high proportion of carbon fibre to accommodate bending and torsion.

Aramid fibre has also found recent application in boat-building because of its stiffness and impact resistance. In view of the poor properties in compression, it is understandable that hybrids of glass and aramids are preferred in a wide range of leisure and working boats. The weight saving is still considerable over all-glass constructions.[5]

We may summarise by saying that glass, carbon and aramid, separately or as hybrids, now play a central part in the boat-building industry. The products range from rowing and sailing dinghies, large and small yachts, powered cabin cruisers and high-speed motor-boats, lifeboats, police and coastguard patrol boats, in-shore and deep-sea fishing vessels and large mine detectors and mine sweepers for the British Navy, and many others (Hall and Hall 1984).

Figure 7.13 shows the twin-keel survey vessel *Blom Surveyor*, which is typical of modern practice, while Figure 7.14 shows the hull of a mine sweeper of the Hoddesden class.

7.5 SPACE STRUCTURES

The rocket power which launches spacecraft into the earth's orbit and beyond, is achieved by burning fuel at an enormous rate. Most of the weight at lift-off is due to the fuel, and the remainder is for structure, auxiliary engines and the useful payload. Thus in space technology there is every incentive for weight saving by the widest use of composite materials.

While some fuels and oxidants (peroxide or nitric acid) will require metal or metal-lined tanks of great resistance to corrosion, the solid propellants are normally enclosed in filament-wound bodies (Krings and Scharring-hausen 1985, and Shankar 1980), the size of which puts them among the largest structures made from composites.

The giant bodies of American and European practice are wound from glass fibre and epoxy resin. Indeed, the development of S-glass, which is somewhat stronger, somewhat stiffer and less dense that E-glass derives directly from the drive for greater structural efficiency.

It was failure of the rubber seals between sections, in cold weather, during the last abortive flight of the Shuttle, which has led to the examination and testing of filament-wound carbon/epoxy cases for the solid propellant.

More readily accepted for structural parts in carbon fibre are the nose fairings (on Ariane 4 (Eymard 1985) and 5, for example), the interstage structures and various other components of the satellites which are placed in orbit as part of the payload.

The introduction of the reusable Shuttle provided new scope for space structures. The enormous cargo-bay doors are of sandwich construction with carbon/epoxy/honeycomb materials, and the remote manipulator arm is a favourite design case for the instruction of students.

The manipulator arm is a 15 m (50 ft)-long articulated arm for handling and placing items in the payload bay (Dunbar, Robertson and Kerrison 1980). It has the full range of movements of a human arm, comprising the equivalent of the shoulder joint, elbow and wrist joint and a payload-grappling mechanism which replaces the human hand. The mechanical arm comprises two 5.4 m (18 ft)-long booms, connected with pivot points, and made of carbon/epoxy laminate. A television camera monitors the movement, and the operator gives instructions to the electric motors supplying power.

In order to work with precision, the deflections under load must be extremely small. This leads to the choice of very high-modulus carbon/epoxy material GY70/934 in an orientation which is primarily unidirectional. However, it is recognised that all such unidirectional materials are lower in shear strength and stiffness; since the manipulator arm, like the human wrist, needs a high torsional stiffness, it is necessary to add a proportion of $\pm 45^\circ$ fibres as well.

Both upper and lower arms have internal carbon/epoxy stiffening rings, while an outer buffer layer, comprising a covering of aramid laminate bonded to a sandwich core of Nomex honeycomb, protected the structural tube from impact damage. The current weights are 24.7 kg (54.4 lb) for the upper arm and 23.8 kg (52.4 lb) for the lower.

An important requirement for structures operating in the space environment is the need for a very low coefficient of thermal expansion. The temperature normally ranges from -200°C to $+80^\circ$C (-328° to $+176^\circ$F), though for purposes of earth-bound testing a range from -150°C to $+80^\circ$C (-238° to 176°F) is specified.

Apart from the main structures already mentioned, components such as antennae (aerials) reflectors, panels for arrays of cells producing solar power, mirror carriers, optical telescopes and their platforms and benches, all depend on a precise knowledge of thermal coefficients.

Yates and his team at Salford University have measured these very carefully with a wide range of composite materials and have looked at the influence of different fibres/resins/volume fractions and orientations on the expansion coefficients over the temperatures encountered in space and up to 350°C (662°F) (Yates 1977).

Other workers in Europe (Galipienso and Dell'Amico 1985) and the USA have checked particular structures to verify the comprehensive theories now established.

Of great importance is the possibility of blending fibres having different coefficients of expansion (such as high- and medium-modulus carbon or high-modulus carbon with glass) in order to obtain a coefficient of almost zero.

The design of a typical carbon/epoxy structure for the Japanese domestic communications satellite CS3 has been described in some detail by Kawashima, Inoue and Seko (Kawashima, Inoue and Seko 1985). The satellite has a capacity of 6,000 simultaneous telephone calls (circuits) with an operational life in excess of 7 years.

Figure 7.15 shows the structural configuration adopted. There are two equipment panels separated by struts. The upper panel carrying mission subsystems weighs 110 kg (242.5 lb), while the lower panel carries bus equipment and weighs 260 kg (573.2 lb).

The massive thrust tube rises through central holes cut in the upper and lower panels. Inside the thrust tube is installed the apogee kick

Figure 7.15 *Design of a typical carbon/epoxy structure for the Japanese domestic communications satellite, CS3*

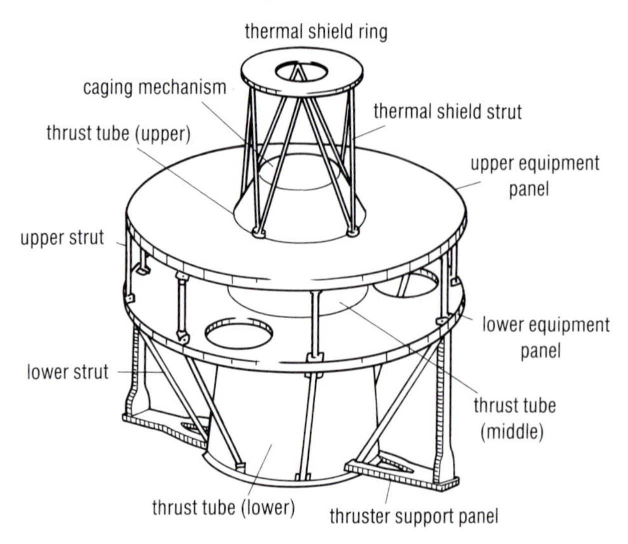

Source Kawashima, Inoue and Seko 1986

motor. The bottom portion of the central thrust tube bears on aluminium rings which form the rocket interface plane and transmit the satellite loads to the rocket.

Above the top of the upper equipment panel is a thermal shield ring, which is joined to the panel by a cage consisting of further struts. Apart from the interface rings, the only other aluminium components are the upper and lower equipment panels. Here the alloy is necessary because of the heat conductivity requirements, while the instruments are working. Except as stated, the entire satellite structure is composed of carbon-epoxy material. The use of carbon/epoxy struts for joining together and creating load-bearing structures (as in the Japanese CS3) is very widespread (Franz and Laube 1985). The subject of joints has been pursued in several research programmes because of its fundamental importance (Stöffler and Wurlinger 1981, and Garilotti, Johnson and Cwiertny 1980).

Very often the tubes must be provided with end-caps, either in light alloy or in titanium, through which the joints are realised. Again the choice of metal is dictated by the thermal expansion required.

Marconi has developed Europe's first remote sensing satellite – the ERS-1 – for ocean observation, and has a major role in putting military hardware into space (Figure 7.16). And, now that a decision has been taken by Europe to join with the USA in the 'Columbia' space station project, there is a revival of interest in the construction of very large space

Figure 7.16 *Skynet 4 military satellite is primarily an aluminium honeycomb and skin structure. The antenna systems clustered on top of the spacecraft are constructed from carbon fibre/epoxy materials while the special and spot beam reflectors are made of ultra-high modulus GY70 skins with aluminium honeycomb. This choice of materials gives an optimum trade-off of stiffness, strength and thermal stability*

COURTESY MARCONI SPACE SYSTEMS LTD

structures. Various suggestions have been put forward of which the following should be mentioned:

(i) The complete tubes, already hinged together, are packed tightly before launch. After entering orbit they are erected and locked together permanently, rather like an umbrella. A number of neat erection mechanisms have been devised for this purpose.
(ii) A soft fibrous array, impregnated with resin but as yet uncured, is packed tightly into a container. It can be drawn from the container and stretched fully and at the same time irradiated strongly with UV light to cure into a rigid network. As an alternative, some members only of the framework are already cured, but the junctions are tied with precatalysed uncured resin, already in position on the fibrous ties. Again, after erection the resin needs irradiation to lock the structure securely. A thin pultruded rod of circular cross-section which has been closely coiled to the minimum radius will spring back strongly to take the straightened form.
(iii) An array of pultruded rods is made with a thermoplastics matrix. The rods can be hot-welded together quite rapidly (about 15 seconds in the NASA experiments) since the carbon fibres conduct heat along the grain without difficulty.
(iv) Finally, long planks of composite made with a thermoplastics matrix can be stored in a compact roll and taken into orbit. This was described in Chapter 2. The thickness is chosen so that the plank is readily rolled into a drum, while the width is chosen so that the beam welder, as described by NASA, can bend and then weld it into a rigid beam of triangular section.

In these various ways, large platforms, immense solar arrays and large reflectors will pave the way for civil engineering projects in space.

7.6 APPLICATIONS IN SURGERY

Conventional glass-reinforced plastics have been used for many years as splints of a minor nature. They have sometimes been moulded directly off the patient's body, eg for a dislocated thumb or retracted top finger joint, or to support a broken forearm, using a grease parting agent in conjunction with an acetate lacquer, and using the body heat to effect quick gelation of the matrix resin. Polyester resins are used since amine-hardened epoxy may cause dermatitis.

With more complicated orthoses, a plaster cast is made first, so that the composite can be moulded off it without the patient's presence being needed.

Artificial legs are moulded in glass/polyester and filled with polyurethane foam to stabilise the thin shell. They are used both in UK and in Europe to replace metal.

Often, glass cloth can be bonded directly to the outside of a plaster cast, to make it more resistant to abrasion and impact loads. For example, young children with the lower leg in plaster are soon very boisterous, and can cause disintegration of an unreinforced plaster heel within a few days.

With the introduction of carbon fibre and carbon-fibre-reinforced plastics (CFRP), more spectacular advances have been made in several ways, as follows.

7.6.1 Frames for power-assisted arms

The thalidomide tragedy of the early 1960s created a sudden demand for lightweight harnesses for the attachment of power-assisted arms.

A common technique, having taken an accurate cast, is to hot-form strips of closed-cell polyethylene foam over the cast. Shallow grooves are machined into these strips previously, so that carbon and resin can be laid directly into the grooves and allowed to cure. The result is a rigid lightweight frame which is comfortable to wear because of the cushion of PE foam over the framework.

Since the growing child needs a larger frame from time to time in order to accommodate longer power-assisted arms which are in proportion, this method of moulding provides a quick and convenient method of renewal.

7.6.2 Artificial lungs

The original 'iron lung' was a necessary but cumbersome apparatus, and there have been many efforts to devise a portable version, so that the patient can sit or move around for a few hours outside the chamber.

One of the most successful designs involves a lightweight shell or 'cuirasse' made of hybrid glass/carbon/polyester laminate, shaped rather like the back of a tortoise and having around the rim a deep flexible seal conforming to the patient's chest. From the domed centre of the cuirasse – approximately above the sternum – an armoured hose is attached so that a portable pump can extract and admit air. The changes in pressure within the cuirasse act on the diaphragm to allow respiration at the normal rate.

Figure 7.17 *ERS-1 is a scientific development satellite featuring a number of new composite technologies due to the special nature of the payload: the Active Microwave Instrument (AMI). This is a high power radar instrument. RF energy is passed through metallised CFRP waveguides to slotted waveguide antennae. These antennae are flat structures and irradiate the RF energy through slots. The large SAR antenna is likewise made from metallised CFRP waveguides*

COURTESY MARCONI SPACE SYSTEMS LTD

7.6.3 Artificial limbs

The firm of Chas A Blatchford and Sons Ltd has developed a modular system for artificial legs which consists of a series of carbon fibre/epoxy components, in standard stock sizes, which can be joined at the knee and ankle to suit the patient (Shorter 1986).

Use of the structural tubing in this manner (known as the Endalite

Figure 7.18 *Artificial legs using carbon fibre/epoxy components. Their strength, stiffness, and low density make carbon fibre components ideal in orthopaedics*

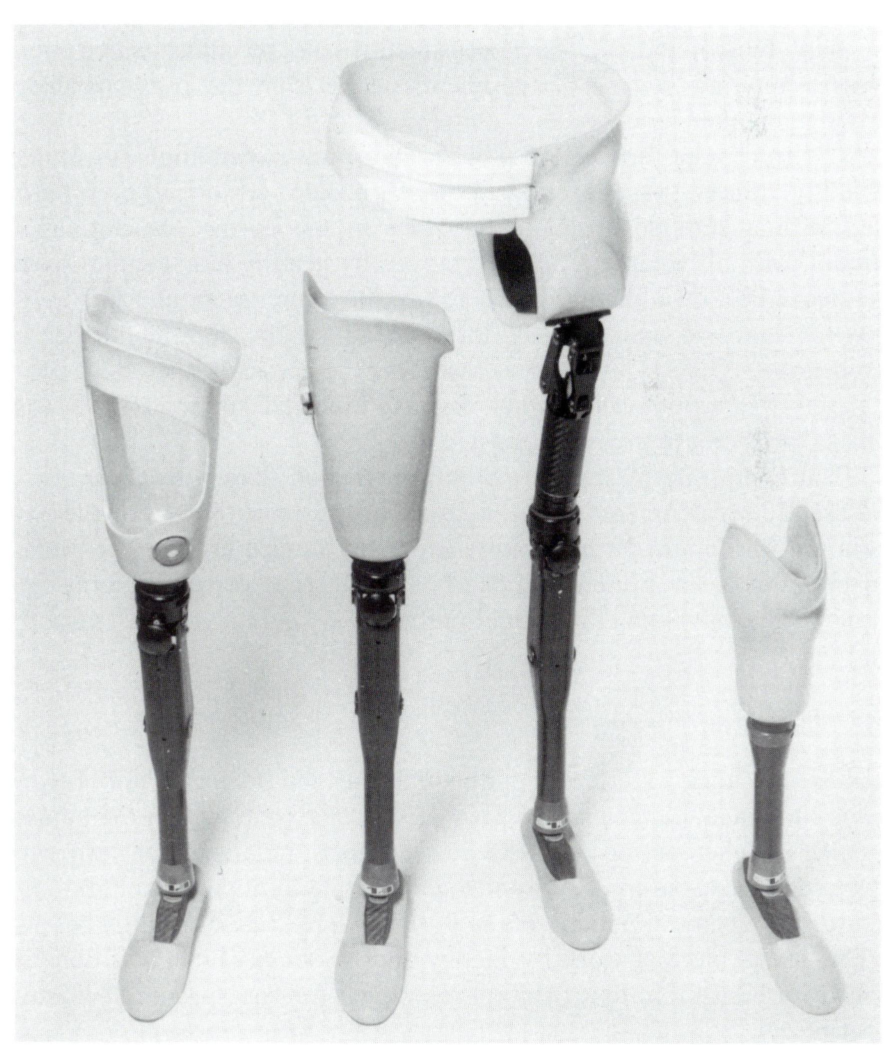

COURTESY CHAS A. BLATCHFORD

system because it provides an endo-skeleton), together with an outer soft covering for cosmetic effect, has proved outstanding in terms of compression and bending strength and in torsion. The fatigue testing shows that several lifetimes of cyclic stress are exceeded. There is a considerable saving in weight, thus helping the older and weaker amputees.

7.6.4 Bone surgery

The same company and several others now produce and market carbon fibre/epoxy plates which are employed in bone surgery to replace the expensive titanium plates previously used.

The use of CFRP within the body depends on having materials which are fully bio-compatible. The occasional tissue reactions which were sometimes found in earlier experiments derive from the epoxy/hardener formulations then in use.

As Scheer found (Scheer 1981) in an important paper, aliphatic amines generally induced tissue reactions. Where a purified epoxy had been hardened with a suitable aromatic hardener in the correct stoichiometric amount, with all volatile residues removed by heating in a vacuum oven, and cleaned before implantation, there were no tissue reactions.

With regard to epoxy resins, those based on the diglycidyl ether of Bisphenol-A (DGBA) and those based on the glycidyl ether of a polyphenol (GPP) were excellent when diamino diphenyl sulphone (DDS) was used as the aromatic cross-linking agent.

When composite plates are made from carbon fibre with these resins and used to repair broken bones, a layer of connective tissue (polymerised fibrin and thrombin) forms rapidly across the surface of the plate. Many red blood cells are present but there is an absence of white blood corpuscles, which is a sign of incompatibility.

7.6.5 The brain

Because of their inertness, individual carbon filaments of 8 μm (0.3×10^{-3} inch) diameter have been used as electrodes within the human brain. It is sometimes necessary to go even finer in diameter, and workers have used oxidation in air or in nitrogen dioxide to obtain a uniform and progressive 'chemical milling' of the fibre down to 2 μm or slightly below.

This useful piece of work by Jacoby and Shorrock (1980) was done in order to produce thin fibres suitable as targets for laser-produced plasma experiments.

Figure 7.19 *Tows of carbon fibre can be used for the repair of the Achilles tendon. The growth of new tissue is also speeded considerably. The photograph of sheep's tendon shows that the freshly repaired tissue* (*right*) *is somewhat larger than the original tendon* (*left*)

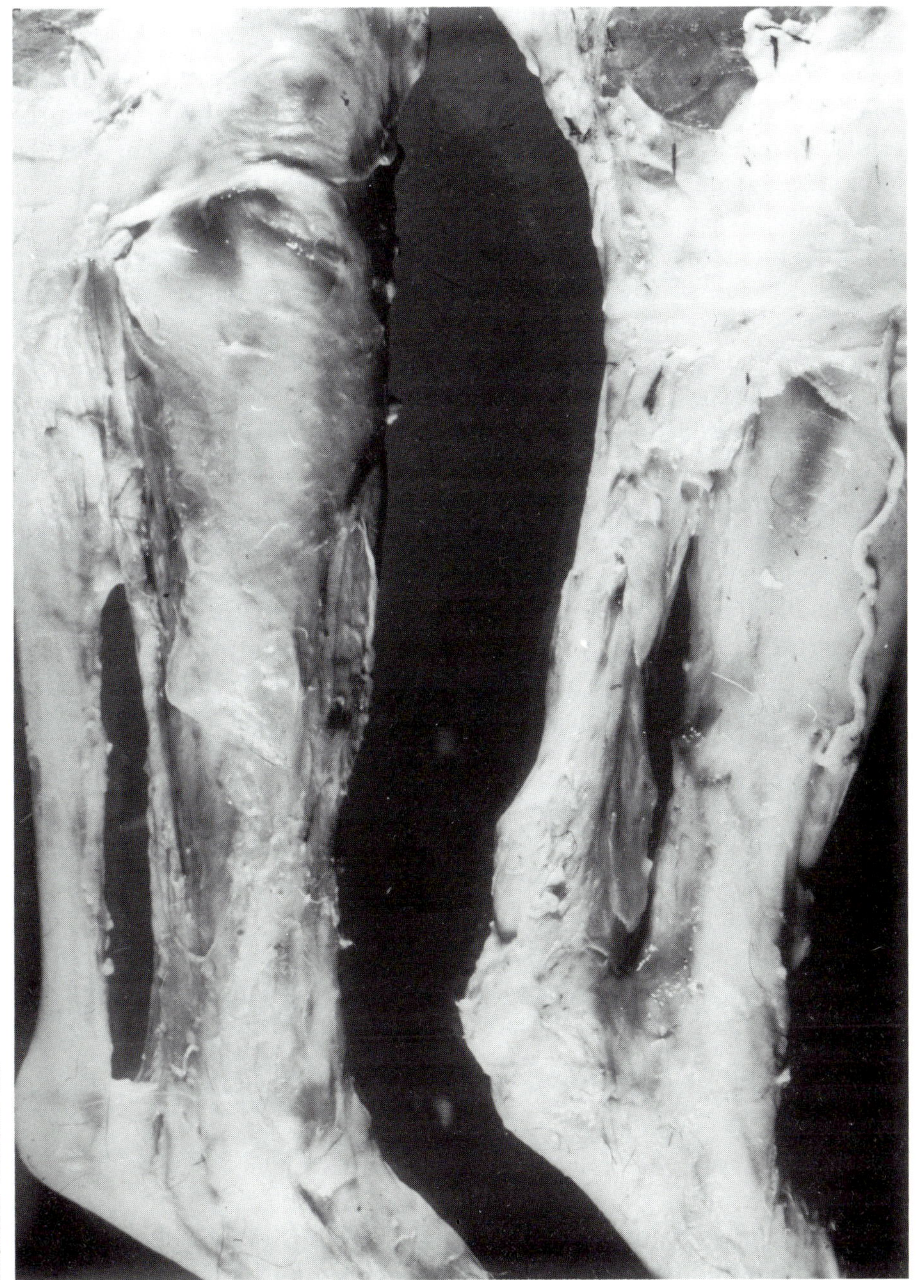

COURTESY McKIBBIN 1983

7.6.6 Ligaments

Carbon fibres based on polyacrylonitrile are known to be biocompatible, and this has led to one of the most remarkable uses in surgery, ie the repair of damaged ligaments.

Early and successful experiments were done by Jenkins on knee ligaments using four tows of 10,000 filaments each plaited together (Picquet 1981). This was used to join the patient's knee joint together by drilling holes in both tibia and femur and passing the plaited string through the resulting channels. Having adjusted the correct movement or 'play' in the joint, the fibre was knotted and secured at the exit holes.

Another surgeon has used a slightly different arrangement of fibre for repairing the Achilles tendon which joins the heel bone to the calf muscle.

A loop of several tows is passed through holes in both heel and tendon end and is 'tuned' by twisting the loop (by means of a small sterile stick inserted in the loop like a tourniquet) until the correct tension has been established. It has been found that this tension promotes more rapid healing (McKibbin 1983).

Elderly patients in particular, for whom tearing the Achilles tendon is a serious matter, normally requiring 12 or more weeks of convalescence, find they are able to resume exercises (even old-time dancing) in 6–8 weeks.

Animal experiments show that, in the presence of carbon fibre, new tissue grows from the ends of the loop towards the centre. The final stages of repair show that the carbon is gradually removed via the lymph glands to leave the new tendon somewhat larger than it was before (Figure 7.19).

7.6.7 Dentures

In the allied field of dentistry, papers of carbon fibre have been used to reinforce the acrylic resins used in dentures. For example, British Army dentists at Aldershot have saturated several layers of carbon fibre paper with monomer before 'flasking' or moulding. This strengthens and stiffens the thinnest part of the denture, across the roof of the mouth (normally an unseen area) in order to resist more easily the shocks of normal use.

7.7 APPLICATIONS IN SPORTS EQUIPMENT

One of the first practical steps to take with a strong stiff fibre is to impregnate a number of strands and draw them into a long tube, acting as a mould, where they are allowed to cure. This leads directly to the development of parallel rods of solid cross-section (soon to be done much better by the pultrusion technique) and then to solid tapered rods and tapered tubular rods.

Early experiments at RAE in conjunction with Hardy Bros (quality rod makers to the world) and with Richard Walker, the famous fisherman and writer on angling subjects, showed that the prototype rods had an extraordinary performance.

7.7.1 Fishing rods

Fly rods, acting as beams, threw the flies considerably further than either split bamboo or glass fibre. These new rods were thinner and lighter, had excellent fatigue resistance, by working always within the straight-line portion of the stress-strain curve, were perfectly elastic, and all the energy stored on the back cast was returned during the forward throw.

As the design was improved, manufacture was simplied by the development of fine unidirectional and bidirectional fabrics, so that the rolling technique with preimpregnated fabrics could be performed, as already practised with glass.

Starting from these modest beginnings, a vast number of different rods have evolved, ranging from the most delicate of fly rods to the more powerful spinning rods, reservoir rods and salmon rods. Beach casting rods and boat rods have been produced by other companies. These are often hybrid carbon/glass constructions.

A particular case of elegant exploitation of the high specific stiffness of CFRP is in the design of the so-called ‘roach pole’. Originally this was a one-piece cantilever of hollow bamboo with a fixed line, hook and float, about 9 metres (nearly 10 yards) in length. The idea is to raise the rod over any bank-side obstacles and place the bait accurately towards centre stream. This method of fishing was and is a special favourite with French anglers.

The old-fashioned pole has been given a new lease of life by making 2- or 3-piece carbon fibre rods as a small fraction of their former weight, making them much more comfortable to use.

Somewhere in the archives of the Materials Department at the RAE, there is a letter from Richard Walker in which he says that the development of carbon fibre for fishing rods ‘saved the British tackle industry’. The same remarks could doubtless be made with respect to the North American, Swedish, Scottish and Japanese efforts in this field.

7.7.2 Golf clubs

Because of the ease of producing rods and tubes, an enormous effort has gone into the development of the carbon fibre golf shaft. The popular theory is that, by having a light, stiff shaft, a higher proportion of the weight can be put into the head, which is accelerated faster and whose energy is returned elastically on impact. Early trials indicated an increased distance of the drive by more than 20–30 metres (25–35 yards).

At any rate, several companies had been dismayed by fatigue failure in light-alloy shafts, while the golfing public were becoming conditioned to the expense of costly stainless steel shafts. Many golfers, including top professionals, took eagerly to carbon-shafted clubs, so that for a few years it helped to sustain the demand for carbon fibre before the aeronautical applications had taken off.

7.7.3 Bows

There have been many attempts, going back to the Asiatic and Turkish horsemen, to improve on the compound bow, which used dissimilar materials (sinew and horn) for the tensile and compressive faces of the bow (Hardy 1976).

With the advent of glass fibre, laminations of glass and wood had greatly improved performance. Aramid fibres are poor in compression and have not proved a success, but there have been several designs involving carbon fibre.

In view of the large distortions involved in stringing a bow, the consensus is for a high-strain, low-modulus carbon on the tension side.

In order to avoid bending the arrow round the stock on release, there have been many ingenious designs put forward, including a moulded carbon stock having a central hole. The increased stiffness of lightweight carbon tubing makes this a desirable improvement over light alloy for the arrows themselves.

7.7.4 Racquets and bats

A large number of designs for tennis racquets have been produced which incorporate carbon fibre. There are two main types: (*a*) frames are injection-moulded with a thermoplastic moulding compound with a high percentage of carbon fibres. The fibrous reinforcement is in short lengths; (*b*) continuous carbon fibres are moulded around a central foamed core, usually of polyurethane material in order to stabilise and stiffen the primary load-bearing hollow frame.

The Dunlop Company has pioneered the first approach, while the Belgian firm of Donnay has promoted the second. As a fabrication technique, with general application, it is hard to beat for directness and simplicity.

Strips of polyurethane foam are inserted into the centre of the hollow braid length of carbon fibre. After smoothing down the braid (which decreases in diameter when the length is increased) the assembly is impregnated with a solution of hot-setting epoxy resin, dried, precured and hot-formed into the racquet shape before placing in the mould for final cure. The two short straight arms are bound together rigidly before inserting and bonding into the hand-grip.

Similar methods are used for squash rackets, although the amount of reinforcement needs to be increased substantially. The patent literature records the efforts of many inventors to improve the stringing, to enlarge the 'sweet-spot', to decrease weight, improve the weight distribution, alter the shape etc, since there is a large international market to be reached.

Table tennis is a popular indoor sport, and the traditional plywood bat is considerably improved by making a carbon/plywood sandwich construction.

7.7.5 Skis

Skis and ski poles are outlets for composite materials on an increasing scale. The technical problem with skis is to tailor the reinforcement (skins over a wooden core) so that the flexural properties and the torsional stiffness, required in turn, are satisfied. The American company K2 uses braiding technology to manufacture composite alpine skis (Figure 7.20), while the Finnish company of Excel OY has built up a considerable business making ski poles. Lapinleiumu (1985) has described the growth from nil in 1973 to an average of 1.5 million units by 1982–83.

A number of lay-ups are possible. The all-carbon tube is the most efficient. Where a heavier and tougher unit is required, a carbon/glass hybrid is still superior to the best aluminium tube, while for outstanding impact resistance a carbon tube with a Kevlar (aramid) outer layer gives a better result at the minimum increase in weight. The author speculates that the same manufacturing process can be used for hockey sticks, ice hockey sticks and windsurfing masts.

Before leaving the fascinating subject of sports equipment, one should mention that the use of composites for the upgrading of wood is under-exploited. The racing oar is one of the few successful examples, along with double-bladed paddles for canoes.

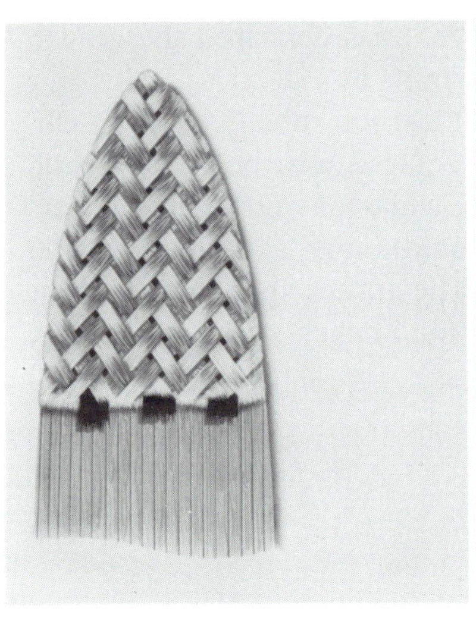

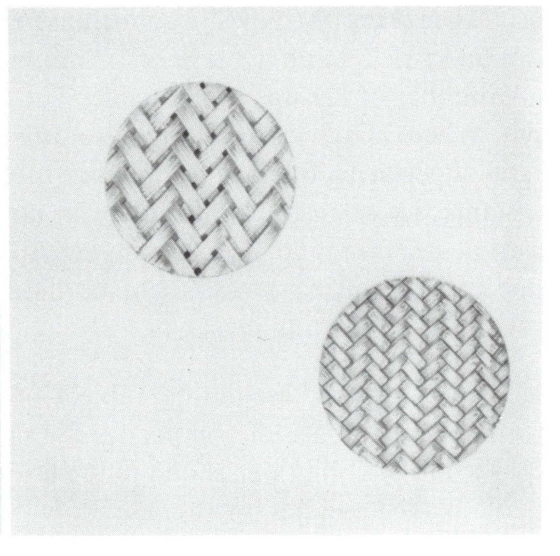

COURTESY K2 CORPORATION, WASHINGTON, USA

Figure 7.20 *K2 TRIAXIAL alpine snow skis use fibreglass, carbon and Kevlar strands which are interlocked or braided under tension around a laminated wood core. This mixing of fibres is of great significance in ski design because it allows the most efficient use of composite fibres*

There has been speculation from time to time as to whether those traditional sporting symbols of American and British life, the baseball bat and the cricket bat, respectively, would be improved by composites. Granted that there are now 'springs' of carbon fibre/epoxy laminate in the handles of top-grade cricket bats, it is unlikely to progress further; while the rules of the game and plentiful supplies of hickory will preserve the status quo of the great American game.

7.8 MECHANICAL ENGINEERING

The applications to general mechanical engineering have been rather slow to materialise. This is understandable for the following reasons:

(i) Composite materials are far costlier than the general range of metals.
(ii) The techniques of fabrication are considerably outside the expertise of the average production engineer.
(iii) There are pitfalls in designing with anisotropic materials.
(iv) Normally, metals work very well, and it is only when driven by known drawbacks such as corrosion, poor resistance to wear, excessive weight etc that there is incentive to try something new.
(v) The expenses of research and development are high and liable to rise even higher as the new consumer protection legislation forces manufacturers into careful field trials in order to furnish proof that their products are safe and 'fit for the purpose'.

With these words of explanation for the rate of progress in mechanical engineering, one should nevertheless list those specific areas where application has been made already:

(i) Friction and wear: The coefficients of friction and the rates of wear were determined at an early date at Farnborough on blends of carbon fibre with a wide range of polymers.

Figure 7.21, from the work of Lancaster (1966) gives typical results. It will be noted that wear rate (amount removed in milligrams per centimetre travelled per kilogram of load applied) is registered on a logarithmic scale, on pairs of specimens with and without fibre added.

Figure 7.21 *Effect of carbon fibres on the wear rate of polymers. The incorporation of carbon fibre can reduce the wear rate by a thousand-fold in many cases*

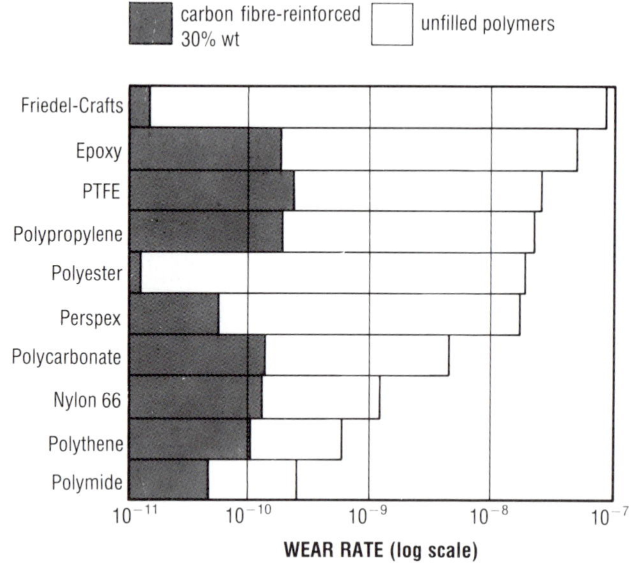

Source Lancaster 1966 (adapted with permission of the Controller of Her Majesty's Stationary Office)

Clearly, a whole new class of dry self-lubricating bearing materials has been unearthed, with wide application.

For example, Figure 7.22 shows a schematic diagram of an oil-free compressor, used for pumping sensitive fluids such as vegetable oils,

Figure 7.22 *Oil-free rotary compressor (with CFRP blades). These compressors are required in the pumping of milk, wine etc, which must be taint-free*

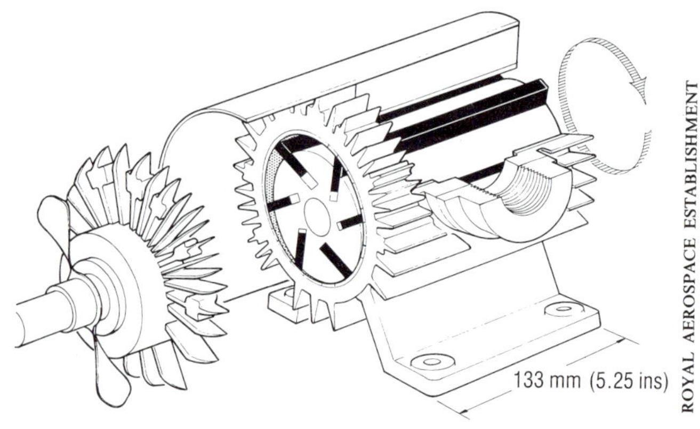

blood and milk. For best wear resistance there should be $\pm 45°$ fibres as well as $0°$ fibres present.

In the same way, gear wheels can be cast (using nylon 6 monomer as the matrix) or injection-moulded (nylon 6.6, polyacetal etc) as the working parts of fishing reels which do not suffer from corrosion in a salt-water environment.

A large number of small injection-moulded parts made with carbon or carbon/glass hybrid moulding granules have appeared which make use of the electrical conductivity of the material, or the very low coefficient of expansion, or both.

Camera bodies for 'compact' 35mm photography are an example of the very accurate parts which are now produced by injection-moulding, which were formerly made in machined metal.

(ii) Centrifugal forces: A fly-wheel, rotating at high speed, can store, and return, a large amount of energy. The potential advantages are considerable and the design of such devices has been studied in some detail. Christensen and Wu (1978) of the Lawrence Livermore Laboratory, concluded that equal reinforcement in both radial and circumferential directions was required.

Worthington (1978), of the Central Electricity Generating Board, examined the requirements for hoops and rings in the design of new types of future generator. There are advantages in CFRP because of the high specfic strength and stiffness and good fatigue resistance.

Potter (1985) has described the production of a high-speed flywheel (with protecting case) which was installed recently in a passenger bus, where much energy is normally lost in stopping and starting. With energy stored, the bus can be started off with flywheel assistance, and the fuel saving is considerable.

7.9 VEHICLES

The application of carbon fibre to racing car bodies was demonstrated dramatically with the hybrid GT40 Ford which won the 24 hour endurance race of Le Mans in 1968 and 1969 (Phillips 1969). Since then the amount of composite has increased steadily and the gradual change in design philosophy has been summarised neatly by Clarke (1985).

At first there was a simple fairing. Then the driver's seat was incorporated, along with side tanks. The final stage is a three-dimensional shell, reinforced in the necessary places for attachment of engine, wheel mounts, tanks, steering controls, bulkhead and instrumentation – in other words, body and chassis have become united (Figure 7.23).

Figure 7.23 *Advanced composites are being used in the latest racing car bodies. Top: is the all-carbon chassis, in an integral monocoque design, by ACT Ltd. Bottom: The TWR Jaguar XJR-6, the first-ever sports racing car to be composed of varying thicknesses of a strong and light combination of carbon fibre and Kevlar (the Du Pont fibre)*

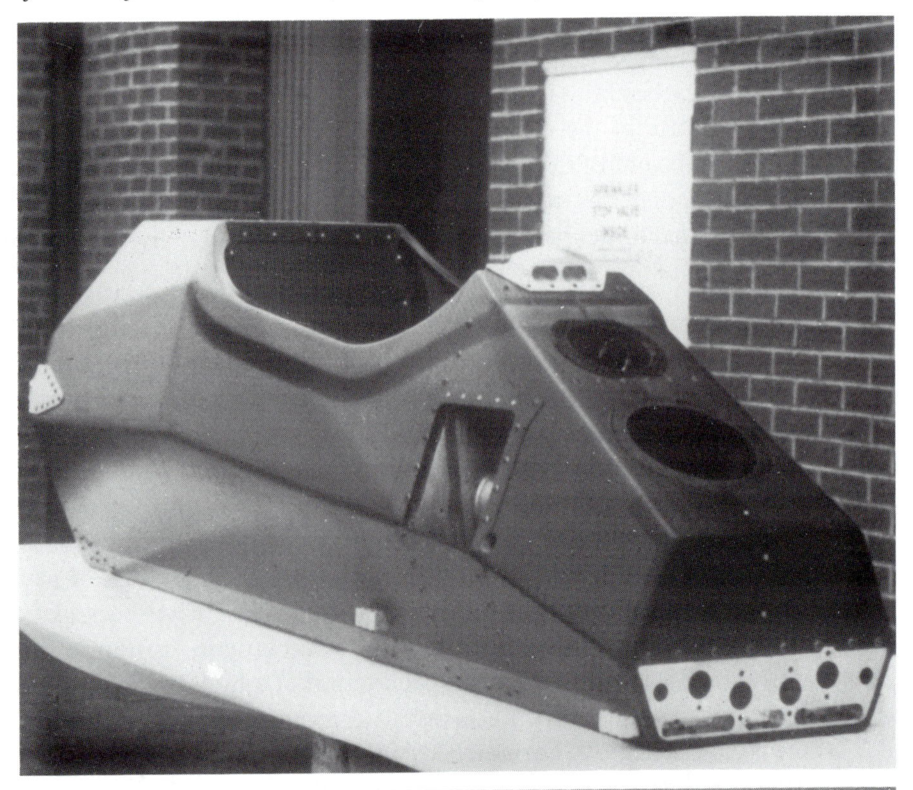

COURTESY TOM WALKINSHAW RACING LTD

Figure 7.24 *Leaf springs are increasingly being made of carbon fibre, owing to its low weight, excellent fatigue life and lack of corrosion*

COURTESY GKN TECHNOLOGY

Two other developments have considerable general importance:

(i) The first is convincing proof, both analytical and practical, of the advantages of the composite drive shaft. This has been examined carefully in Germany, the USA, Japan and the UK. A small decrease in the price of carbon fibre should clinch it (Ruegg 1981).
(ii) The second is the use of laminated leaf springs to replace steel in both private cars and heavy lorries (Kliger and Yates 1980). Fatigue life is now fully adequate, and one company at least (GKN) is in production for commercial road transport.

Mention should also be made here of a grand 'tour-de-force' by the American Ford company, who built a prototype saloon car almost entirely in carbon fibre to establish feasibility (Dharan 1978). Weight savings of roughly 565 kg (1,246 lb) were possible on an original vehicle weight of 1,700 kg (3,748 lb), which would in turn lead to considerable fuel economy (Beardmore, Harwood and Horton 1980).

7.10 MUSICAL INSTRUMENTS

At the London College of Furniture and elsewhere there has been much interest in the use of advanced composites in designing musical instruments.

In the case of stringed instruments such as the guitar and lute, the tension in the strings is often sufficient to distort the neck until they become unplayable. There are many possible designs to stiffen and strengthen the neck, but one of the simplest is to machine narrow grooves in the wooden neck and to bond into these grooves lengths of unidirectional carbon-reinforced pultrusion. A section 3 mm square (an eighth of a square inch) would be suitable in most cases.

Sounding boards are another possibility where laminated ply or foamed plastics as the core, bonded to thin skins of carbon or hybrid, can replace the very expensive wood laminations (Haines and Chang 1975). Guitar and other bodies have received attention, and reproducibility of sound quality is an important advantage of the moulded construction.

Lastly, a violin of high concert quality has been made almost entirely of carbon fibre/epoxy material (John 1980). If such violins could be mass-produced using the best moulding techniques, they would be very popular. With some fine tuning, perhaps they could rival the few remaining examples of the old violin makers at a fraction of the prices obtained at auction (Figure 7.25).

Figure 7.25 *Carbon fibre/epoxy violin. The use of CFRP gives a violin of full concert quality. There are now many such applications in musical instruments*

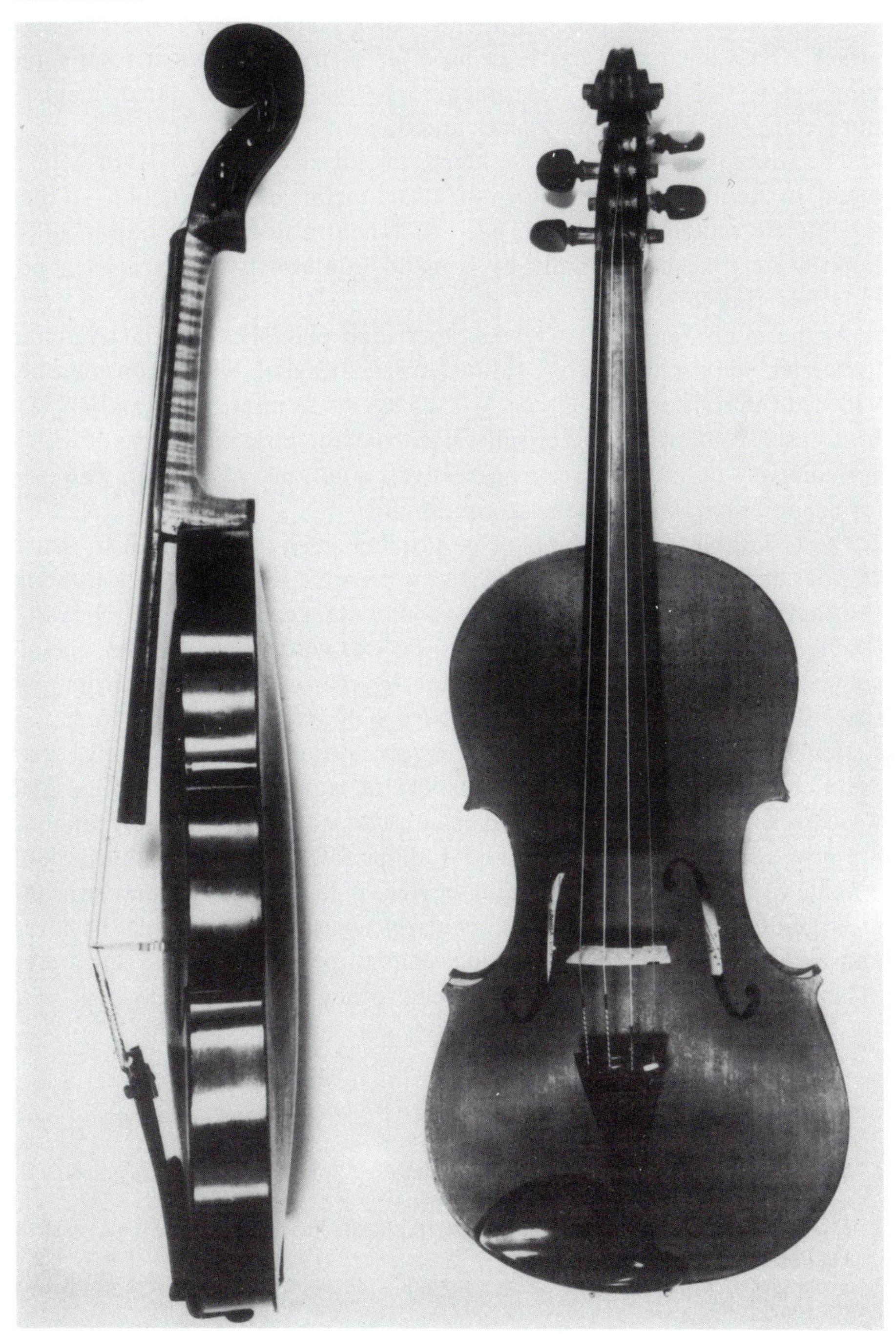

7.11 CHEMICAL PLANT

The chemical engineer faces a constant battle against metal corrosion and is always alert to the possible use of other materials to resist chemical attack. Glass-lined steel has been used for many years, but it took some time before the industry was prepared to risk complete tanks in glass fibre, complete with auxiliary pipes and ducts.

The strength of some of these items was lower than the steel or Monel metal, or stainless steel that were the usual metals of construction, so that a complete redesign was often needed. When in doubt, the non-metallic tanks were braced externally by a metal framework to share the load. However, this corrodes.

At the same time, matrix resins other than phenolics became available (polyester, vinyl-ester, epoxy, furane, Friedel-Crafts), which considerably widened the resistance to attack by such agents as dilute acids and alkalis, fruit juices, tomato and other sauces, petroleum crude oils (some of which containing sulphur are very aggressive), solutions of sodium chloride (brine) and many other chemical compounds.

By a suitable combination of reinforcing fibre (asbestos, glass, silica, carbon) and the correct matrix, it is now possible to offer good long-term resistance to many aggressive environments encountered in industry (Phillips and Judd 1969). Thus, over the past two or three decades, chemical plant makers have transformed themselves from metal fabricators into specialist companies with a wide knowledge of materials.

Items such as pumps, reaction vessels, storage vessels for oils, feed grain, wine etc, pipes for the transport of water, chemical sludge and waste, sewage, and for the handling of crude oil installations on tankers, are now commonplace (Green and Phillips 1978). Because of their light weight, which is less than 1/4 that of steel pipe, filament-wound sections in large-diameter glass fibre are regularly transported by mule, by lorry and by helicopter to inaccessible locations to pump away mud and water in landslides and to provide portable emergency water supplies.

NOTES

1 Weight savings for aircraft using Kevlar. Booklet from Du Pont de Nemours, Wilmington, Delaware
2 Mylar – the tradename for polyetheleneterephthalate film, produced and marketed by Du Pont
3 Turbopropeller technology. Brochure no DR/118/2/4000/TL/DF. Dowty-Rotol Ltd, Gloucester, England

4 Modern developments in materials applicable to yacht construction. Advisory Committee for Yacht Research. Proceedings of symposium held in April 1969. SUYR Report no 26, Dept of Aeronautics and Astronautics, University of Southampton, England

5 Hand lay-up laminates with Kevlar 49 aramid fibre. Brochure no E39608, Du Pont de Nemours SA, Geneva, Switzerland

8 The Quality Control and Non-Destructive Evaluation of Composite Materials and Components

P R Teagle

President
Qualcorp, Aerospace Systems
Chatsworth
California

8.1 INTRODUCTION

The proliferation of manufacturing organisations undertaking the production of composite components is reaching vast proportions. In addition to established manufacturers in a number of industries acquiring composite skills, we see the emergence of a large number of specialist composite manufacturing companies. There are many reasons for this rapid growth. The three most widely put forward are:

(i) The ability to enhance the functional or environmental performance of an existing component by replacing it, or supplementing it, with a composite equivalent. This makes use of such benefits as improved fatigue performance and reduced weight

(ii) The ability to manufacture structures which could not be produced with any material other than the high-stiffness, high-strength carbon composites. This category includes many applications in reciprocating machinery as used in the textile and agricultural machinery industries

(iii) Reduced manufacturing costs. It has been shown that once a composite culture has been instilled into a company, benefits can be realised at every stage from the initial design through to final production. It is possible to reduce manufacturing times by whole orders of magnitude.

In the first two areas, improvements in metallic materials are likely, in the long term, to reduce these advantages; lithium-based aluminium alloys are beginning to gain some ground in these applications. However, it is currently inconceivable that metallic material manufacturing would ever be able to achieve the reduced manufacturing costs realised with composites.

One major area of concern, however, is that composite structures present unique problems with respect to the verification of their 'as-manufactured' and long-term fitness for use. The problems arise because material of extreme anisotropy is produced by a process which is complex in nature, involving programmed variations in temperature and pressure, intricate lay-up of the required layers of prepreg, often manually, and with a large number of variables and opportunities for deviation from specified procedures. Furthermore, component manufacturers are unavoidably moving into the realms of raw-material production. In addition to their established role of fabrication, they perform the final processes of composite-material manufacture by virtue of the fact that the material and the component are produced in the same operation.

The measurement, control and reduction of production variability is one of the fundamental objectives of the quality organisation within manufacturing companies. It can be seen therefore, that quality assurance has a vital role to play in realising the full potential of composite technology.

This role is considered in general terms in this chapter. Discussion will centre on the quality control (QC) principles applied at the various stages of the manufacturing process. The significant points, resulting from the peculiarities of composites, are explained. The key stages discussed are:

(i) Material quality approval: the assurance that the materials provided by the prepreg manufacturers are as specified and to the required standard. In addition to detailed visual examination, the material quality approval checks described below consist of agreed mechanical tests which are supplemented by a range of physical and chemical analysis techniques. Additionally, non-destructive testing (NDT) methods are applied to the mechanical test samples for two principal reasons: (a) to determine the uniformity of quality of these samples and hence identify any local anomalies which may invalidate the mechanical test results; (b) to help form an NDT database for evaluation of the components when in production.

(ii) Manufacturing quality control: to ensure that the correct manufacturing procedures have been adhered to, thus making it possible to achieve designed form, fit and function. This covers a range of activities including personnel certification, documentation and process control, together with process verification which is normally conducted on test coupons cured with a component.

(iii) Fabricated component inspection: to ensure that the fitness for use of the component is achieved. The fabricated component is inspected using a variety of NDT techniques in an attempt to identify any production-induced defects. The types of inspection techniques used fall broadly into two categories: (a) those used to verify that the composite laminate has been correctly consolidated and cured; (b) those used to determine that the structure produced from the laminates is free from structurally significant defects.

8.2 MATERIAL QUALITY APPROVAL

It is common amongst component manufacturers to assume that their QC responsibility begins with the prepreg. However, the material producers' responsibilities are obviously extended when pre-cured laminate is supplied, as is the case with thermoplastic matrix materials. Material quality approval therefore takes the form of receipt checks, carried out on the as-received prepreg, followed by a number of similar revalidation or 're-life' tests, at specified intervals while the material is stored in the uncured condition.

8.2.1 Receipt checks

The extent and type of batch receipt checking varies greatly. Visual examination and other simple tests are usually carried out on a sampling basis to determine:

- freedom from unflagged cosmetic defects such as splits, gaps, whorls and edge straightness deviation
- fibre distribution
- surface tack
- resin/fibre content
- volatile content
- flow characteristics
- resin solids content

Some manufacturers limit further prepreg evaluation to tests aimed at determining the gelation characteristics of a resin sample. This is to acquire information necessary to make adjustments to the cure cycle. It is the norm, however, to conduct mechanical tests on samples produced to standard procedures, as agreed with the supplier. The usual tests are:

(*a*) *Unidirectional tape*:

(i) flexural strength to indicate fibre properties
(ii) room temperature interlaminar shear strength (ILSS) to assess degree of adhesion between fibre and resin
(iii) elevated temperature ILSS to indicate resin reactivity
(iv) thickness and volume fraction measurement to determine effective consolidation

(*b*) *Woven cloth*: Room temperature tests are carried out to determine the following characteristics:

(i) tensile strength/modulus
(ii) compression strength/modulus
(iii) short beam shear.

In an attempt to reduce the number of mechanical tests necessary to qualify 'as-received' material and 'as-cured' coupon specimens, a number of analytical chemistry techniques have been widely adopted as an alternative. It is usual to evaluate the following properties of the received prepreg and sections cut from small cured samples to characterise the material and assess its compliance with standards:

- chemical formulation
- resin viscosity and its variation as a function of time and temperature
- resin reactivity and temperatures of reaction
- degree of cure of the resin in the test sample
- fibre/resin volume fraction.

The techniques used for monitoring these properties are, respectively,

(i) infra-red spectrophotometry, and
high-performance liquid chromatography
(ii) temperature-programmable cone and plate viscometry
(iii) differential scanning calorimetry
(iv) differential scanning calorimetry, and
dynamic mechanical analysis
(v) acid digestion.

The effective curing and consolidation of a component is evaluated by monitoring the degree of cure of the processed resin. The techniques used are as described in (iv) above. These clearly are destructive and require samples to be removed from simultaneously cured coupon specimens.

For major producers, the simple assessment of prepreg properties against fixed acceptance standards has been superseded by statistical analysis of batch test results. Prepreg manufacturers are thus informed of undesirable trends in their product before they reach a level which would cause rejection.

8.2.2 Revalidation

Having satisfied themselves that the as-received material meets approved quality standards, the component manufacturers are required to ensure that it remains within these standards until cured into components.

The time and temperature of storage, in approved refrigeration conditions, is logged – together with the time at ambient temperature while material is being removed from the batch. When a predetermined period has lapsed, it is necessary to re-life the prepreg. This usually involves a repetition of the mechanical and thermal analysis tests described above. It is now becoming common practice in the USA to input the material storage parameters into the cure cycle control instrumentation. This is to adjust the cure cycle automatically to compensate for the prepreg condition.

8.3 MANUFACTURING QUALITY CONTROL

The QC effort described above is aimed at ensuring that the prepreg to be cured into a component is of an approved standard. It is now necessary to control the manufacture to ensure that any process variables remain within acceptable limits. This implies that strict, monitored control must be exercised over personnel, documentation and processes.

8.3.1 Personnel

For the majority of producers, the QC procedures involved bear only the slightest resemblance to those previously established for metallic components. As a result, it has been shown necessary to re-educate all personnel involved in CFRP production. Some manufacturers have extensive formal training for their operators, highlighting the problems associated with composite production. Others operate an approved operator scheme based on written and oral examinations. Even though relatively high levels of automation can be brought into prepreg tailoring and component lay-up, the operator is still regarded as being the greatest potential source of quality variation.

8.3.2 Documentation

The range of QC documentation employed by the majority of manufacturers follows the same general pattern:

(i) Component drawing: This is the controlling document giving component geometry, including lay-up detail. It also calls up all relevant QC documentation.
(ii) Process specifications: These contain the general requirements for all composite component manufacture.
(iii) Technique sheets: These contain requirements relating to particular materials or components.
(iv) Planning sheets: These travel with the component, giving details of all operations to be carried out. On completion of an operation, the card must be stamped by an approved inspector.
(v) Component history cards: These travel with the component and record all relevant process details and QC test results. All entries must be stamped by an approved inspector.

In addition to the process parameters, these manufacturing documents define the QC procedures together with the acceptance limits. It is usual to incorporate these in the technique sheets, although in some organisations they may be found elsewhere.

8.3.3 Process control

Having ensured that the prepreg available to the production personnel is satisfactory, that adequate documentation exists to define the component and control its production, and that the personnel are conversant with the general procedures required, the task of QC is now to monitor the production of the component. In particular, the following must be determined:

- the type of prepreg is as specified
- the laminate geometry is correct, ie number, direction and sequence of plies
- the moulding tool components are correct, ie bleed packs, caul plates, vacuum bags etc
- the cure cycle is as specified

The first three are normally achieved by establishing a closely monitored production sequence with frequent visual inspections. It is normal for inspectors to examine the components 'kits' before, during and after assembly. The counting and re-recording of kit remnants, such as backing paper and polythene film, is an important aspect of the latter.

The conformity of the cure cycle with laid down standards is usually demonstrated by permanent records produced by the control instrumentation (see Chapter 2, Section 2.3.3). The correlation between the response of these instruments and the cure conditions over the component surface is previously established with component calibration trials.

8.3.4 Process verification

Although nothing can be done to affect the quality of a cured laminate, postcure QC procedures are necessary to confirm that the specified process has been successfully performed. To this end it is common practice to include one or more coupon specimens in each component batch during cure. Unidirectional 2 mm (0.08 inch)-thick specimens, for ILSS and flexure tests, are produced almost universally. Some manufacturers include additional coupons which are more representative of the component being cured. Nevertheless, it is generally accepted that the purpose of the coupon specimens is to confirm that the cure cycle is correct, and not to represent qualitatively the component being produced. The tests performed on the coupon specimens are as follows:

(i) Flexural tests: Although these results are influenced by several factors, they give an indication of the tensile and compressive properties of the fibre. Hence it is possible to confirm that the fibres are of the correct type.
(ii) Room-temperature ILSS: This test essentially gives a measure of the cohesive shear strength of the resin and the adhesive strength of the fibre to resin interface. Hence it is possible to confirm adequate fibre surface treatment and the absence of voids.
(iii) Elevated-temperature ILSS: Satisfactory elevated-temperature ILSS tests confirm that the resin is correctly formulated and adequately cured.
(iv) Fibre volume fraction: The fibre volume fraction is determined to aid in the calculation of flexural strength. It gives little indication of the success of the cure in its own right, although major process anomalies can be indicated by inadequate resin bleed.

8.4 FABRICATED COMPONENT INSPECTION

The inspection applied to the fabricated component varies considerably according to its purpose, geometry and method of production. In addition to the usual visual and dimensional checks, it is often possible to provide a section of laminate cut from the component for destructive evaluation.

As the size and shape are governed by the component, it is often not possible to carry out the standard coupon mechanical tests. Normally some form of compression test is devised as this gives an indication of the performance of both resin and fibre. However, the principal fabricated component inspection is performed by subjecting the component to a range of non-destructive tests. The types of test used and the reasons for their selection are discussed in the remainder of this chapter. A summary of the applicability of the methods is given in Table 8.1. In this section, the techniques used in production are discussed.

Table 8.1 Non-destructive tests applied to CFRP laminates and structures

Technique	Defect or property evaluated														
	Material									Structure					
	Void content	Fibre content	Resin content	Fibre orientation	Material homogeneity	Mechanical properties	Delaminations	Fatigue damage	Contamination	Delaminations	Disbonds	Cracks	Debris	Condition of internal	components
Radiographic															
low-kV, high-definition				•	•				•				•	•	•
radio-opaque penetrants	•														
radio-opaque tracers				•											
gamma		•	•												
Ultrasonic															
attenuation measurements	•	•				•	•	•			•	•			•
velocity measurement	•	•		•	•	•									
goniometry				•	•	•	•								
spectroscopy	•				•	•	•	•							
resonance					•	•	•				•	•			
holography											•	•			
Optical															
optical holography											•	•			
speckle photography											•	•	•		
laser interferometry											•	•			
Others															
acoustic emission						•		•			•	•	•		
eddy currents		•		•	•	•		•					•		
microwaves	•						•								
mechanical impedance											•	•		•	•
Thermography											•	•		•	•

8.4.1 Conventional ultrasonics

This involves interrogation of the materials or structure using ultrasound. Usually for composites a frequency range from 1 to 10 MHz is used. Both continuous-wave and pulse techniques are used, the latter being by far the more common. There are two fundamentally different techniques of ultrasonic inspection: through-transmission and pulse echo. The ultrasonic principles behind these techniques have remained unaltered since the earliest application of ultrasonic technology. The major advances achieved in recent years have resulted from the digital control, acquisition and analysis power now available to the average manufacturing organisation. Few fundamentally new techniques have emerged from the laboratories. Those novel applications most likely to be adopted for widespread use are discussed briefly in Section 8.4.2. The through-transmission and pulse-echo techniques are described as follows:

1 Through-transmission

This technique involves the measurement of variations in a reference signal. Modulations to this reference signal, caused by the variability in the propagation of the sound waves as they pass through the medium being inspected, are interpreted in terms of the structure and properties of that medium. A number of features of the transmitted ultrasonic pulse are measured. The most common is pulse amplitude. Also of interest, and increasingly finding their way into production environments, are pulse frequency content monitoring, known as ultrasonic spectroscopy, and pulse velocity measurement. The reference signal used takes one of many forms:

(*a*) *Pitch-catch*: This most simple case is achieved by transmitting sound directly from one transducer to another. The reference signal is that received with nothing but the coupling medium between the transducers. The sample is then introduced and variations in the ultrasonic response monitored. The most common method of applying this technique to the inspection of aerospace components is with squirter or jet probes. Here the ultrasound is coupled to the component through columns of water. The digitally controlled jets are usually scanned over the panel using scanning equipment with data being acquired automatically for real-time and post test analysis and display. Typical equipment is shown in Figure 8.1.

(*b*) *Reflector plate*: Another common method uses the signal reflected from a standard reflector, such as a piece of glass, as the reference. This technique, widely used in the inspection of solid laminates, has a

Figure 8.1a *Automated ultrasonic scanning equipment is used to detect faults in carbon fibre composite laminates and sandwich panels. Shown here, the through transmission, water jet system being utilised to inspect a Harrier GR MK V carbon fibre composite tailplane*

COURTESY BRITISH AEROSPACE

number of practical advantages over the previous method. Only one transducer is required, the pulse being transmitted and received by the same transducer. This gives considerable potential saving on scanning equipment as the requirement to keep the transducers co-axially aligned is eliminated. Multiple transmission through thin laminates can be monitored to increase the effective signal/noise ratio. It has the disadvantage that for highly attenuative samples, particularly honeycomb panels, it is not possible for the sound to pass twice through the

Figure 8.1b *Computer-aided ultrasonic inspection of carbon fibre composite aircraft structures*

1 A beam of energy, vibrating at high frequency, is generated by the transmitter probe through a jet of water on to one side of the composite. The energy passes through the composite, and is detected at the other side by a receiver probe, again via a jet of water. The water acts as a coupling agent.

2 The amount of energy detected by the receiving probe is affected by the quality of composite through which it has passed. Voids and porosity will reduce the amount of transmitted energy, and large laminar faults will completely block transmission.

3 The received ultrasonic energy is electronically amplified and presented as a voltage, which is proportional to the quantity of received energy. This voltage is displayed on an oscilloscope, the time base of which indicates the time taken from ultrasonic transmission to its reception.
An electronic gate, which can be set in time captures the received voltage and passes it to the computer.

4 The motorised test rig drives the ultrasonic water jet coupled probes back and forth along the test item. At each end of the travel the probes are indexed, up or down, to generate a raster scan. This raster scan is controlled by the computer, in length and index, to cover the pre-determined shape of the test item.

5 Point-by-point data files are generated by the computer and stored on hard discs. These files comprise X and Y scan co-ordinates, and for each co-ordinate, which usually represents one square millimetre, a block of digitised data representing the voltage of the received signal, is stored.

6 The computer system then generates a picture of the test item with the digitised ultrasonic transmission values converted into colour bands. The range, width, and colour code, of these bands are selected by operator interaction to optimise the sensitivity and quality of the picture presented for viewing, and operator interpretation.
From this picture the quality of the item and its acceptance, or non-acceptance, are determined.

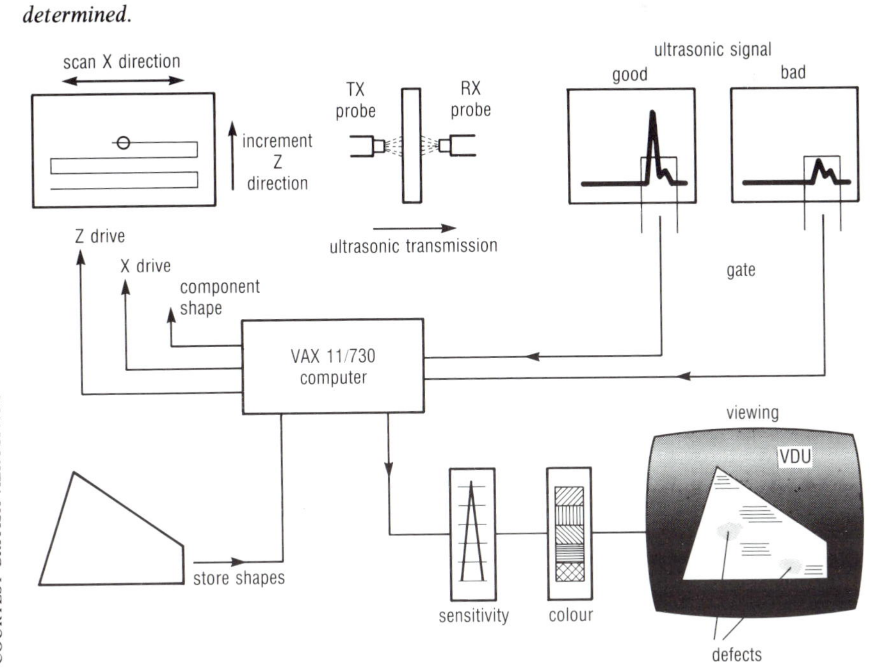

COURTESY BRITISH AEROSPACE

structure. This technique has applications in coupon specimen testing and is the principal inspection method for thermoplastic matrix material, which is supplied to the component manufacture in the form of pre-cured, pre-tested sheets.

(c) *Back-wall echo*: This through-transmission technique involves the monitoring of the signal reflected from the back surface of the item being tested. This has the added practical advantage of no auxiliary reflector being required. It has the disadvantage that the variation in ultrasonic response due to changes in the condition of the back surface are superimposed on response changes due to component structural variation. Back-wall echo techniques are used where it is not convenient to position an auxiliary reflector. In particular, complex geometry components can be inspected in a fully programmable contour-following immersion tank using this technique. Ultrasonic pulse echo testing can also be performed on similar equipment.

2 Pulse echo

The second ultrasonic technique is referred to as pulse echo. For this we do not observe how the specimen under test modulates a standard reference signal. Instead we monitor the reflections from discrete discontinuities within the laminated structure. A single transducer is used, resulting in the advantages described in (ii) above. Again the pulse features monitored are amplitude, frequency content and time of flight. In this case the latter is used for defect depth determination rather than velocity measurement. One of the major advantages of these single-transducer techniques is that access is usually required from one side of the component only. This has great significance for in-service inspection.

The data acquisition, analysis and display is common to all types of ultrasonic test. A typical graphics display giving the plan position of the transducer together with the amplitude at that position is shown in Figure 8.2. This is normally termed a 'C-scan'. The different colours are operator selectable to represent various amplitude thresholds. Once acquired, the data can be manipulated to aid the analysis and determination of the structural significance of the ultrasonic response. It is possible to vary the thresholds and increase or reduce the area being displayed. It is also possible to use a wide range of digital image enhancement and analysis techniques. In the latest generation of ultrasonic inspection systems complex contour following, using high-speed curvilinear motion control, is performed. Motion control, through-transmission testing, pulse-echo testing, data acquisition, analysis and presentation are all performed simultaneously, in real time.

Figure 8.2 *Ultrasonic C-scan with liquid immersion tank*

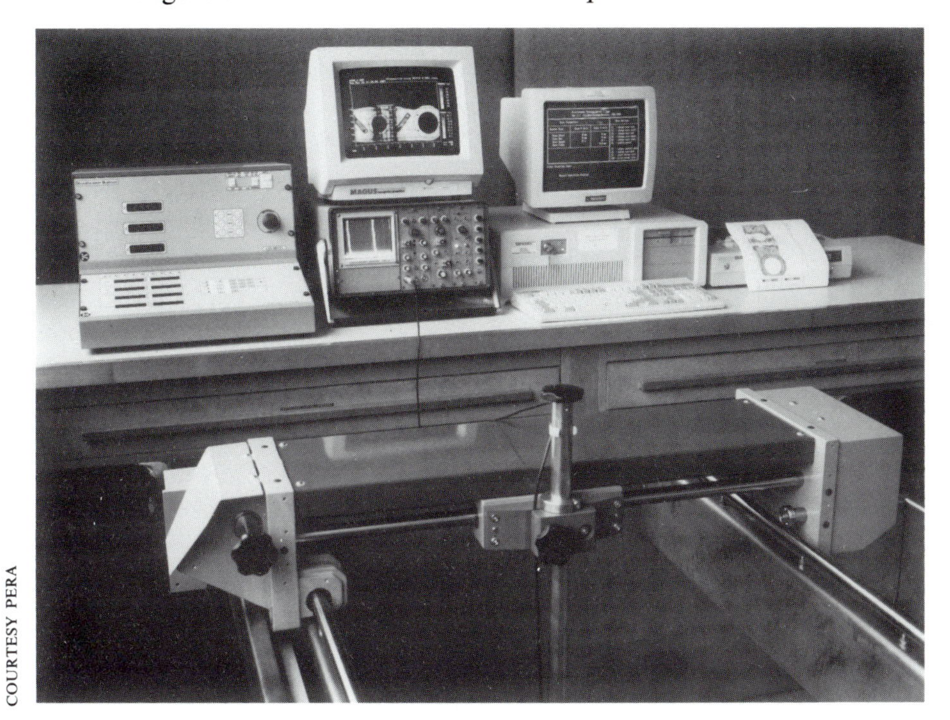

COURTESY PERA

8.4.2 Emergent ultrasonics techniques

A number of novel ultrasonics techniques are now finding increased use in the production environment. This list is by no means exhaustive; it contains some techniques which are known to have made a contribution to production testing in some companies, but are not yet widely used:

1 Dry-coupled roller probes
Dry-coupled ultrasonic transducers employing a silicon rubber contact face are widely available. The rotary wheel types allow automatic scanning. Recent developments (Dickson 1988) have provided multi-element assemblies which increase the speed of inspection. Pulse-echo wheel probes have also been developed. The major advantage of this product is that it is capable of inspecting a large area rapidly with little or no contamination to the surface. A wheel probe is shown in Figure 8.3.

2 Air-coupled probes
There were obviously enormous benefits to be obtained if the advantages of non-contact and dry coupling could be achieved simultaneously. Accordingly, considerable effort was directed towards the production of a

Figure 8.3 *Multi-element dry coupled wheel probes*

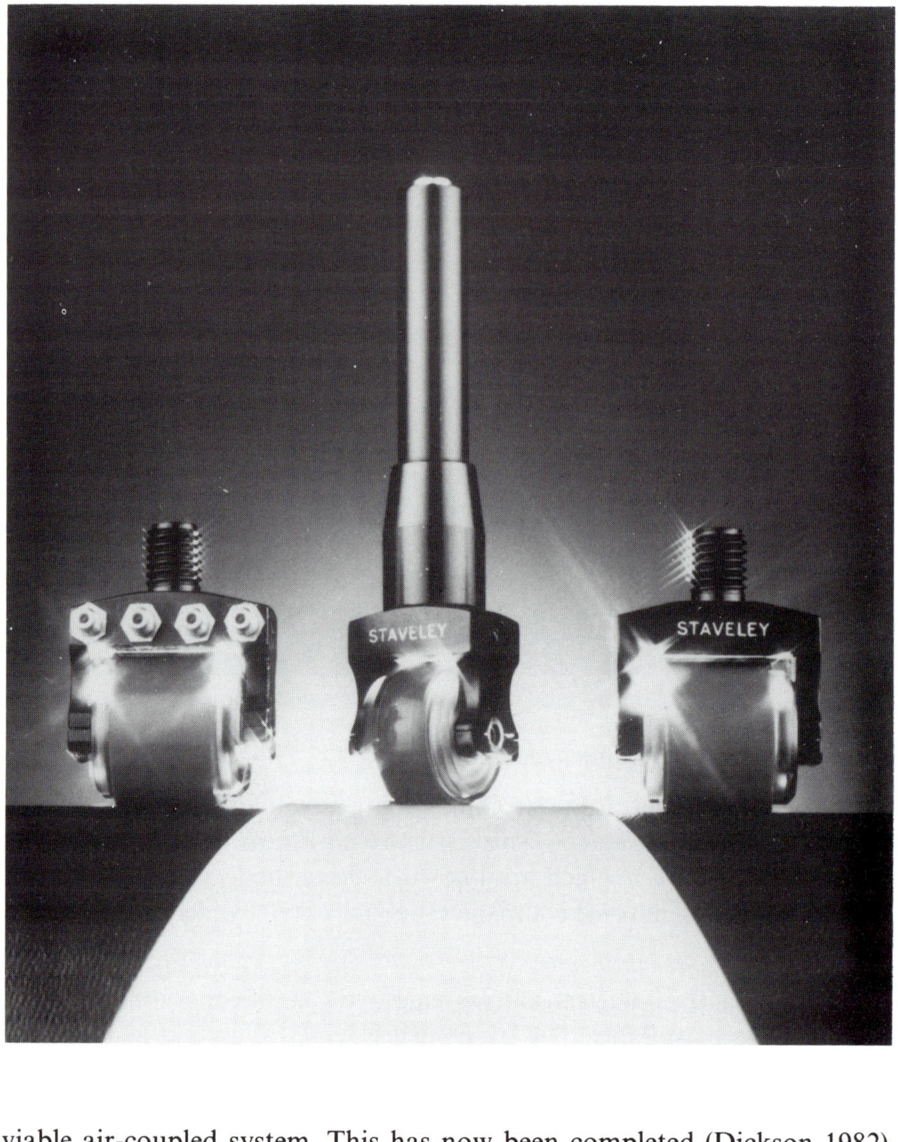

COURTESY STAVELEY INDUSTRIES PLC

viable air-coupled system. This has now been completed (Dickson 1982). This recent development involved improvements in both the transducer and ultrasonic pulse generator designs to enable transmission of pulsed ultrasound at moderately high frequencies across large air gaps.

3 Ultrasonic spectroscopy

As described in Section 8.4, ultrasonic spectroscopy involves the measurement of the frequency content of an ultrasonic pulse. Features of this

Figure 8.4 *Typical ultrasonic spectroscopy results*

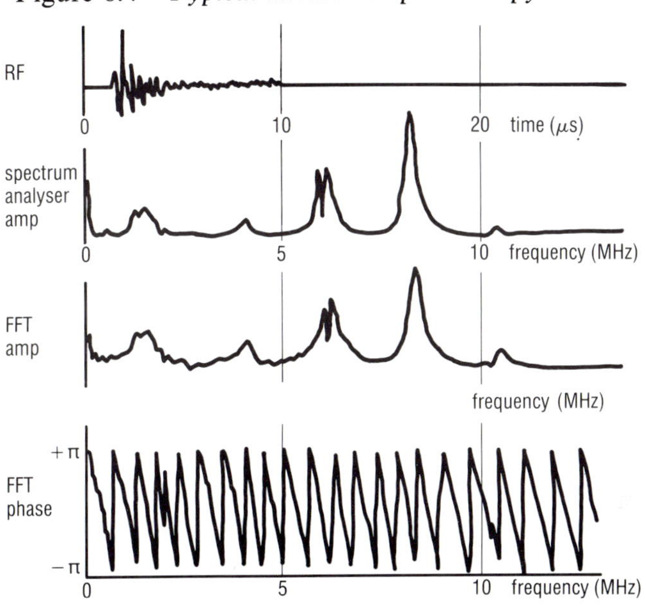

spectrum are then correlated with material characteristics. All spectroscopy is now performed digitally. This requires that a digitised time-domain signal, or A-scan, be acquired. This then becomes the real component of a complex matrix which is transformed into frequency-domain information using a fast-Fourier transform (FFT). One advantage of this method is that the imaging component of the complex matrix can be used to give phase information. These potentially useful data are irretrievable by analogue methods. The rapid development of high-speed digitising boards has meant that signals with frequencies up to 10 MHz can now be amply accommodated by an IBM PC plug-in board. Typical results are shown in Figure 8.4.

The testing of CFRP is likely to develop along the lines of current research. Interference and scattering techniques will be extended, but diffraction techniques, for predicting reflector geometries, will have little application. Conceivably, both velocity measurements techniques, as described above, and attenuation measurements techniques, as described below, will be performed with spectral analysis equipment. For velocity measurement in thin laminates, this technique has already been demonstrated to be superior to conventional methods. The ability to determine the relationship between attenuation and frequency from one spectrum has possibilities as a rapid production inspection technique.

4 Ultrasonic velocity measurement

There are numerous advantages to be gained from being able to measure the elastic constants of a composite laminate non-destructively. Mechanical tests (ILSS and flexural strength etc) described earlier are indirect methods of measuring the constants or their effect, and are generally time-consuming and expensive. In addition, the relationship between these measured mechanical strengths and the elastic constants are complex, as they were devised for isotropic materials. Hence calculations of the structural behaviour of a laminate using results of established mechanical tests is difficult.

Ultrasonic velocity measurement has been shown to be a viable method of directly determining the dynamic elastic constants, truly reflecting the anisotropic nature of the laminates. As an illustration, for unidirectional laminates, the five independent elastic constants could be measured, as opposed to a single inter-laminar shear strength. Monitoring these five constants will facilitate the separation of the effects of the fibre-to-matrix bond, the fibre and resin volume fractions and the void contents.

8.4.3 Radiography

The range of radiographic techniques applied to composites is vast. Both conventional X- and gamma-ray techniques are widely employed using equipment identical to that used in the inspection of metallic aircraft structures. Such techniques are able to detect cracks parallel to the radiation beam, damage to and presence of internal components, eg honeycomb core, and the presence of most types of debris and contaminant. In addition, lack of bond due to the absence of adhesive can usually be identified. Disbonds and cracks normal to the radiation beam cannot be detected as there is no loss of material. Stereography provides the most sensitive technique for characterising damage throughout the volume of a composite (Drummond, Brinkerloff and Tedrow 1980). The addition of radio-opaque penetrants and filament tracers has significantly increased the effectiveness of conventional techniques. This has resulted in viable production and in-service methods of determining ply orientation, stacking sequence, presence of butt joints and the extent of de-lamination caused by impact damage. X-ray microscopy techniques have been employed with some success. Recent advances combining micro-fluoroscopy with modern video processing techniques have resulted in a valuable laboratory tool which is being adapted for production inspection (Burch 1982).

More recently, real-time, filmless techniques with digital image enhancement capabilities have greatly extended the application of radiography. A clear example of the benefits of digital techniques is given by computerised tomography. The wealth of sensor manipulation, data acquisition, reduction and enhancement facilities used for ultrasonic inspection are now being evaluated for use with real-time X-ray systems. For the majority of producers, however, conventional, low-kilovolt, beryllium-windowed radiographic equipment is the mainstay.

8.4.4 Acoustic emission

All structures will tend to exist at their lowest energy level. If they are stressed, mechanically or thermally, they attempt to dissipate some of the increase in strain energy by moving towards their previous state. This relaxation is achieved by deformation of the structure. The deformation initiates at microscopic discontinuities and the strain energy is released in several forms, one of which is by acoustic- or stress-wave emissions.

Various methods are used to detect and process the emissions to obtain information about the microscopic failures. These are present in composites, usually in the form of resin fracture, fibre fracture or failure of resin-to-fibre bond.

The stress waves emitted during deformation are usually detected by one or more piezoelectric transducers. These convert the stress wave into a voltage pulse of the type shown in Figure 8.5. This pulse is typical of an undamped piezoelectric transducer and displays the characteristic sinusoidal decay.

Figure 8.5 *Stress-wave/voltage pulse conversion by an undamped piezoelectric transducer*

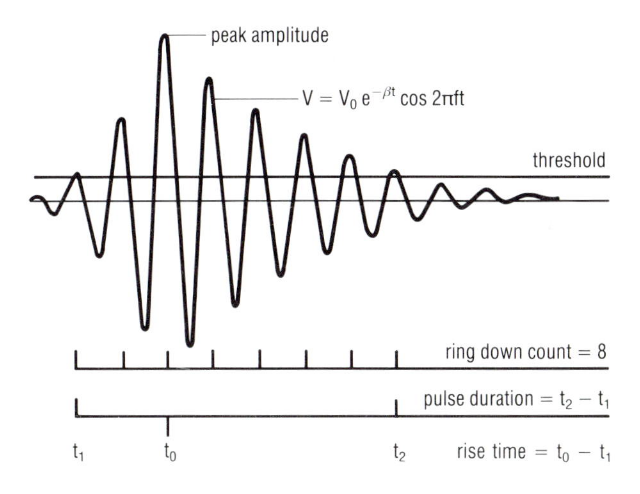

Variations in the quantified features of the pulse are then correlated with changes in the structural properties of the component being examined. This correlation is usually determined empirically during a series of mechanical tests on similar items.

The simplest feature to quantify is the number of times the received signal exceeds a threshold amplitude. This is called ring-down counting, and Figure 8.5 shows that the ring-down count for our typical pulse is 8. Obviously the count achieved is determined by the amplification of the signal and the level selected for the threshold. Although both are arbitrary, attempts to standardise are being made in order to aid comparison of data.

The stress-wave pulse is regarded as being produced by a single physical event, such as a matrix crack initiation in composites, or dislocation migration on a metal lattice. A useful alternative form of analysis therefore is to quantify the pulse as a single event. This is known as event counting. Effectively, all activity between the first and last ring-down count is called a single event.

A further commonly quantified and analysed feature of the pulse is the peak amplitude. The amplitude sorting of emissions is able to give an indication of the damage mechanisms generating the stress waves. Likely amplitude-characterised mechanisms include microcracking of the matrix, fibre fracture, failure at the matrix fibre board, and major discontinuities such as an interlaminar void and foreign bodies.

The pulse feature which is attracting most attention currently is the duration. This is the interval between the first and last ring-down counts. It has been found that medium-amplitude, long-duration pulses are characteristic of delaminations, particularly in pressure vessels and pipes. Standards have been agreed in the procedure for monitoring pulse duration, and such standard tests are now used as the basis of the manufacturing quality control of certain types of pressure vessel.

A less widely used method of analysis is frequency analysis or spectroscopy, similar to that described earlier for ultrasonics. Considerable work has been carried out attempting to correlate variations in the emission frequency spectrum with structurally significant anomalies. Although there have been several successful laboratory applications, related to damage mechanism identification, there are as yet no viable production techniques.

Finally, one of the most valuable characteristics of the pulse to be quantified is its differential arrival time at a number of transducers. This is used to give the location of the defect by intersecting loci. Although there are a great number of complications in the propagation of sound in a

Table 8.2 Applications of acoustic emission

Pulse feature	Anomally evaluated
counts	damage severity
events	damage severity
amplitude	damage mechanism
duration	delamination – quality
frequency	damage mechanism
arrival time	location

highly attenuative, anisotropic, heterogeneous material, this application is extremely successful. A summary of the applicability of the various techniques is given in Table 8.2.

It can be seen that the practical applications of acoustic emission are essentially concerned with the use of sophisticated electronic equipment to quantify, analyse, display and evaluate as many features as possible in a stress-wave pulse. The evaluation generally takes the form of deductions based on an empirical correlation between the measured feature and known physical anomalies.

An interesting development has been to extend the application of the acoustic emission equipment to evaluate how the material passively modifies an introduced pulse of ultrasound rather than the active internally generated stress wave. The technique, called acoustic ultrasonics, has the advantage that it is truly non-destructive as the component does not need to be stressed.

In addition to providing information about the local failure mechanisms, acoustic emission is used to monitor structures during service as a means of incipient failure detection; it is applied during proof tests to identify flaws which could induce failure later. On-board continuous monitoring systems are now being developed. The disadvantage of the technique is that it is not essentially non-destructive and, at the moment, it is still regarded as a laboratory tool.

8.4.5 Mechanical impedance testing

There are a number of commercial bond test instruments which excite a structure with comparatively low-frequency mechanical vibrations and measure the structural response. Although the responses monitored vary between instruments, the operating fundamentals are similar – many of them were spawned by the same research team at MIT in the late 1950s. Peizo electric excitation is widely used, although other methods are employed. Because of the frequency of operation, a couplant is usually required.

Nearly all these instruments require different probes for different materials and structures. The responses monitored include resonant-frequency shift, received signal amplitude and transmitted/received signal phase difference.

The performance of the instruments varies from structure to structure. Some are better at non-metallics and others at metallics; some are better at skin-to-skin bond and others at honeycomb components. Equipment employing a single dry-coupled probe has the advantage that it can be scanned rapidly in an appropriate manipulating device to provide a C-scan facsimile recording (Cawley 1987).

8.4.6 Thermography

Thermography has been used in the inspection of aerospace structures for many years. Several methods are employed to detect variations in the surface temperature resulting from the difference in thermal properties between good and flawed areas. All airframe inspection involves passive techniques, where a heat source or cooling has to be used to obtain thermal information.

The most commonly used methods of thermal monitoring fall into two categories, ie thermographic and thermometric. Both categories include contact and non-contact methods. Examples are given as follows:

- Thermographic contact: liquid crystals, treated paper, paints, and phosphors
- Thermographic non-contact: infra-red scanners, cameras, film
- Thermometric contact: thermocouples, thermopiles, thermistors
- Thermometric non-contact: radiometers, pyrometers.

All of these methods have been used for inspection of composites but there are several major disadvantages common to all thermal techniques. Firstly, a suitable heat/cold source must be provided, and the establishment of consistent temperature gradients is difficult. This in turn leads to difficulty in interpretation of results. Secondly, highly anisotropic materials with different thermal properties in different directions can cause problems. The difficulty with metallic structures, with high thermal conductivity, is that they are only able to maintain the temperature gradient around flaws for short periods.

The advent of video-compatible infra-red cameras and improved thermal sources has now enhanced this technique considerably. Many of the variables associated with traditional techniques can now be eliminated by using digital signal processing algorithms to extract information from noise. The technique has been further enhanced by applying non-stationary pulsed heat sources such as lasers and high-powered photographic flash tubes (Cawley 1987).

8.5 THE FUTURE

For production NDT development resources are currently being directed towards goals which lie in two distinctly different areas. The first, of more immediate benefit to production quality control, is to achieve improvement in the organisation and application of existing techniques. The second is the adaptation of fundamental developments in the laboratory for production use.

8.5.1 Improved application

In this area, the future of quality control and the non-destructive testing of composite components lies, as with many other manufacturing technologies, in advanced automation. Technology is available today to allow simultaneous automatic metrology, ultrasonic inspection and radiographic inspection unattended by a human operator. The data used to control these tests systems will be accessed using manufacturing automation protocol and data highways linked to the computer-aided-design facilities of the manufacturer. The concept of an NDT section will disappear because, within one plant, automated NDT equipment will be incorporated into a number of manufacturing cells. The rate of throughput of components will be the critical parameter for each of these cells. This arises because of the need to keep the 'in-process' storage, of components awaiting test, to a minimum. Hence, the emphasis on NDT equipment will move from cost to speed.

8.5.2 New techniques

It is beyond the scope of this chapter to discuss, in detail, current development activity in NDT. In addition to the fundamental development of new techniques, considerable success will be achieved in enhancing the performance of established methods. Two areas of research with considerable potential benefit are:

- the development of hybrid inspection systems combining many techniques simultaneously

- the development of large-area, rapid-evaluation techniques, such as acoustic emission, to the point where they have resolution and sensitivity equivalent to the scanned point evaluation techniques, such as ultrasonics.

It can be said that the range of production NDT methods is enormous and growing. No one technique is able to solve all the inspection requirements. Each new development seems only to add an additional weapon.

The Mathematics of Design

9 Design of Composites

Leonard Hollaway PhD MSc(Eng) CEng FICE MIStrutE

Professor in Composite Structures
University of Surrey

9.1 INTRODUCTION

The design of structures and structural components manufactured from composite materials for vessels, tanks, columns, tubes etc would initially involve the selection of the most suitable materials, fibre orientation and manufacturing techniques for the composite; each item will have a particular influence upon the mechanical properties of the material and the strength and stiffness of the final component.

The distinction between the design of reinforced plastics and that of metals is that the composite materials for particular applications are manufactured at the time of fabrication of the component, and therefore the design stage must commence with an integrated decision on the choice of the manufacturing technique.

Initially the design procedures for composite fibre-matrix materials followed closely those used for metals. However, these procedures proved to be inaccurate and now composite theory is used. It would be of value at this point to illustrate the difference between the ductile material and the polymer characteristics. A typical stress/strain curve for a ductile metal is shown in Figure 9.1. It exhibits elastic behaviour at small strains of less than 0.2%, followed by yield and plastic flow.

Design procedures for metals are based upon linear, isotropic analysis when calculating the stress and strain distribution within a component of small elastic strains. The parameters of deflection, stiffness, buckling etc depend upon the elastic constants for the material (eg the modulus of elasticity, modulus of rigidity and Poisson's ratio), which are usually isotropic and are independent of time, temperature and rate of loading. Failure would probably occur by yielding.

The specification for standard materials is given in the British Standard Specifications, and the mechanical properties will show little scatter because of the homogeneous material structure and the ductility, which reduces the influence of stress concentrations. The elastic moduli of metals are generally high; consequently the deformations and strains are small and the designs are based upon the limitation of yield stress. In addition, stress concentrations within a limited region of a metal component could be relieved by local yielding of the material.

The polymer material characteristics, on the other hand, are very different. The mechanical behaviour of polymers is viscoelastic; consequently the stiffness and strength properties, the frequency and rate of loading are all time-dependent. Although the fibre component in a composite material has a stabilising influence on the mechanical properties, it is still time-dependent, but the dependence will vary with the volume percentage and orientation of the fibres. Figure 9.2 shows a typical stress/strain curve for three types of polymer, and it will be observed that the failure stresses are much lower than those of metals. In addition, the behaviour of the polymers may be anisotropic, and this anisotropy arises at the time of manufacture of the material when the polymer molecules may be directly orientated. In a composite material, unless a randomly orientated fibre arrangement is used (in which case a quasi-isotropic material is produced), the fibres will be directionally aligned and anisotropy in the material will result. Unlike conventional materials, because of the wide range of material behaviour, plastics and composites have few recognised design procedures for the production and presentation of test data and design curves.

Polymer and composite materials show little ductility compared with metals, and therefore greater emphasis is placed on the stress and strain analysis of components. The polymer components of the composite generally fail in a brittle manner, particularly at high rates of loading or at low temperatures, and the 'yielding' is sometimes accompanied by surface 'whitening' of the material, this phenomenon being generally associated

Figure 9.1 (*left*) *Typical stress-strain curve for a ductile metal*
Figure 9.2 (*right*) *Typical stress-strain curve for three types of polymer*

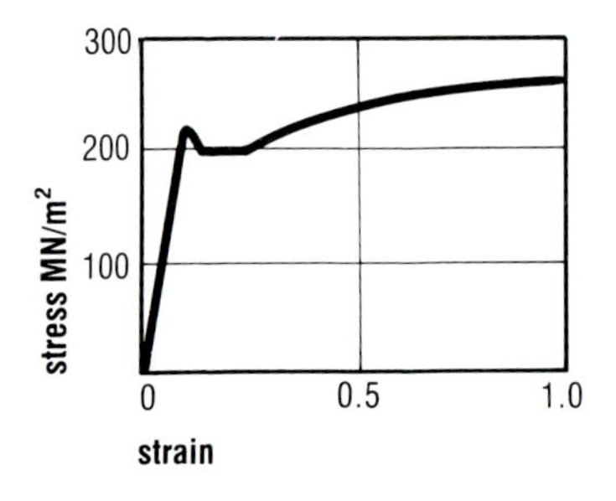

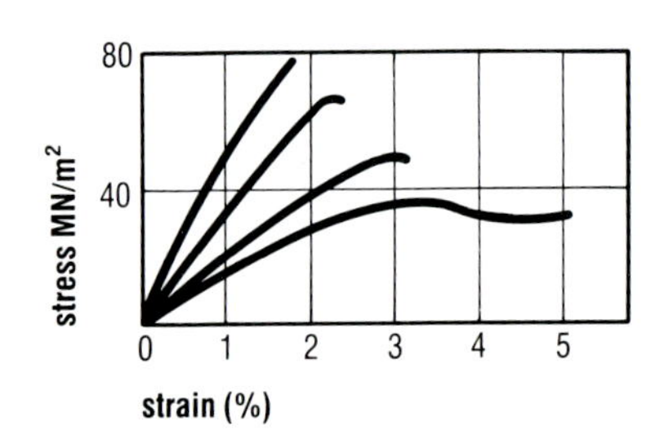

with thermoplastics polymers; consequently, ductility in polymers is not as acceptable as it is in metals. It will be seen that it is vital to have detailed information on the level of the values of stress and deformation of the composite. Therefore, stress concentrations within composite materials, particularly around bonded and mechanical joints, should be avoided by undertaking a detailed investigation and then designing the component accordingly.

Polymers are viscoelastic and therefore the deformations are dependent upon such factors as the time under load and the temperature. The fundamental characteristic features of viscoelastic materials are that they exhibit a time-dependent strain response to a constant stress (this is known as the creep of the material) and a time-dependent stress response to a constant strain (this is known as relaxation). Therefore, as the stress (σ) is a function of strain (ε) and time (t), it can be represented by an equation of the form

$$\sigma = f(\varepsilon, t) \tag{9.1a}$$

This response is a non-linear viscoelastic system, and for design it is often reduced to a linear viscoelastic type as

$$\sigma = \varepsilon f(t) \tag{9.1b}$$

This equation states that for a given interval of time the stress will be directly proportional to time. Figure 9.3 gives the stress/strain behaviour of elastic and viscoelastic materials. In design calculations it is often convenient to use elastic analyses and to modify these where necessary to take into account the more complex nature of the material behaviour. The designer is then able to use the wealth of design procedures that are available for small deflection and strain properties for elastic materials.

Figure 9.3 *Typical stress-strain curves for elastic and viscoelastic materials*

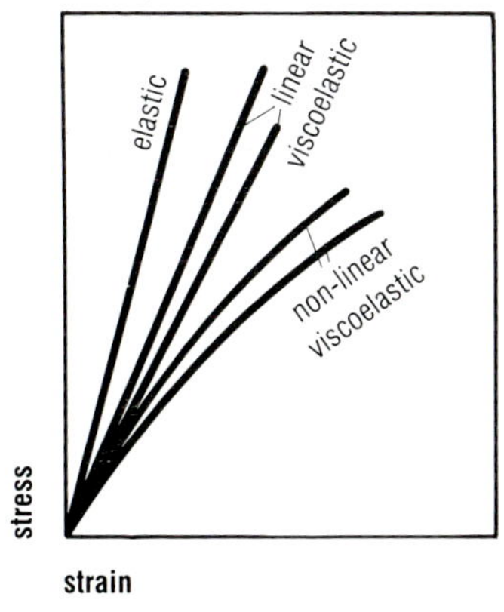

However, composites and filled polymers do exhibit some viscoelastic behaviour, but this will be slight for many loading conditions, particularly short-term ones where the behaviour is essentially linear. If this latter assumption is not acceptable under particular conditions, a pseudo-elastic analysis may be made where the elastic constants are modified to take into account the time- and rate-dependent component of the material. Williams (1980) and Ogorkiewicz (1974) have used the elastic design and the pseudo-elastic design methods, respectively.

As polymer materials have low stiffness but reasonable strength properties and the composite is generally manufactured in thin sheet form, it is likely that the criterion for design will be stiffness. Consequently, it is usual first to compare the maximum deflection with the limiting values imposed on the design and then to calculate the strength of the composite in order to assess the likely failure stresses.

It is possible that other relevant stresses will be required, such as resin cracking stress in sheet moulding compounds (SMC) components or the early stages of stress whitening in thermoplastics polymers.

One of the most important considerations during the design procedure concerns the mechanical properties of the components of the composite and the composite itself. If the properties are obtained from the plastics materials suppliers, only the tensile and flexural moduli of elasticity are provided, together with tensile yield stress and elongation to failure. Poisson's ratio, shear stiffness, the effects of loading rate and the anisotropic properties are rarely given.

It should be understood, however, that the design procedures for polymers and composite materials are only as accurate as the material properties used. There is, currently, considerable concern over the variety of test methods used to determine the basic property values and the anisotropy of the test specimen. There is also concern that some test methods do not relate directly to a material property. They relate to that particular specimen or to any other nominally identical one. In obtaining material data it is advisable to test large-size units as these should represent more closely the actual properties of the mouldings and, in addition, they will incorporate the effects of the processing variables in the results.

9.2 MECHANISM OF REINFORCEMENT IN COMPOSITE MATERIALS

The reinforcement of a low-modulus matrix material with high-modulus fibres uses the plastic flow of the matrix material under stress to transfer the load to the fibre; this results in a high-strength, high-modulus composite. The aim of the combination is to produce a two-phase material in which the primary phase, which determines stiffness, is in the form of particles of high aspect ratio (ie the fibre). The principal constituents which influence the strength and stiffness of the composites are the reinforcing fibres, the matrix and the interface. Each of these individual phases has to perform certain essential functional requirements based upon their mechanical properties so that a system containing them may function satisfactorily as a composite.

9.2.1 Fibres

The fibres in fibre-matrix composites should have the following properties:

- a high modulus of elasticity in order to provide efficient reinforcement
- a high ultimate strength
- low variation of strength between individual fibres
- uniform diameter
- stability and retention of strength during handling and fabrication.

9.2.2 Matrix

The matrix is required to fulfil the following functions:

- to bind together the fibres and to protect their surfaces from damage during handling, fabrication and the service life of the composite
- to disperse the fibres and separate them in order to avoid any catastrophic propagation of cracks and subsequent failure of the composite
- to transfer stresses to the fibres by adhesion and/or friction (when the composite is under load)
- to be chemically and thermally compatible with fibres.

9.2.3 Interface

The interface between the fibre and the matrix is an isotropic transition region exhibiting a graduation of properties. It must provide adequate chemically and physically stable bonding between the fibres and the matrix. Its functional needs vary considerably according to the performance requirements of the composite during its various stages under service conditions.

9.3 FIBRE ARRANGEMENT IN COMPOSITES

Reinforcing fibres in composite materials vary in diameter from 7 μm to 100 μm (0.3 to 4 $\times 10^{-3}$ inches), and they may be in the form of continuous or randomly orientated arrays; the latter would have lengths varying between 3 mm and 50 mm (0.1 and 2 inches).

The properties of short fibre composites are dependent upon the length to diameter ratio of the fibre. This ratio is known as the aspect ratio. The higher the aspect ratio, the greater will be the strength and stiffness of the composite.

In analysing fibre-reinforced matrix materials, the primary aim is to obtain from the properties of the components predictions of the average behaviour of the composite. From the point of view of the mechanical properties of the composite, the area of interest to the designer will be its stiffness and strength, which will be influenced by the following:

- the mechanical properties of the fibre and matrix
- the fibre volume fraction of the composite
- the fibre cross-section
- the fibre orientation within the matrix
- the method of manufacture of the composite.

9.4 MACROMECHANICAL ANALYSIS OF COMPOSITE LAMINATES

During the manufacture of a fibre/matrix composite it is usual to introduce multidirectional reinforcement and in so doing to increase the thickness of the composite to the required value; depending upon its structural requirement, the reinforcement could be either:

- randomly orientated, which for design purposes is assumed to be isotropic
- bidirectionally orientated, which is assumed to be orthotropic
- unidirectionally orientated, which is assumed to be orthotropic.

It will therefore be seen that the properties of the composite lamina (a lamina being a single layer of the composite and the basic module in the material) will depend upon the properties of the component materials, the arrangement of the fibres and the method of manufacture of the composite. The fibres are generally assumed to be linear elastic to failure, and although the resin is linear in the low-stress region, it does exhibit non-linear properties in the failure region. However, the ultimate strain of the brittle fibre is invariably less than that of the 'ductile' matrix and consequently, to justify the assumption of linear elasticity of the matrix, the latter's stress would be relatively low at failure of the composite. A further assumption in the analysis of composite laminates is that there is a complete bond between the fibre and the matrix.

The effect of creep in the matrix (polymer) on the stress distribution within and on the stiffness characteristics on the composite can be reduced to a minimum by ensuring that the fibres are positioned in the most effective way, which for axial forces would be along the line of action of the applied force. In addition, increasing the proportion of glass fibres in the composite implies a larger percentage of the load being taken by the fibres and consequently a lower proportion being carried by the matrix. On the other hand, if the fibre content is low, the effect of creep on the matrix may be considerable at high loads. However, for these composites the applied force is generally sufficiently low to ensure low creep.

The composite material property assumptions which will be made in the following development of the relationships between stresses and the corresponding strains are that:

- the composite material has linear elastic properties
- the tension and compression characteristics are the same.

9.4.1 Isotropic laminae

Composites of strong, stiff fibres in a polymer matrix need to be understood in order that design procedures may be developed. The mechanical properties of the composites are clearly controlled by their constituent properties and by the microstructural configurations. It is therefore necessary to be able to predict properties under varying conditions.

If fibres are randomly orientated in a composite material, such as a chopped strand mat laminate, it is reasonable to assume that, on a macroscopic scale, the material properties in the plane of the laminate will be the same in all directions, and hence the composite will be subjected to a state of stress as shown in Figure 9.4.

Figure 9.4 *General stress systems*

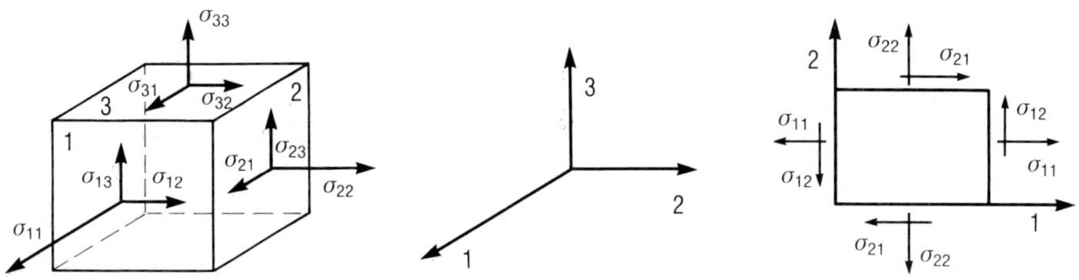

(*a*) three-dimensional stress system (*b*) co-ordinate axes (*c*) element subjected to plane stress

Hooke's law relationships are:

$$\sigma_{11} = \{E/(1 - \nu^2)\}[\varepsilon_{11} + \nu\varepsilon_{22}]$$
$$\sigma_{22} = \{E/(1 - \nu^2)\}[\varepsilon_{22} + \nu\varepsilon_{11}]$$
$$\sigma_{12} = \{E/(1 + \nu)2\}[\varepsilon_{12}] \qquad (9.2a)$$

In matrix form,

$$\begin{bmatrix} \sigma_{11} \\ \sigma_{22} \\ \sigma_{12} \end{bmatrix} = \begin{bmatrix} Q_{11} & Q_{12} & Q_{13} \\ Q_{21} & Q_{22} & Q_{23} \\ Q_{31} & Q_{32} & Q_{33} \end{bmatrix} \begin{bmatrix} \varepsilon_{11} \\ \varepsilon_{22} \\ \varepsilon_{12} \end{bmatrix}$$
$$[\boldsymbol{\sigma}] = [\boldsymbol{Q}][\boldsymbol{\varepsilon}] \qquad (9.2b)$$

where $Q_{11} = Q_{22} = E/(1 - \nu^2) \qquad Q_{13} = Q_{23} = Q_{31} = Q_{32} = 0$

$$Q_{12} = Q_{21} = E\nu/(1 - \nu^2)$$
$$Q_{33} = E/2(1 + \nu) = G$$

The modulus of rigidity (G) is generally independent of the modulus of elasticity (E) and Poisson's ratio (ν), but in the case of an isotropic material the modulus of rigidity may be expressed as a function of the other two elastic properties, as given in equation of Q_{33}. The above equations imply that only two elastic constants are required to characterise an isotropic material under plane stress.

A corresponding set of equations to those of Equation 9.2*a* relate the strains to stresses as

$$\begin{bmatrix} \varepsilon_{11} \\ \varepsilon_{22} \\ \varepsilon_{12} \end{bmatrix} = \begin{bmatrix} S_{11} & S_{12} & S_{13} \\ S_{21} & S_{22} & S_{23} \\ S_{31} & S_{32} & S_{33} \end{bmatrix} \begin{bmatrix} \sigma_{11} \\ \sigma_{22} \\ \sigma_{12} \end{bmatrix}$$
$$[\boldsymbol{\varepsilon}] = [\boldsymbol{S}][\boldsymbol{\sigma}] \qquad (9.3)$$

where $$S_{11} = S_{22} = \frac{1}{E}$$

$$S_{12} = S_{21} = -\frac{\nu}{E}$$

$$S_{33} = \frac{1}{G}$$

The square matrix $[\boldsymbol{Q}]$ given in equation 9.2*b* describes the relationship between the stress and strain vectors. This matrix is known as the material matrix and its components are a function of the elastic properties of the material. The square matrix $[\boldsymbol{S}]$ given in Equation 9.3 is the inversion of matrix $[\boldsymbol{Q}]$ and is known as the compliance matrix.

9.4.2 Orthotropic laminae

If fibres are unidirectionally or bidirectionally aligned in a composite material, high strengths and stiffnesses in the fibre direction will result; these mechanical properties will vary with the volume of fibres used. Consequently, in the unidirectionally aligned fibre array, high strength and stiffness will be attained in the direction of the fibres compared with those values at right angles to the fibres. In the composite, which consists of bidirectional reinforcement, equal strength and stiffness values will result in these two orthogonal directions. Such composites, which exhibit material symmetry about two axes at right angles to each other, are known as orthotropic and their properties are far more complex than those of the isotropic material.

The Hooke's law relationships are:

$$\begin{aligned}\sigma_{11} &= Q_{11}\varepsilon_{11} + Q_{12}\varepsilon_{22}\\ \sigma_{22} &= Q_{12}\varepsilon_{11} + Q_{22}\varepsilon_{22}\\ \sigma_{12} &= Q_{33}\varepsilon_{12}\end{aligned} \tag{9.4a}$$

or in matrix form

$$\begin{bmatrix}\sigma_{11}\\ \sigma_{22}\\ \sigma_{12}\end{bmatrix} = \begin{bmatrix}Q_{11} & Q_{12} & Q_{13}\\ Q_{21} & Q_{22} & Q_{23}\\ Q_{31} & Q_{32} & Q_{33}\end{bmatrix}\begin{bmatrix}\varepsilon_{11}\\ \varepsilon_{22}\\ \varepsilon_{12}\end{bmatrix}$$

$$[\boldsymbol{\sigma}] = [\boldsymbol{Q}][\boldsymbol{\varepsilon}] \tag{9.4b}$$

where
$$\begin{aligned}Q_{11} &= E_{11}/(1 - \nu_{12}\nu_{21})\\ Q_{22} &= E_{22}/(1 - \nu_{12}\nu_{21})\\ Q_{12} &= Q_{21} = \nu_{21}E_{11}/(1 - \nu_{21}\nu_{12}) = \nu_{12}E_{22}/(1 - \nu_{12}\nu_{21})\\ Q_{33} &= G_{12}\end{aligned}$$

The modulus of rigidity, in the orthotropic case, is defined with respect to the directions in which the stresses and distortions occur. In this case G is independent of the values of E and ν.

The composite which has orthotropic properties is completely defined by four independent elastic constants E_{11}, E_{22}, ν_{12} and G_{12}.

The corresponding set of equations to those in Equation 9.4*b* which relate strains to stresses are:

$$\begin{bmatrix} \varepsilon_{11} \\ \varepsilon_{22} \\ \varepsilon_{12} \end{bmatrix} = \begin{bmatrix} S_{11} & S_{12} & S_{13} \\ S_{21} & S_{22} & S_{23} \\ S_{31} & S_{32} & S_{33} \end{bmatrix} \begin{bmatrix} \sigma_{11} \\ \sigma_{22} \\ \sigma_{12} \end{bmatrix}$$

$$[\varepsilon] = [S][\sigma] \tag{9.5}$$

where

$$S_{11} = \frac{1}{E_{11}}$$

$$S_{22} = \frac{1}{E_{22}}$$

$$S_{33} = \frac{1}{G_{12}}$$

$$S_{12} = -\frac{\nu_{21}}{E_{22}} = -\frac{\nu_{12}}{E_{11}}$$

Poisson's ratio notation requires some explanation. The ν_{12} refers to the strain produced in the 2-direction when the line of action of the load application is in the 1-direction (see Figure 9.4).

It should be realised that Poisson's ratio for orthotropic materials can be greater than the maximum of 0.5 that is inherent in isotropic materials.

To be able to form the relationships between the stresses and strains relative to a set of arbitrary axes, orientated by an angle θ to the principal directions of the material, it is necessary to obtain the relationship between the stresses defined by the two sets of axes and then to obtain a similar relationship between the corresponding strains.

1 Orthotropic lamina – arbitrary orientation

The relationships defined in Equation 9.4 are related to the principal directions of the material. Consequently, because of material symmetry, the effects of the normal stress are independent of those of the shearing stresses and hence the total effects can be obtained by superposition.

If the lamina principal axes (1, 2) do not coincide with the reference axes (x, y) at some arbitrary orientation θ to them (shown in Figure 9.5 *a*), then the above constitutive relationship for each individual lamina must be transformed to the reference axes. Therefore, firstly the relationship between the stresses defined with respect to the two sets of axes in a lamina is determined, and then a similar relationship between the strains is obtained.

Figure 9.5 *Orientation of orthotropic lamina about reference axes*
(*a*) *Lamina principal axes* (*1, 2*) *at* θ *orientation to reference axes* (x. y),
(*b*) *Element under stress in axes* (x, y) *and* (*1, 2*)

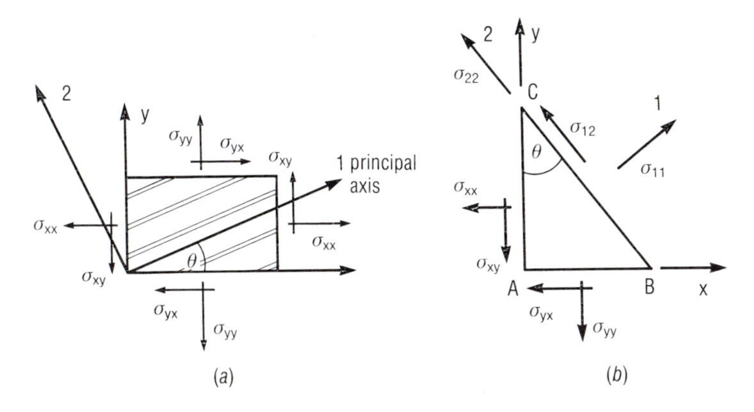

2 Stress relationships
By resolving the forces in direction 1 of Figure 9.5*b*, the equilibrium equation will be

$$\sigma_{11} BC - \sigma_{xx} AC \cos\theta - \sigma_{yy} AB \sin\theta - \sigma_{xy} AC \sin\theta - \sigma_{yx} AB \cos\theta = 0$$

Now $\quad AC = BC \cos\theta \quad$ and $\quad AB = BC \sin\theta$

and $\quad \sigma_{xy} = \sigma_{yx}$

therefore $\quad \sigma_{11} = \sigma_{xx}\cos^2\theta + \sigma_{yy}\sin^2\theta + \sigma_{xy}\, 2\sin\theta\cos\theta$

and, resolving in direction 2,

$$\sigma_{12} = -\sigma_{xx}\sin\theta\cos\theta + \sigma_{yy}\sin\theta\cos\theta + \sigma_{xy}(\cos^2\theta - \sin^2\theta)$$

Similarly $\quad \sigma_{22} = \sigma_{xx}\sin^2\theta + \sigma_{yy}\cos^2\theta - \sigma_{xy}\, 2\sin\theta\cos\theta$

Therefore the following relationships in matrix form equate the stresses in the principal (1, 2) and the reference axes (x, y) as follows:

$$\begin{bmatrix}\sigma_{11}\\ \sigma_{22}\\ \sigma_{12}\end{bmatrix} = [\boldsymbol{T}]\begin{bmatrix}\sigma_{xx}\\ \sigma_{yy}\\ \sigma_{xy}\end{bmatrix} \tag{9.6}$$

or

$$\begin{bmatrix}\sigma_{xx}\\ \sigma_{yy}\\ \sigma_{xy}\end{bmatrix} = [\boldsymbol{T}]^{-1}\begin{bmatrix}\sigma_{11}\\ \sigma_{22}\\ \sigma_{12}\end{bmatrix} \tag{9.7}$$

The transformation matrix $\boldsymbol{T}$ is

$$[\boldsymbol{T}] = \begin{bmatrix} m^2 & n^2 & 2mn \\ n^2 & m^2 & -2mn \\ -mn & mn & m^2 - n^2 \end{bmatrix} \tag{9.8a}$$

and

$$[\boldsymbol{T}]^{-1} = \begin{bmatrix} m^2 & n^2 & -2mn \\ n^2 & m_2 & 2mn \\ nm & -nm & (m^2 - n^2) \end{bmatrix} = \text{inverse } [\boldsymbol{T}] \tag{9.8b}$$

where

$$m = \cos\theta$$

$$n = \sin\theta$$

3 Strain relationships

Consider an element OFAE (Figure 9.6) to be under strains of ε_{xx}, ε_{yy}, ε_{xy}, where the line 0–1 makes an angle of θ with the line 0–X. It is necessary to obtain a relationship between the strains allied to the axes 1–2 and to those allied to the axes x–y of the system.

The normal and shear strains in the x–y directions are:

$$\varepsilon_{xx} = \frac{\partial u}{\partial x}$$

$$\varepsilon_{yy} = \frac{\partial v}{\partial y}$$

$$\varepsilon_{xy} = \frac{\partial u}{\partial y} + \frac{\partial v}{\partial x} \tag{9.9}$$

Also, from Figure 9.6 it can be seen that, under strain, the line O–A of element OFAE is deformed to the line O–C. Therefore the displacement can be expressed in either the x–y or the 1–2 coordinate directions.

The coordinate position of C relative to A is (AB, BC) and (AD, DC) in the x–y and 1–2 coordinate axes, respectively. Therefore:

$$AB = \frac{\partial u}{\partial x}\,\delta x + \frac{\partial u}{\partial y}\,\delta y$$

$$BC = \frac{\partial v}{\partial x}\,\delta x + \frac{\partial v}{\partial y}\,\delta y \tag{9.10}$$

Also, from the transformation relationship in Figure 9.6,

$$\begin{bmatrix} AD \\ CD \end{bmatrix} = \begin{bmatrix} m & n \\ -n & m \end{bmatrix} \begin{bmatrix} AB \\ BC \end{bmatrix}$$

Figure 9.6 *Deformation pattern of an element under normal and shear strains. Rigid body displacements have been eliminated*

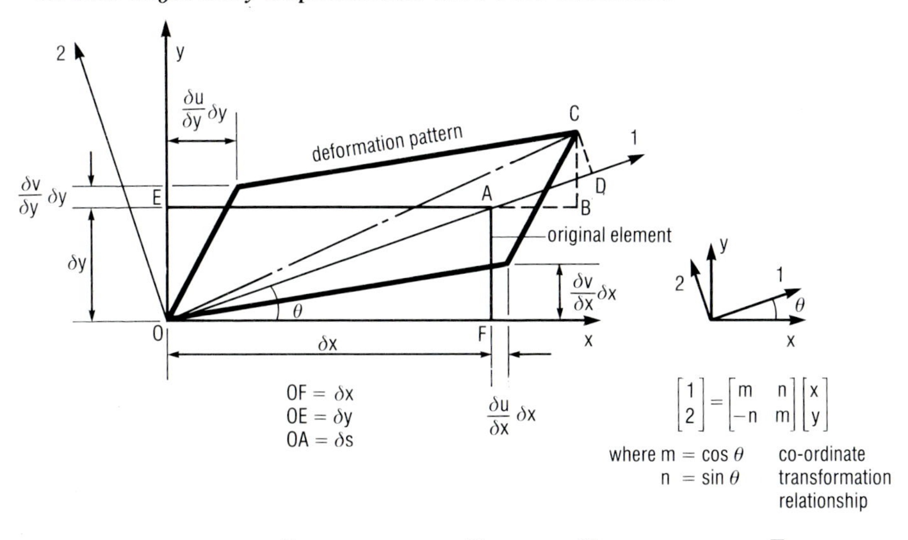

so $$AD = \cos\theta\left[\frac{\partial u}{\partial x}\,\delta x + \frac{\partial u}{\partial y}\,\delta y\right] + \sin\theta\left[\frac{\partial v}{\partial x}\,\delta x + \frac{\partial v}{\partial u}\,\delta y\right]$$

and $$CD = -\sin\theta\left[\frac{\partial u}{\partial x}\,\delta x + \frac{\partial u}{\partial y}\,\delta y\right] + \cos\theta\left[\frac{\partial v}{\partial x}\,\delta x + \frac{\partial v}{\partial y}\,\delta y\right]$$

now $$\varepsilon_{11} = \frac{AD}{OA} = \cos\theta\left[\frac{\partial u}{\partial x}\frac{\delta x}{\delta s} + \frac{\partial u}{\partial y}\frac{\delta y}{\delta s}\right] + \sin\theta\left[\frac{\partial v}{\partial x}\frac{\delta x}{\delta s} + \frac{\partial v}{\partial y}\frac{\delta y}{\delta s}\right]$$

where $OA = \delta s$

and $$\frac{\delta x}{\delta s} = \cos\theta, \qquad \frac{\delta y}{\delta s} = \sin\theta$$

Therefore $$\varepsilon_{11} = \frac{\partial u}{\partial x}\cos^2\theta + \frac{\partial v}{\partial y}\sin^2\theta + \left[\frac{\partial u}{\partial y} + \frac{\partial v}{\partial x}\right]\sin\theta\cos\theta \qquad (9.11a)$$

From equation (9.9), substituting the strain values into Equation 9.11*a* gives

$$\varepsilon_{11} = \varepsilon_{xx}m^2 + \varepsilon_{yy}n^2 + \frac{\varepsilon_{xy}}{2}\,2nm \qquad (9.11b)$$

Similarly, it can be shown that

$$\varepsilon_{22} = \varepsilon_{xx}n^2 + \varepsilon_{yy}m^2 - \frac{\varepsilon_{xy}}{2}\,2nm \qquad (9.12)$$

and $$\frac{\varepsilon_{12}}{2} = -\varepsilon_{xx}nm + \varepsilon_{yy}nm + \frac{\varepsilon_{xy}}{2}(m^2 - n^2) \qquad (9.13)$$

Therefore the following relationships in matrix form ally the stresses in the principal axes (1, 2) and the reference axes (x, y):

$$\begin{bmatrix} \varepsilon_{11} \\ \varepsilon_{22} \\ \varepsilon_{12}/2 \end{bmatrix} = [\boldsymbol{T}] \begin{bmatrix} \varepsilon_{xx} \\ \varepsilon_{yy} \\ \varepsilon_{xy}/2 \end{bmatrix} \tag{9.14}$$

where $[\boldsymbol{T}]$ is the transform matrix given in Equation 9.8

or

$$\begin{bmatrix} \varepsilon_{xx} \\ \varepsilon_{yy} \\ \varepsilon_{xy}/2 \end{bmatrix} = |\boldsymbol{T}|^{-1} \begin{bmatrix} \varepsilon_{11} \\ \varepsilon_{22} \\ \varepsilon_{12}/2 \end{bmatrix} \tag{9.15}$$

Therefore the stress and strain transformation equations are identical. (Compare Equations 9.6 and 9.7 with Equations 9.14 and 9.15, respectively.)

4 Stress–strain relationship

The relationships between the stress and strain in the axes x–y in terms of the material properties relative to the principal axes 1–2 may be determined as follows:

$$\begin{bmatrix} \varepsilon_{11} \\ \varepsilon_{22} \\ \varepsilon_{12} \end{bmatrix} = [\boldsymbol{S}] \begin{bmatrix} \sigma_{11} \\ \sigma_{22} \\ \sigma_{12} \end{bmatrix} \tag{9.16}$$

and

$$\begin{bmatrix} \sigma_{11} \\ \sigma_{22} \\ \sigma_{12} \end{bmatrix} = [\boldsymbol{Q}] \begin{bmatrix} \varepsilon_{11} \\ \varepsilon_{22} \\ \varepsilon_{12} \end{bmatrix} \tag{9.17}$$

and from Equations 9.7 and 9.17

$$\begin{bmatrix} \sigma_{xx} \\ \sigma_{yy} \\ \sigma_{xy} \end{bmatrix} = [\boldsymbol{T}]^{-1} [\boldsymbol{Q}] \begin{bmatrix} \varepsilon_{11} \\ \varepsilon_{22} \\ \varepsilon_{12} \end{bmatrix} \tag{9.18}$$

and from Equations 9.6 and 9.16

$$\begin{bmatrix} \varepsilon_{11} \\ \varepsilon_{22} \\ \varepsilon_{12} \end{bmatrix} = [\boldsymbol{S}][\boldsymbol{T}] \begin{bmatrix} \sigma_{xx} \\ \sigma_{yy} \\ \sigma_{xy} \end{bmatrix} \tag{9.19}$$

In order to transform Equations 9.14 and 9.15 from half shear strain to the full shear strain it is necessary to pre- and post-multiply the $[\boldsymbol{T}]$

matrix by a matrix $[\boldsymbol{R}]$ and the inverse of $[\boldsymbol{R}]$, respectively, where $[\boldsymbol{R}]$ is

$$\begin{bmatrix} 1 & 0 & 0 \\ 0 & 1 & 0 \\ 0 & 0 & 2 \end{bmatrix}$$

Therefore Equations 9.14 and 9.15 become

$$\begin{bmatrix} \varepsilon_{11} \\ \varepsilon_{22} \\ \varepsilon_{12} \end{bmatrix} = [\boldsymbol{R}][\boldsymbol{T}][\boldsymbol{R}]^{-1} \begin{bmatrix} \varepsilon_{xx} \\ \varepsilon_{yy} \\ \varepsilon_{xy} \end{bmatrix} \tag{9.20a}$$

and

$$\begin{bmatrix} \varepsilon_{xx} \\ \varepsilon_{yy} \\ \varepsilon_{xy} \end{bmatrix} = [\boldsymbol{R}][\boldsymbol{T}]^{-1}[\boldsymbol{R}]^{-1} \begin{bmatrix} \varepsilon_{11} \\ \varepsilon_{22} \\ \varepsilon_{12} \end{bmatrix} \tag{9.20b}$$

It can be shown that $[\boldsymbol{R}][\boldsymbol{T}][\boldsymbol{R}]^{-1} = [[\boldsymbol{T}]^{-1}]^{\mathrm{T}}$

and $[\boldsymbol{R}][\boldsymbol{T}]^{-1}[\boldsymbol{R}]^{-1} = [\boldsymbol{T}]^{\mathrm{T}}$

where

$$[[\boldsymbol{T}]^{-1}]^{\mathrm{T}} = \begin{bmatrix} m^2 & n^2 & nm \\ n^2 & m^2 & -nm \\ -2mn & 2mn & (m^2 - n^2) \end{bmatrix} \tag{9.21a}$$

and

$$[\boldsymbol{T}]^{\mathrm{T}} = \begin{bmatrix} m^2 & n & -nm \\ n^2 & m & nm \\ 2nm & -2nm & (m^2 - n^2) \end{bmatrix} \tag{9.21b}$$

so Equations 9.20*a* and 9.20*b* can be written as

$$\begin{bmatrix} \varepsilon_{11} \\ \varepsilon_{22} \\ \varepsilon_{12} \end{bmatrix} = [[\boldsymbol{T}]^{-1}]^{\mathrm{T}} \begin{bmatrix} \varepsilon_{xx} \\ \varepsilon_{yy} \\ \varepsilon_{xy} \end{bmatrix} \tag{9.22}$$

and

$$\begin{bmatrix} \varepsilon_{xx} \\ \varepsilon_{yy} \\ \varepsilon_{xy} \end{bmatrix} = [\boldsymbol{T}]^{\mathrm{T}} \begin{bmatrix} \varepsilon_{11} \\ \varepsilon_{22} \\ \varepsilon_{12} \end{bmatrix} \tag{9.23}$$

Substituting Equation 9.22 into Equation 9.18 will give the stress components in the reference axes x, y in terms of the strain components in that axis:

$$\begin{bmatrix} \sigma_{xx} \\ \sigma_{yy} \\ \sigma_{xy} \end{bmatrix} = [\boldsymbol{T}]^{-1}[\boldsymbol{Q}][[\boldsymbol{T}]^{-1}]^{\mathrm{T}} \begin{bmatrix} \varepsilon_{xx} \\ \varepsilon_{yy} \\ \varepsilon_{xy} \end{bmatrix} \tag{9.24}$$

or

$$\begin{bmatrix} \sigma_{xx} \\ \sigma_{yy} \\ \sigma_{xy} \end{bmatrix} = [\bar{Q}] \begin{bmatrix} \varepsilon_{xx} \\ \varepsilon_{yy} \\ \varepsilon_{xy} \end{bmatrix}$$

The matrix $[\bar{Q}]$ is a 3×3 material matrix where the components are:

$$\bar{Q}_{11} = Q_{11}m^4 + Q_{22}n^4 + 2(Q_{12} + 2Q_{33})n^2m^2$$

$$\bar{Q}_{12} = \bar{Q}_{21} = (Q_{11} + Q_{22} - 4Q_{33})n^2m^2 + Q_{12}(n^4 + m^4)$$

$$\bar{Q}_{13} = \bar{Q}_{31} = (Q_{11} - Q_{12} - 2Q_{33})nm^3 + (Q_{12} - Q_{22} + 2Q_{33})n^3m$$

$$\bar{Q}_{22} = Q_{11}n^4 + Q_{22}m^4 + 2(Q_{12} + 2Q_{33})n^2m^2$$

$$\bar{Q}_{23} = \bar{Q}_{32} = (Q_{11} - Q_{12} - 2Q_{33})n^3m + (Q_{12} - Q_{22} + 2Q_{33})nm^3$$

$$\bar{Q}_{33} = (Q_{11} + Q_{22} - 2Q_{12} - 2Q_{33})n^2m^2 + Q_{33}(n^4 + m^4)$$

where m and n have been defined in Equation 9.8 and Q_{11}, Q_{22}, Q_{12}, Q_{21} and Q_{33} have been defined in Equation 9.4*b*.

An equivalent expression for strain components in the reference axes x, y in terms of the stress components in that axis can be obtained by substituting Equation 9.19 into Equation 9.23. A compliance matrix will then describe the relationship between these two components as

$$\begin{bmatrix} \varepsilon_{xx} \\ \varepsilon_{yy} \\ \varepsilon_{xy} \end{bmatrix} = [\bar{S}] \begin{bmatrix} \sigma_{xx} \\ \sigma_{yy} \\ \sigma_{xy} \end{bmatrix} \tag{9.25}$$

The matrix $[\bar{S}]$ is a 3×3 compliance matrix where the components are:

$$\bar{S}_{11} = S_{11}m^4 + S_{22}n^4 + (2S_{12} + S_{33})n^2m^2$$

$$\bar{S}_{12} = \bar{S}_{21} = S_{12}(n^4 + m^4) + (S_{11} + S_{22} - S_{33})n^2m^2$$

$$\bar{S}_{13} = \bar{S}_{31} = (2S_{11} - 2S_{12} - S_{33})nm^3 - (2S_{22} - S_{12} - S_{33})n^3m$$

$$\bar{S}_{23} = \bar{S}_{32} = (2S_{11} - 2S_{12} - S_{33})n^3m - (2S_{22} - 2S_{12} - S_{33})nm^3$$

$$\bar{S}_{22} = S_{11}n^4 + S_{22}m^4 + (2S_{12} + S_{33})n^2m^2$$

$$\bar{S}_{33} = 2(2S_{11} + 2S_{22} - 4S_{12} - S_{33})n^2m^2 + S_{33}(n^4 + m^4)$$

where m and n have been defined in Equation 9.8 and S_{11}, S_{22}, S_{12}, S_{21} and S_{33} have been defined in Equation 9.5.

5 Engineering constants relationships

Using the approach outlined above it is possible to obtain expressions for the elastic properties E_{xx}, E_{yy}, G_{xy} and ν_{xy}, corresponding to the x–y axis, in terms of the elastic constants in the 1–2 axis. The equations are:

$$\frac{1}{E_{xx}} = \frac{1}{E_{11}} m^4 + \left(\frac{1}{G_{12}} - \frac{2\nu_{12}}{E_{11}}\right) n^2 m^2 + \frac{1}{E_{22}} n^4 \tag{9.26a}$$

$$\frac{1}{E_{yy}} = \frac{1}{E_{11}} n^4 + \left(\frac{1}{G_{12}} - \frac{2\nu_{12}}{E_{11}}\right) n^2 m^2 + \frac{1}{E_{11}} m^4 \tag{9.26b}$$

$$\frac{1}{G_{xy}} = 2\left(\frac{2}{E_{11}} + \frac{2}{E_{22}} + \frac{4\nu_{12}}{E_{11}} - \frac{1}{G_{12}}\right) n^2 m^2 + \frac{1}{G_{12}} (n^4 + m^4) \tag{9.26c}$$

$$\nu_{xy} = E_{xx}\left[\frac{\nu_{12}}{E_{11}} (n^4 + m^4) - \frac{1}{E_{11}} + \frac{1}{E_{22}} - \frac{1}{G_{12}}\right] n^2 m^2 \tag{9.26d}$$

Thus if E_{11}, E_{22}, G_{12} and ν_{12} are known, the elastic properties at any angle to the x–y axis can be calculated.

9.4.3 The strength characteristics and failure criteria of composite laminae

In the preceding section, the stiffness relationships in terms of stress and strain have been presented for isotropic and orthotropic materials. It is now necessary to have an understanding of the ultimate strengths of the laminae to enable a complete characterisation of the composite material to be made.

The ultimate strength behaviour of composite systems may be different in tension and compression and the characteristics of the failure mode will be highly dependent upon the component materials. Therefore, it is not possible to undertake a systematic development of the strength characteristics as was the case for the stiffness relationships; consequently, a series of failure criteria for composite materials will be given.

In isotropic materials it is usual to equate a combined stress situation to a basic strength value derived from a uniaxial tension or compression test. Similarly, in the case of orthotropic materials the strength values could be related to the principal material axes.

1 Strength theories for isotropic material

When a tensile load is applied to a specimen in a uniaxial test it is possible for failure in the specimen to be initiated by either an ultimate tensile stress or a shear stress, because a tensile stress of σ (the maximum principal stress in this type of test) on the specimen produces a maximum shear stress value of $\sigma/2$. Failure can, therefore, be defined as a departure from the linear stress–strain relationship (ie a yield of the material) or an actual fracture of the specimen. Therefore, the failure theories are related

Figure 9.7 *Isotropic and orthotropic materials under normal stresses (a) Element of isotropic material under three principal stresses* $\sigma_{xx} > \sigma_{yy} > \sigma_{zz}$, *(b) Orthotropic material under normal stress*

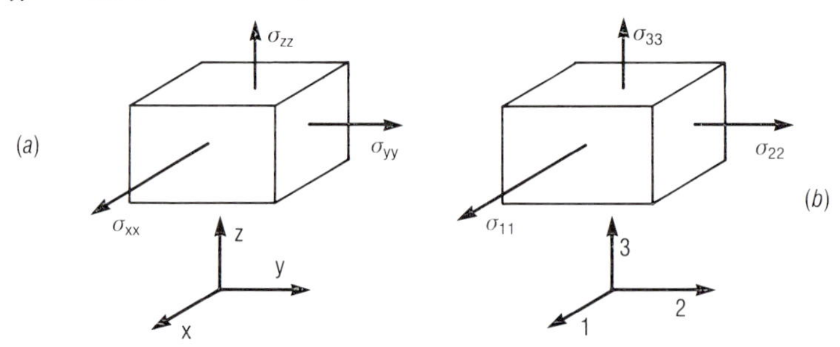

to the applied tensile or compressive stresses that cause failure in uniaxial tests, irrespective of whether it was a normal or a shear stress failure:

(*a*) *Maximum principal stress theory*: If the $\sigma_{xx} > \sigma_{yy} > \sigma_{zz}$ in an isotropic material where σ_{xx}, σ_{yy} and σ_{zz} are the principal stresses acting on an element, as shown in Figure 9.7*a*, then failure will occur when

$$\sigma_{xx} = \sigma^* \tag{9.27a}$$

where σ^* is the failure stress in a uniaxial tensile test. Also, if σ_{zz} is the minimum principal stress and is compressive, then

$$\sigma_{zz} = \sigma^* \tag{9.27b}$$

where σ^* is the failure stress in a uniaxial compressive test.

(*b*) *Maximum principal strain theory*: This theory, introduced by St Venant, assumes that failure occurs when the strain reaches its ultimate value; thus, if ε_{xx} is the maximum principal tensile strain, then

$$\varepsilon_{xx} = \varepsilon^* \tag{9.28a}$$

where ε^* is the tensile strain at failure of a uniaxial specimen or, in terms of stress,

$$\frac{1}{E}(\sigma_{xx} - \nu\sigma_{yy} - \nu\sigma_{zz}) = \frac{\sigma^*}{E}$$

where ν is Poisson's ratio, or

$$\sigma_{xx} - \nu\sigma_{yy} - \nu\sigma_{zz} = \sigma^* \tag{9.28b}$$

Similarly, for a compressive strain,

$$\varepsilon_{zz} = \varepsilon^*$$

or

$$\sigma_{zz} - \nu\sigma_{xx} - \nu\sigma_{yy} = \sigma^* \tag{9.28c}$$

The above two theories consider failure due to normal stresses and ignore shear stresses, irrespective of their magnitude. The theories, therefore, are relevant only to the failure of brittle materials under tension.

(*c*) *Maximum shearing stress theory*: This theory states that failure occurs when the maximum shearing stress in the lamina equals the maximum shearing stress in a uniaxial tensile test at failure.

Considering Figure 9.7*a*, the maximum and minimum principal stresses are σ_{xx} and σ_{zz} and the maximum shear stress is $(\sigma_{xx} - \sigma_{zz})/2$, and this is equal to $\sigma^*/2$, so

$$\sigma_{xx} - \sigma_{zz} = \sigma^* \tag{9.29}$$

Because this theory assumes that the shear stresses are lower than the tensile stresses, it is particularly relevant to ductile materials.

(*d*) *The total strain energy theory*: The above theories express the failure criteria as either limiting stress or limiting strain; the total strain energy theory attempts to combine the two.

The work done on a linear elastic uniaxial tensile specimen which is extended by an amount u by the external force P is

$$\text{work done} = \tfrac{1}{2}Pu = \text{strain energy stored in specimen}$$

$$\text{strain energy } U = \tfrac{1}{2}(P/A)(u/l)$$

$$= \tfrac{1}{2}\sigma\varepsilon$$

where A = uniform cross-section, l = length of specimen, σ = stress in specimen and ε = strain in specimen; therefore

$$U = (\tfrac{1}{2}E)\sigma^2$$

At failure

$$U^* = (\tfrac{1}{2}E)\sigma^{*2} \tag{9.30}$$

From Figure 9.7*a*

$$U = \tfrac{1}{2}\sigma_{xx}\varepsilon_{xx} + \tfrac{1}{2}\sigma_{yy}\varepsilon_{yy} + \tfrac{1}{2}\sigma_{zz}\varepsilon_{zz} \tag{9.31}$$

but

$$\varepsilon_{xx} = 1/E(\sigma_{xx} - \sigma_{yy}\nu - \sigma_{zz}\nu)$$

$$\varepsilon_{yy} = 1/E(\sigma_{yy} - \sigma_{xx}\nu - \sigma_{zz}\nu)$$

$$\varepsilon_{zz} = 1/E(\sigma_{zz} - \sigma_{xx}\nu - \sigma_{yy}\nu)$$

Equation 9.31 becomes

$$U = \tfrac{1}{2}E(\sigma_{xx}^2 + \sigma_{yy}^2 + \sigma_{zz}^2) - \nu/2E(2\sigma_{xx}\sigma_{yy} + 2\sigma_{yy}\sigma_{zz} + 2\sigma_{zz}\sigma_{xx}) \tag{9.32}$$

Equating Equations 9.30 and 9.32,

$$\sigma^{*2} = \sigma_{xx}^2 + \sigma_{yy}^2 + \sigma_{zz}^2 - 2\nu(\sigma_{xx}\sigma_{yy} + \sigma_{yy}\sigma_{zz} + \sigma_{zz}\sigma_{xx}) \tag{9.33}$$

Assuming lamina theory, Equation 9.33 becomes

$$\sigma^{*2} = \sigma_{xx}^2 + \sigma_{yy}^2 - 2\sigma_{xx}\sigma_{yy}\nu \tag{9.34}$$

This theory applies in particular to brittle materials where the ultimate tensile stress is less than the ultimate shear stress.

(*e*) *Deviation strain energy theory*: This theory is known as the von Mises criterion, and in it the principal stresses σ_{xx}, σ_{yy} and σ_{zz} can be expressed as the sum of two components (which are shown schematically in Figure 9.8) namely

$$\sigma_{xx} = \sigma_{vv} + \bar{\sigma}_{xx}$$

$$\sigma_{yy} = \sigma_{vv} + \bar{\sigma}_{yy}$$

$$\sigma_{zz} = \sigma_{vv} + \bar{\sigma}_{zz} \tag{9.35}$$

Figure 9.8 *Volume and deviation stress system*

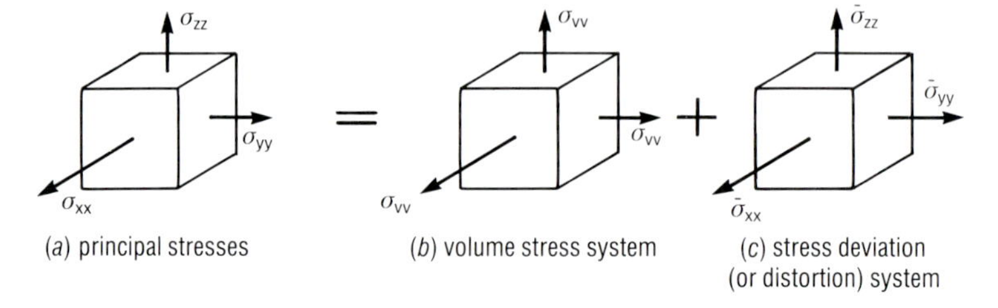

where σ_{vv} is a hydrostatic stress and causes only a change in volume. $\bar{\sigma}_{xx}$, $\bar{\sigma}_{yy}$ and $\bar{\sigma}_{zz}$ are stresses causing a distortion of the body. Both components of stress act at a point in a body.

The hydrostatic stress components produced are equal in magnitude and are consistent in the three directions, thus producing equal strains in these directions. Consequently, the material undergoes a change in volume but not in shape. The deviation stress components will cause the material to change its shape but not its volume.

As the hydrostatic stress component alone does not cause appreciable deformation in materials, it is usual to subtract it from the actual stresses when developing criteria for yielding and to assume that the remaining stress deviation component produces yielding.

As the assumption is that the deviation components produce zero volume change, then the sum of the strains due to these components must also be zero. Therefore

$$\bar{\varepsilon}_{xx} + \bar{\varepsilon}_{yy} + \bar{\varepsilon}_{zz} = 0 \tag{9.36}$$

In terms of stress,

$$(1 - 2\nu)/E(\bar{\sigma}_{xx} + \bar{\sigma}_{yy} + \bar{\sigma}_{zz}) = 0 \tag{9.37}$$

ie

$$\bar{\sigma}_{xx} + \bar{\sigma}_{yy} + \bar{\sigma}_{zz} = 0$$

as $(1 - 2\nu)/E$ cannot be zero. Therefore, from Equation 9.35,

$$\sigma_{xx} + \sigma_{yy} + \sigma_{zz} = 3\sigma_v$$

or

$$\sigma_{vv} = \tfrac{1}{3}(\sigma_{xx} + \sigma_{yy} + \sigma_{zz}) \tag{9.38}$$

and this stress will produce a strain

$$\varepsilon_{vv} = [(1 - 2\nu)/E]\sigma_{vv} \tag{9.39}$$

The total strain energy (U_T) in the composite per unit volume is equal to the strain energy due to change in volume (U_v) plus strain energy due to change in distortion (U_D):

$$U_T = U_v + U_D$$

$$U_v = \tfrac{1}{2}\sigma_{vv}\varepsilon_{vv} + \tfrac{1}{2}\sigma_{vv}\varepsilon_{vv} + \tfrac{1}{2}\sigma_{vv}\varepsilon_{vv}$$

$$= [3(1 - 2\nu)/2E]\sigma_{vv}^2 \tag{9.40}$$

From Equation 9.32,

$$U_T = \tfrac{1}{2}E\{\sigma_{xx}^2 + \sigma_{yy}^2 + \sigma_{zz}^2\} - \nu/2E\{2\sigma_{xx}\sigma_{yy} + 2\sigma_{yy}\sigma_{zz} + 2\sigma_{zz}\sigma_{xx}\} \tag{9.41}$$

$$U_D = U_T - U_v$$

Therefore from Equations 9.40 and 9.41, and having substituted Equation 9.38 into Equation 9.40,

$$U_D = \tfrac{1}{2}E(\sigma_{xx}^2 + \sigma_{yy}^2 + \sigma_{zz}^2) - \nu/2E(2\sigma_{xx}\sigma_{yy} + 2\sigma_{yy}\sigma_{zz} + 2\sigma_{zz}\sigma_{xx}) - (1 - 2\nu)/6E(\sigma_{xx} + \sigma_{yy} + \sigma_{zz})^2 \tag{9.42}$$

$$= (1 + \nu)/6E[(\sigma_{xx} - \sigma_{yy})^2 + (\sigma_{yy} - \sigma_{zz})^2 + (\sigma_{zz} - \sigma_{xx})^2] \tag{9.43}$$

The distortional strain energy value per unit volume of a lamina at uniaxial tensile failure can be derived from Equation 9.43, where

$$\sigma_{zz} = \sigma_{yy} = 0$$

and

$$U_D = [(1 + \nu)/6E]2\sigma^{*2} \tag{9.44}$$

From Equations 9.43 and 9.44,

$$(\sigma_{xx} - \sigma_{yy})^2 + (\sigma_{yy} - \sigma_{zz})^2 + (\sigma_{zz} - \sigma_{xx})^2 = 2\sigma^{*2} \tag{9.45}$$

and for lamina theory, where $\sigma_{zz} = 0$, Equation 9.45 becomes

$$\sigma_{xx}^2 + \sigma_{yy}^2 - \sigma_{xx}\sigma_{yy} = \sigma^{*2}$$

As with the maximum shear stress theory, the above is most relevant to ductile materials.

It is not immediately obvious which of the above failure criteria are most relevant to composites, as the fibre volume fraction and orientation of the fibres in the polymer will influence their strength and ductility properties. However, the last theory has been applied to quasi-isotropic composites with some success, although the results are only based upon simple fundamental tests.

2 Strength of orthotropic laminae

The theories based upon the strength characteristics of orthotropic materials are necessarily more complicated than the isotropic ones.

The results of the orthotropic material analysis are based upon the uniaxial strength properties in the three principal axes. These tests will determine the modulus of elasticity, Poisson's ratio and the strength characteristics in the direction of the principal axes, and, as stated earlier, will eliminate any coupling effects of shearing and normal strains which would occur if the laminate were tested in any other direction.

Unlike the case of isotropic materials, where the shearing strength can be obtained from the uniaxial test, the orthotropic material properties must be determined in the principal axes' direction from separate tests. This means that independent ultimate uniaxial strengths must be determined in the 1, 2 and 3 directions and a separate test to determine the shear strengths in these directions must also be undertaken. These tests are shown schematically in Figure 9.9.

Figure 9.9 *Critical stress values in the principal material axes*

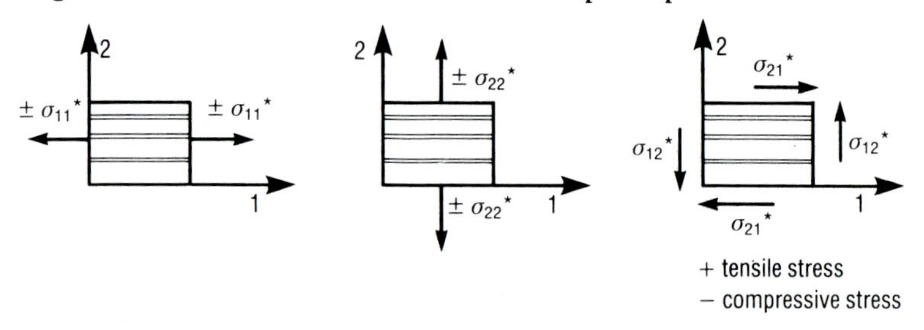

Consequently, the stress conditions in orthotropic materials have to be considered in connection with the normal and shearing components relative to their principal axes:

(*a*) *Maximum stress theory*: The maximum stress theory of failure assumes that failure occurs when the stresses in the principal material axes reach a critical value. There are three possible modes of failure, and the conditions for these are:

$$\sigma_{11} = \sigma_{11}^*$$

$$\sigma_{22} = \sigma_{22}^*$$

$$\sigma_{12} = \sigma_{12}^*$$

where σ_{11}^* is the ultimate tensile or compressive stress in direction 1, σ_{22}^* is the ultimate tensile or compressive stress in direction 2 and σ_{12}^* is the ultimate shear stress acting in plane 1 in direction 2.

If the load were applied to the lamina at an angle θ to the principal axis direction (see Figure 9.5*a*), then by transformation

$$\sigma_{11} = \sigma_{xx} \cos^2 \theta = \sigma_\theta \cos^2 \theta$$

$$\sigma_{22} = \sigma_{xx} \sin^2 \theta = \sigma_\theta \sin^2 \theta$$

$$\sigma_{12} = -\sigma_{xx} \sin \theta \cos \theta = -\sigma_\theta \sin \theta \cos \theta$$

The failure strength predicted by the maximum stress theory would depend upon the relative values of σ_{11}, σ_{22} and σ_{12}, and would therefore be the smallest value of the following:

$$\sigma_\theta = \sigma_{11}^*/\cos^2 \theta$$

$$\sigma_\theta = \sigma_{22}^*/\sin^2 \theta$$

$$\sigma_\theta = \sigma_{12}^*/\sin \theta \cos \theta \tag{9.46}$$

(*b*) *Maximum strain theory*: The maximum strain theory of failure assumes that failure occurs when the stresses in the principal material axes reach a critical value. Here again there are three possible modes of failure:

$$\varepsilon_{11} = \varepsilon_{11}^*$$

$$\varepsilon_{22} = \varepsilon_{22}^*$$

$$\varepsilon_{12} = \varepsilon_{12}^*$$

where ε_{11}^* is the maximum tensile or compressive strain in direction 1, ε_{22}^* is the maximum tensile or compressive strain in direction 2 and ε_{12}^* is the maximum shear strain on plane 1 in direction 2.

(*c*) *Tsai-Hill energy theory*: The Tsai-Hill criterion is based upon the von Mises failure criterion which was originally applied to homogeneous isotropic bodies (Section 9.4.3.1(*e*)). It was then modified by Hill to suit anisotropic bodies and finally applied to composite materials by Tsai.

In the development of the deviational strain energy for the isotropic material (Equation 9.35) it was assumed that the normal stresses causing failure were the principal stresses. If the failure stresses had not been the principal stresses, the lamina's plane of failure would have been subjected to shear stresses in addition to the normal stresses. This would have given rise to additional shear strains.

The additional deviational strain energy, therefore, would be

$$\tfrac{1}{2}\sigma_{xy}\,\varepsilon_{xy} + \tfrac{1}{2}\sigma_{yz}\,\varepsilon_{yz} + \tfrac{1}{2}\sigma_{zx}\,\varepsilon_{zx}$$

and in terms of material values and stress components the shear strain energy would be

$$\{(1 + v)/E\}[\sigma_{xy}^2 + \sigma_{yz}^2 + \sigma_{zx}^2]$$

Therefore the total deviational strain energy is the sum of the components from the normal strain and the shear strain energies:

$$U_D = \{(1 + v)/6E\}[(\sigma_{xx} - \sigma_{yy})^2 + (\sigma_{yy} - \sigma_{zz})^2 + (\sigma_{zz} - \sigma_{xx})^2] + \{(1 + v)/E\}[\sigma_{xy}^2 + \sigma_{yz}^2 + \sigma_{zx}^2] \qquad (9.47)$$

Equating Equation 9.47 to the failure in a uniaxial tensile test in terms of the deviational strain energy,

$$\{(1 + v)/6E\}[2\sigma^{*2}]$$

gives

$$\{(1 + v)/6E\}[2\sigma^{*2}] = \{(1 + v)/6E\}[(\sigma_{xx} - \sigma_{yy})^2 + (\sigma_{yy} - \sigma_{zz})^2 + (\sigma_{zz} - \sigma_{xx})^2] + \{(1 + v)E\}[\sigma_{xy}^2 + \sigma_{yz}^2 + \sigma_{zx}^2]$$

and therefore

$$1 = 1/2\sigma^{*2}[(\sigma_{xx} - \sigma_{yy})^2 + (\sigma_{yy} - \sigma_{zz})^2 + (\sigma_{zz} - \sigma_{xx})] + \{3/\sigma^{*2}\}[\sigma_{xy}^2 + \sigma_{yz}^2 + \sigma_{zx}^2] \qquad (9.48)$$

For orthotropic material where the deviational strain energy equation must be related to the principal axes of the lamina, the yield criterion is given (Hill 1960) as

$$H(\sigma_{11} - \sigma_{22})^2 + G(\sigma_{22} - \sigma_{33})^2 + F(\sigma_{33} - \sigma_{11})^2 + 2N\sigma_{12}^2 + 2M\sigma_{23}^2 + 2L\sigma_{31}^2 = 1$$

which is analogous to Equation 9.48. The parameters F, G, H, L, M and N are Hill's yield strength and are regarded as the failure strengths of the lamina.

If σ_{12} acts on the lamina only, then, at failure,

$$2N = 1/\sigma_{12}^{*2} \tag{9.49}$$

Also, if σ_{11} acts on the lamina only, then

$$F + H = 1/\sigma_{11}^{*} \tag{9.50}$$

and if σ_{22} acts on the lamina only, then

$$G + H = 1/\sigma_{22}^{*2} \tag{9.51}$$

Similarly,

$$F + G = 1/\sigma_{33}^{*2} \tag{9.52}$$

On combining Equations 9.50, 9.51 and 9.52:

$$2H = \frac{1}{\sigma_{11}^{*2}} + \frac{1}{\sigma_{22}^{*2}} - \frac{1}{\sigma_{33}^{*2}}$$

$$2G = \frac{1}{\sigma_{22}^{*2}} + \frac{1}{\sigma_{33}^{*2}} - \frac{1}{\sigma_{11}^{*2}}$$

$$2F = \frac{1}{\sigma_{11}^{*2}} + \frac{1}{\sigma_{33}^{*2}} - \frac{1}{\sigma_{22}^{*2}} \tag{9.53}$$

For a unidirectional lamina (ie orthotropic materials) under plane stress when the fibres are in direction 1, then $\sigma_{33} = \sigma_{13} = \sigma_{23} = 0$, and in addition, from geometrical symmetry, $\sigma_{22}^{*} = \sigma_{33}^{*}$. Thus the failure criterion in terms of the lamina strengths σ_{11}^{*}, σ_{22}^{*} and σ_{12}^{*} becomes

$$\frac{\sigma_{11}^{2}}{\sigma_{11}^{*2}} - \frac{\sigma_{11}\sigma_{22}}{\sigma_{11}^{*2}} + \frac{\sigma_{22}^{2}}{\sigma_{22}^{*2}} + \frac{\sigma_{12}^{2}}{\sigma_{12}^{*2}} = 1 \tag{9.54}$$

The equation describes a failure envelope. Consequently, failure of the lamina will not occur provided that σ_{11}, σ_{22} and σ_{12} all lie inside the failure envelope.

For most composite materials $\sigma_{11}^{*} \gg \sigma_{22}$, so that the second term in Equation 9.54 tends to zero and the simplified equation is

$$\frac{\sigma_{11}^{2}}{\sigma_{11}^{*2}} + \frac{\sigma_{22}^{2}}{\sigma_{22}^{*2}} + \frac{\sigma_{12}^{2}}{\sigma_{12}^{*2}} = 1 \tag{9.55}$$

To enable a prediction to be made of the failure strength in direction θ to the principal axes (see Figure 9.5*a*) on unidirectional laminae, Equation 9.54 can be rearranged to give

$$\sigma_{xx} = \sigma_{\theta} = \left[\frac{\cos^4\theta}{\sigma_{11}^{*2}} + \left(\frac{1}{\sigma_{12}^{*2}} - \frac{1}{\sigma_{11}^{*2}}\right)\sin^2\theta\cos^2\theta + \frac{\sin^4\theta}{\sigma_{22}^{*2}}\right]^{-1/2}$$

Jones (1971) has stated that there is a reasonable agreement between this theory and experimental results for E-glass/epoxy composite, and Hull (1985) has found close agreement between carbon fibre/epoxy resin laminae. The two authors have concluded that this failure criterion is suitable for off-axes tests on unidirectional laminae at various orientations of load applications.

Thus the Tsai-Hill failure theory represents a more realistic criterion for the failure of composite materials than do the previous two theories.

Hull (1985) has also observed that for carbon fibre/epoxy resin laminates, when the above angle θ is in the range $8° < \theta < 12°$, there is relatively little interaction between the longitudinal stress and the shear stress or between the transverse stress and the shear stress. Consequently, tests performed on the above range can be used to determine the intralaminar shear strength; from Equation 9.46 it will be seen that

$$\sigma_{12}^* = \sigma_\theta^* \sin\theta \cos\theta$$

The equivalent relationship for the elastic moduli and shear modulus is given in Equation 9.26.

9.5 MICROMECHANICAL ANALYSIS OF COMPOSITES

9.5.1 The analysis of continuous fibre composites

When the fibres are unidirectionally aligned in the matrix material, the composite will possess orthotropic properties with the greatest stiffness and strength in the direction of the fibres.

A composite with continuous aligned fibres is shown in Figure 9.10, and as its properties will depend upon the fibre volume fraction and the

Figure 9.10 *Uniform, continuous and unidirectional composite*

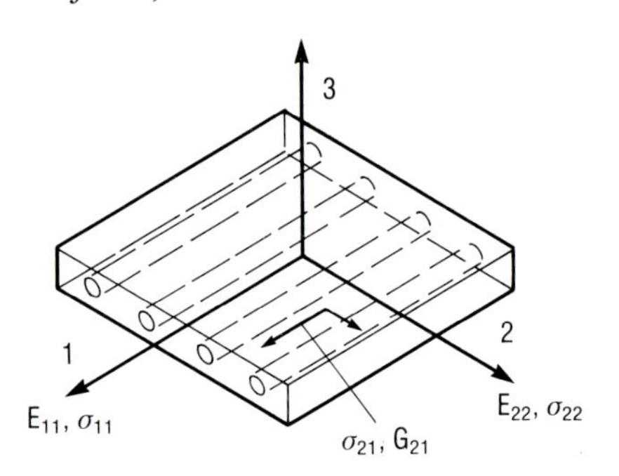

mechanical properties of the component parts, the elastic properties of the composite may be expressed as

$$E = f(E_f, V_f, \nu_f, E_m, V_m, \nu_m)$$

where V is the volume fraction and the suffices f and m refer to fibre and matrix materials, respectively.

1 The longitudinal stiffness E_{11}

To determine the value of the modulus of elasticity E of the composite in the fibre direction, it is assumed that under an axial load the strains in the fibre, matrix and composite in direction 1 are equal, ie

$$\varepsilon_{11} = \varepsilon_m = \varepsilon_f \tag{9.56}$$

The composite resultant axial force P_{11} is shared between the fibre load P_f and the matrix load P_m in the following way:

$$P_{11} = P_m + P_f$$

$$\sigma_{11} A_c = \sigma_m A_m + \sigma_f A_f$$

or

$$\sigma_{11} = \sigma_m V_m + \sigma_f V_f \tag{9.57}$$

where A is the area of the phase, V is the volume fraction of phase ($V_c =$ 1) and σ is the stress in the phase.

Equating Equations 9.56 and 9.57,

$$\varepsilon_{11} E_{11} = \varepsilon_m E_m V_m + \varepsilon_f E_f V_f$$

where E is the modulus of elasticity of phase and E_{11} is the modulus of elasticity of the composite in direction 1. Therefore

$$E_{11} = E_m V_m + E_f V_f \tag{9.58}$$

or

$$E_{11} = (E_f - E_m) V_f + E_m \tag{9.59}$$

Equation 9.59 is the 'law of mixtures' equation and it has been found that experimentally it predicts E_{11} with a fair degree of accuracy.

2 The Poisson's ratio ν_{12}

An expression for ν_{12} may be derived by considering element ABCD in Figure 9.11 under a mean normal stress σ_{11}:

$$\text{Poisson's ratio } \nu_{12} = -\varepsilon_{22}/\varepsilon_{11}$$

ε_{22} is the lateral strain in direction 2 and is a function of two components; these are:

$$\text{fibre component} \quad -\nu_f \varepsilon_{11}$$

$$\text{matrix component} \quad -\nu_m \varepsilon_{11}$$

Figure 9.11 *Lateral strain in element under mean normal stress* σ_{11}

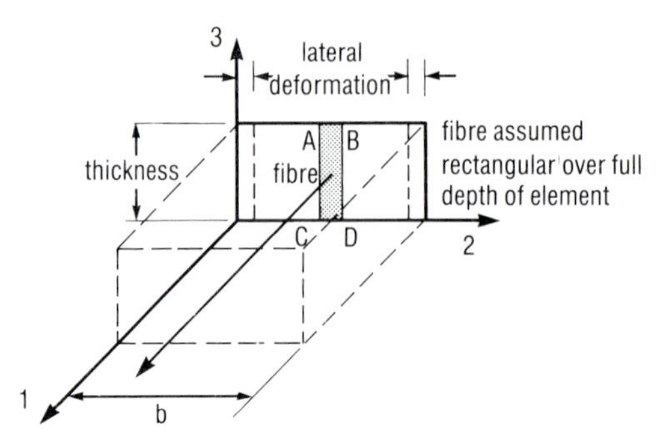

The lateral displacements in the two directions due to these strains are dependent on the geometry of the fibre and on the cross-sectional form of the element. It will be assumed in the following discussion that the fibre is rectangular and extends over the whole thickness of the element, as shown in Figure 9.11, and that the strain effects in direction 3 are negligible in the lamina:

$$\text{lateral displacement of fibre} = -\nu_f\,\varepsilon_{11}(V_f\,b)$$

$$\text{lateral displacement of matrix} = -\nu_m\,\varepsilon_{11}(V_m\,b)$$

Total lateral displacement in direction 2

$$= [-\nu_f\,\varepsilon_{11}\,V_f - \nu_m\,\varepsilon_{11}\,V_m]b$$

where b is the breadth of the element.

Lateral strain in direction 2 $= \varepsilon_{22} = -(\nu_f\,V_f + \nu_m\,V_m)\varepsilon_{11}$

Therefore $$\nu_{12} = -\frac{\varepsilon_{22}}{\varepsilon_{11}}$$

$$\nu_{12} = [\nu_f\,V_f + \nu_m\,V_m] \tag{9.60}$$

Equation 9.60 predicts the value of ν_{12} with a fair degree of accuracy.

3 The modulus of elasticity E_{22}

The modulus of elasticity E_{22} may be determined by considering Figure 9.12, where the assumption of a rectangular fibre extending over the whole thickness of the element is again used. The stiffness of the fibre is large compared with that of the matrix; consequently, when the composite is under a stress σ_{22} the displacement differential in the constituents introduces varying values of shearing stresses in the plane 1–2 which are especially pronounced near the fibre-matrix interface.

The approximate determination of E_{22} will be presented, in which Poisson's ratio is neglected and therefore σ_{22} will be considered constant throughout the element.

Considering the fibre/matrix element in Figure 9.12 under a constant stress σ_{22}, the transverse strains induced in the fibre and matrix are, respectively,

$$\varepsilon_f = \frac{\sigma_{22}}{E_f}$$

and

$$\varepsilon_m = \frac{\sigma_{22}}{E_m} \tag{9.61}$$

The transverse displacements in the fibre and matrix are $\varepsilon_f(V_f b)$ and $\varepsilon_m(V_m b)$, respectively. Therefore the total transverse displacement is

$$b\varepsilon_{22} = (\varepsilon_f V_f b) + (\varepsilon_m V_m b) \tag{9.62}$$

Combining Equations 9.61 and 9.62,

$$b\varepsilon_{22} = \sigma_{22}\left(\frac{V_f}{E_f} + \frac{V_m}{E_m}\right)b$$

but

$$E_{22} = \frac{\sigma_{22}}{\varepsilon_{22}}$$

hence

$$E_{22} = \frac{E_f E_m}{E_m V_f + E_f V_m} \tag{9.63}$$

Equation 9.63 may also be expressed in the form

$$\frac{1}{E_{22}} = \frac{V_f}{E_f} + \frac{V_m}{E_m}$$

$$= \frac{V_f}{E_f} + \frac{1 - V_f}{E_m} \tag{9.64}$$

4 The modulus of rigidity G_{12}

The derivation of G_{12} is analogous to that for E_{22}, and Figure 9.13 shows an element considered to be subjected to uniform shearing and the complimentary shearing stresses along its boundary.

The full derivation of G_{12} will not be given here, but it can be shown that

$$G_{12} = (G_f G_m)/(G_m V_f + G_f V_m) \tag{9.65}$$

Figure 9.12 (*left*) *Element under normal stress* σ_{22}
Figure 9.13 (*right*) *Element under uniform shearing stress* σ_{12}

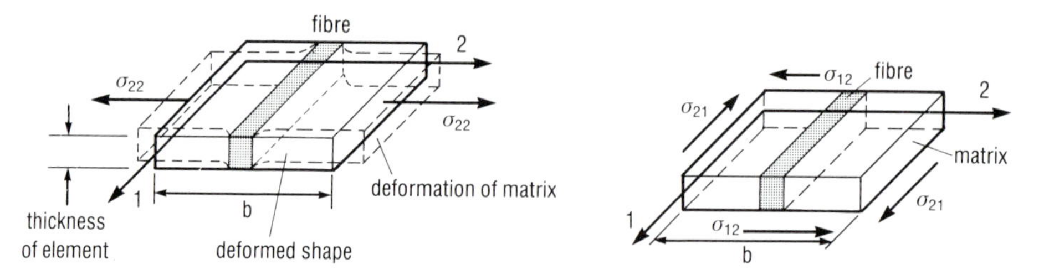

5 *Strength characteristics*

(*a*) *Uniaxial tension*: From the preceding section it will be seen that, to obtain high stresses in fibre-reinforced polymers and therefore to use high-strength reinforcement most efficiently, it is necessary for the fibre modulus to be much greater than the matrix modulus. In addition, the volume fraction of fibres in the composite is in direct proportion to the composite load.

The excellent strengths and strength to weight ratios achieved by fibre-reinforced matrix materials are as a result of the high strength of the fibres, which are imparted to the composite. It is worth mentioning that the modular ratio (E_f/E_m) for glass fibre-reinforced polymers is approximately 20 and at only 10% by volume of this fibre the glass assumes 70% of the total load.

The law of mixtures (Equation 9.59) may be modified to approximate the composite stress σ_{11} in terms of fibre volume fraction V_f and the stresses induced in the matrix σ_m and fibre σ_f. The equation becomes

$$\sigma_{11} = \sigma_f V_f + \sigma_m(1 - V_f) \tag{9.66}$$

The ultimate strain in the fibres will be less than that in the matrix (typical of a carbon fibre/epoxy matrix system) because of the brittle nature of the former, with the result that the fibres will fail first. Also, if the required ultimate composite stress after the failure of the fibres is greater than that of the matrix, complete failure of the composite will occur.

At failure of the composite, Equation 9.66 becomes

$$\sigma_{11}^* = \sigma_f^* V_f + \sigma_m'(1 - V_f) \tag{9.67}$$

where σ_f^* is the ultimate failure stress of fibre and σ_m' is the stress in the matrix at failure of the fibre.

It must be said, however, that the stress relationship is an oversimplification and, although in some cases it is a good approximation, it takes no account of the statistical scatter in fibre strengths and the influence of the microstructural defects. The effect of these quantities may be included in the Equation 9.66, if it is modified to

$$\sigma_{11} = \beta\sigma_{\beta} V_f + \sigma_m(1 - V_f) \tag{9.68}$$

where σ_{β} is the fibre bundle strength and β is the matrix efficiency factor, usually between 1 and 2.

Consequently, it may be seen that the composite strength is principally dependent upon fibre strength and volume fraction, since the contribution from the matrix is small.

(*b*) *Transverse tensile strength*: The transverse strength of a unidirectional fibre array is a difficult quantity to determine. It is influenced by many parameters, including the properties of the fibre, the matrix, the interface strength, the presence of voids and the internal stresses and strains set up by the interaction between the above parameters. Contrary to general expectations, the transverse strength is usually less than the strength of the matrix material, which implies that fibres have a negative reinforcing effect.

(*c*) *Uniaxial compressive strength*: The mechanism of failure of a continuous unidirectional fibre composite subjected to a compressive load is a difficult quantity to estimate and highly complex to analyse. The failure load appears to be controlled by the buckling of the fibre, and it is assumed that each fibre acts independently of its neighbours. The problem, therefore, may be treated similarly to that of the buckling of slender columns on an elastic foundation. The two basic modes of buckling which can occur in the simplified model, shown in Figure 9.14, are

Figure 9.14 *Schematic representation of the two basic modes of buckling in unidirectional laminate*

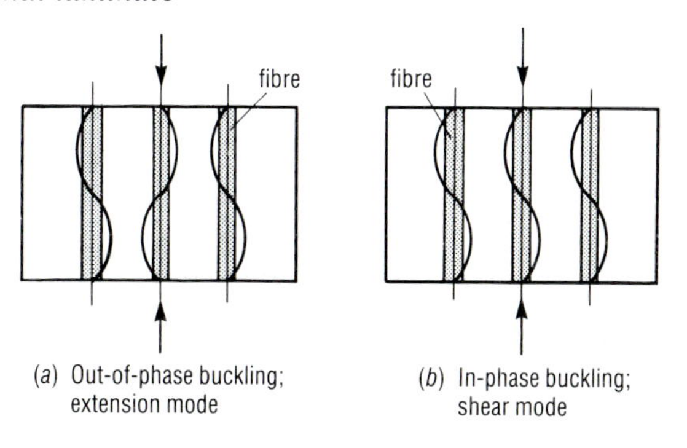

that the adjacent fibres will buckle with equal wavelengths at the ultimate load, either

(i) Out-of-phase with each other, thus producing an extensional mode; this mode is associated with low fibre volume fractions
    or
(ii) In-phase with each other, thus producing a shear mode; this mode is associated with high fibre volume fractions.

The critical compressive stresses are given by

$$\sigma_c = G_m/(1 - V_f) \tag{9.69}$$

for shear-mode, in-phase buckling, and

$$\sigma_c = 2V_f[E_f E_m V_f/3(1 - V_f)]^{1/2} \tag{9.70}$$

for extension-mode, out-of-phase buckling.

In the practical range of V_f the shear mode tends to dominate and the compressive strength as highly dependent upon the shear modulus of the matrix.

Figure 9.15 *Diagrammatic view of shear failure under a longitudinal compressive load*

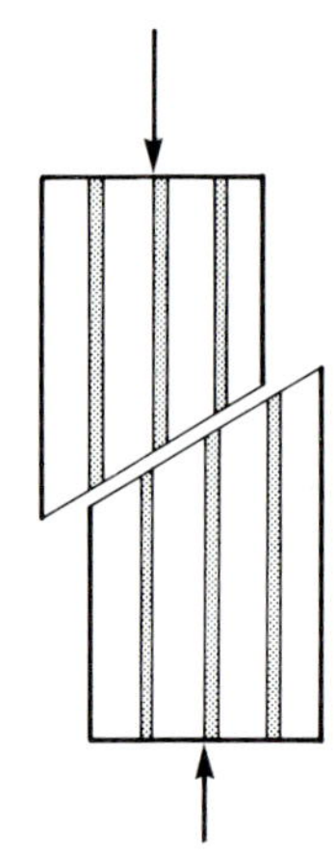

(*d*) *Shear strength*: There appears to be little investigative work undertaken into determining the ultimate shearing strength of composite materials. Ewins and Ham (1973) have suggested that the shear stresses developed in a lamina may cause a shear mode failure under a compressive load, as shown in Figure 9.15. The shear stress would then be

$$\sigma_s = \sigma_{11} \sin\theta \cos\theta$$

which is maximised when $\theta = 45$, ie

$$\sigma_s = \tfrac{1}{2}\sigma_{11}$$

However, it would be expected that this value would be similar to the corresponding matrix strength or the fibre-matrix interface strength, and the ultimate strength would be the smaller of these two values.

9.5.2 Analysis of discontinuous fibres

Discontinuous fibre-reinforced polymers are composites in which the fibres are randomly orientated and have a length to diameter (l/d) ratio, known as the aspect ratio, varying between 100 and 5,000. The ultimate strength and modulus of short-fibre-reinforced composites can approach the values for continuous-fibre composites, provided that the short filaments can be aligned unidirectionally and that their length is much greater than the critical length (l_c) required for shear-stress transfer. The fibres, however, are invariably randomly dispersed in the matrix, producing a composite possessing isotropic properties.

1 Stress–strain relationship for discontinuous fibres

When a composite containing uniaxially aligned discontinuous fibres is stressed in tension parallel to the fibre direction as shown in Figure 9.16, there is a portion at the end of each finite length which is stressed to less than the maximum stress of a continuous fibre. This length, known as the transfer length, depends upon the maximum fibre stress σ_f and the effect-

Figure 9.16 *Schematic representation of discontinuous fibre/matrix composite. Average fibre tensile stress $\bar{\sigma}_{fav}$($\bar{\sigma}_{fl}$ etc) along the fibre length*

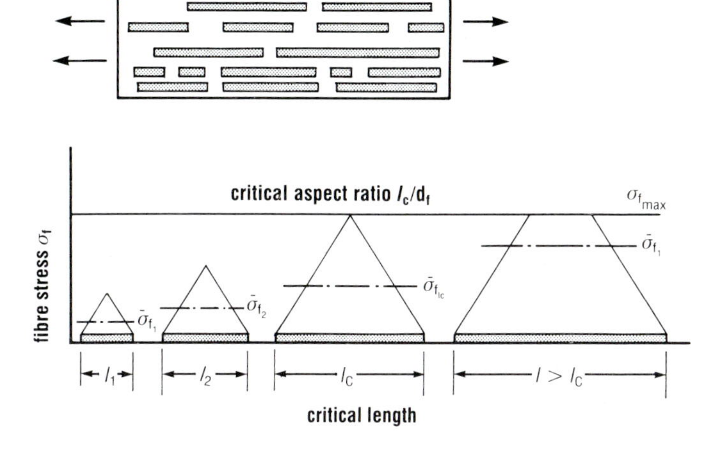

ive shear strength of the fibre-matrix interface σ_{xy},and is given by

$$l_1 = \frac{\sigma_f d_f}{4\sigma_{xy}}$$

where d_f is the diameter of the fibre and σ_f is the maximum stress in the fibre for a fibre of length l_1.

The critical transfer length over which the fibre stress is decreased from $\sigma_{f\max}$ to zero is referred to as l_c. If a quantity α is introduced and defined as the ratio of the area under the stress distribution curve over the length $\frac{1}{2}l_c$ to the area of the rectangle represented by the product

$$\sigma_{f\max} \times \tfrac{1}{2}l_c$$

then the fibre end of the length l_c may be considered as supporting a reduced average stress $\alpha \times \sigma_{f\max}$ or it may be reduced to an effective length of $\frac{1}{2}\alpha l_c$ and subjected to a stress of $\sigma_{f\max}$.

By considering Figure 9.16, the value of the average axial stress σ_{fav} for fibres with lengths not less than l_c may be deduced from the relationship

$$\sigma_{f\max}(l - l_c) + \tfrac{1}{2}\sigma_{f\max} l_c = \sigma_{fav} l$$

or

$$\sigma_{fav} = \sigma_{f\max}(1 - l_c/2l) \qquad (9.71)$$

The equivalent value for a length of fibre less than the critical length is

$$\sigma_{fav} = \sigma_{f\max} l/2l_c \qquad (9.72)$$

It must be emphasised that these equations are only approximations, but the derivations can be useful in predicting the elastic properties of discontinuous fibre composites.

2 Strength characteristics for discontinuous fibres

Ultimate tensile strength for unidirectional aligned fibres

As the variability of the composition of discontinuous fibre composites is much greater than that for continuous fibre composites, the strength predictions for the former can only be approximate.

From Equations 9.57 and 9.71, and assuming that V_f is large, then the ultimate tensile strength of the composite becomes

$$\sigma_c^* = \sigma_{fav}^* V_f + \sigma_{mf} V_m$$

where σ_c^* is the tensile failure in the composite, σ_{fav}^* is the average tensile stress at failure of the fibre, σ_{mf} is the matrix tensile stress at failure of the fibre

or

$$\sigma_c^* = \sigma_f^*\left(1 - \frac{l_c}{2l}\right)V_f + \sigma_{mf} V_m$$

where σ_f^* is maximum ultimate tensile stress of the fibre.

9.6 CHARACTERISTICS OF LAMINATED COMPOSITES

In the preceding sections, attention was directed towards single lamina composites. However, in practice, composite materials rarely exist as a single lamina, but will be fabricated from a number of laminae bonded together. If the separate laminae have been manufactured by utilising randomly orientated fibres in a matrix, the resulting composite will also possess similar in-plane properties and these will be quasi-isotropic. If, however, the separate laminae possess orthotropic properties by virtue of the orientation of the fibres in the matrix, then the resulting composite will have properties depending upon the final arrangement of each independent lamina within the composite. Figure 9.17 illustrates a laminated construction.

Figure 9.17 *An isotropic and an orthotropic laminae (a) Isotropic laminate construction: three individual laminae with randomly oriented fibres in each layer, (b) Orthotropic laminate construction: three individual laminae with unidirectional fibres at 0° and 90° orientation*

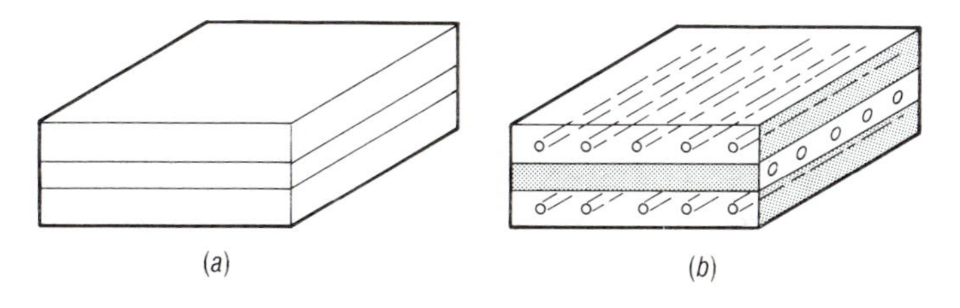

In forming composites, various laminae are brought together during the manufacturing process and consequently there would generally be a thin layer of resin, which does not contain any fibres, between each successive layer of laminae.

If the composite were subsequently to be placed in a flexural situation, shear stresses would be developed through its cross-section and the limiting values of these stresses would depend upon the shear strength of the matrix; these latter values are generally low. Delamination of the adjacent laminae would result; this is one of the main causes of failure of composite materials under flexural loading.

In the preceding sections, only in-plane stresses were considered because the thickness of the lamina was too small to support loading. However, if the thickness of the composite is increased by adding more of the individual laminae, then the material is able to support both in-plane and flexural forces simultaneously. Further, if the composite is manufactured from randomly orientated fibres, a uniform thickness and quasi-isotropic properties are produced. If, however, the thickness is produced from several laminae whose properties vary in given directions, it may not be uniform throughout the composite and this may give rise to a complex stress situation in which the in-plane forces will cause bending moments to develop. This action has been discussed in Agarwal and Broutman (1980), and Broutman and Krock (1967).

9.6.1 Behaviour of composite (laminated) beams

The theory of composite plates may be developed by considering firstly a composite beam, which is a special case in that a plate spans in one direction only. The theory may be developed by considering different laminae stacking arrangements within the cross-section of the beam.

The general equation for symmetrically laminated (isotropic) beams is

$$N_{xx} = Et \frac{du_0}{dx}$$

and

$$M_{xx} = E \frac{t^3}{12} \frac{d^2w}{dx^2}$$

where N is the resultant axial force and M is the bending moment per unit width of beam.

These are the standard equations for axial and flexural effects in a rectangular beam of unit width.

The general equations for the multilayered symmetric cross-sectional beam, as shown in Figure 9.18*a*, would be of similar form to the above but more complex. The stiffness expression for the in-plane forces gives a 'rule of mixtures' result in which each separate component of the equation represents the stiffness of the respective composite section. In addition, the stiffness expression for the flexural forces is obtained from the separate second moments of area for the individual sections. In the above two cases, the deflections produced by the in-plane and flexural forces are uncoupled.

For a multilayered asymmetric cross-sectioned beam, as shown in Figure 9.18*b*, in-plane and flexural expressions are produced similar to those for the symmetric sectioned beam, but the deformation from these two loading components are coupled because the separate applications of in-plane and flexural forces produce both axial and flexural deformations.

Therefore, the general equation for composite (laminated) beams with symmetrical and asymmetrical cross-sections may be expressed in the

Figure 9.18 *Cross sections of laminated beams (a) Variation of stress, strain and modulus of elasticity of symmetric beam, (b) Variation of stress, strain and modulus of elasticity of asymmetric beam*

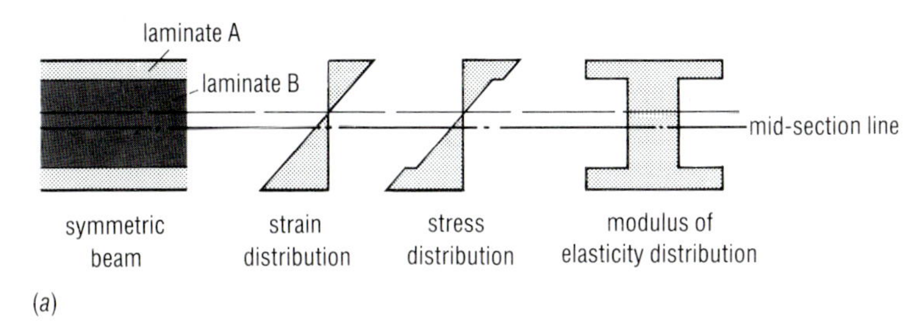

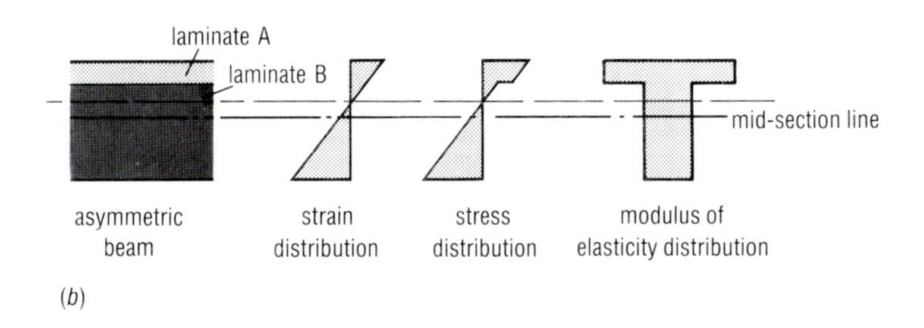

form

$$N = A\frac{du_0}{dx} + B\frac{d^2w}{dx^2} \tag{9.73a}$$

$$M = B\frac{du_0}{dx} + C\frac{d^2w}{dx^2} \tag{9.73b}$$

or

$$\begin{bmatrix} N \\ M \end{bmatrix} = \begin{bmatrix} A & B \\ B & C \end{bmatrix}\begin{bmatrix} du_0/dx \\ d^2w/dx^2 \end{bmatrix}$$

For symmetrical cross-sections, the coefficient B becomes zero but it will be non-zero for asymmetrical composite cross-sectioned beams.

9.6.2 Behaviour of composite (laminated) plates

The general case of the two-dimensional plate theory may be developed by considering a rectangular element with sides dx and dy as shown in Figure 9.19. The important section in these plates is the mid-plane as this will determine the symmetric or asymmetric nature of the plate.

In composite plates, axial deformations can take place in both x and y directions, and also in-plane shear distortions can develop. Consequently, the displacements δu, δv in the x and y directions, respectively, can be

Figure 9.19 *Plate element under load (a) Rectangular plate element of thickness t, (b) Stress distribution in isotropic plate*

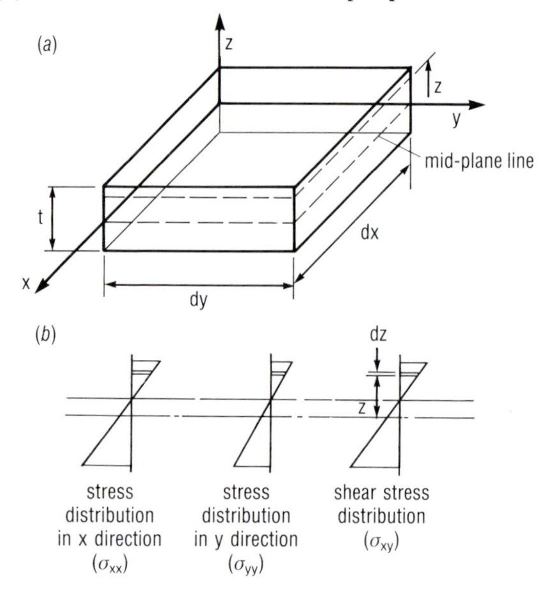

expressed as

$$\delta u = \delta u_0 - z\,\frac{\delta w}{\delta x}$$

$$\delta v = \delta v_0 - z\,\frac{\delta w}{\delta y} \tag{9.74}$$

The normal and shearing strains can be expressed as:

$$\varepsilon_{xx} = \partial u/\partial x = \partial u_0/\partial x - z\partial^2 w/\partial x^2 \tag{9.75a}$$

$$\varepsilon_{yy} = \partial v/\partial y = \partial v_0/\partial y - z\partial^2 w/\partial y^2 \tag{9.75b}$$

$$\varepsilon_{xy} = \partial u/\partial y + \partial v/\partial x = \partial u_0/\partial y + \partial v_0/\partial x - 2z\partial^2 w/\partial x\partial y \tag{9.75c}$$

Partial differentials have been used because u, v and w are functions of both x and y,

or

$$\varepsilon_0 = \begin{bmatrix} \partial u_0/\partial x \\ \partial v_0/\partial y \\ \partial u_0/\partial y + \partial v_0/\partial x \end{bmatrix} \tag{9.76}$$

$$\chi = \begin{bmatrix} \partial^2 w/\partial x^2 \\ \partial^2 w/\partial y^2 \\ 2\partial^2 w/\partial x\partial y \end{bmatrix} \tag{9.77}$$

or

$$[\varepsilon] = [\varepsilon_0] - z[\chi] \tag{9.78}$$

1 Isotropic composite plates

It has been shown (Timoshenko and Woinowsky-Krieger 1959) that for isotropic plates the resultant in-plane forces, namely two axial (N_{xx}, N_{yy}) and one shear (N_{xy}), and moments, namely two bending moments (M_x, M_y) and one torque (M_{xy}), may be obtained by integrating across the thickness of the plate, giving:

$$N_{xx} = \int_{-t/2}^{t/2} \sigma_{xx}\, dz = Et/(1-v^2)\left[\frac{\partial u_0}{\partial x} + v\frac{\partial v_0}{\partial y}\right]$$

$$N_{yy} = \int_{-t/2}^{t/2} \sigma_{yy}\, dz = Et/(1-v^2)\left[\frac{\partial v_0}{\partial y} + v\frac{\partial u_0}{\partial x}\right] \tag{9.79}$$

$$N_{xy} = \int_{-t/2}^{t/2} \sigma_{xy}\, dz = Et/2(1+v)\left[\frac{\partial u_0}{\partial y} + \frac{\partial v_0}{\partial x}\right]$$

and

$$M_{xx} = -\int_{-t/2}^{t/2} \sigma_{xx}\, z\, dz = Et^3/12(1-v^2)\left[\frac{\partial^2 w}{\partial x^2} + v\frac{\partial^2 w}{\partial y^2}\right]$$

$$M_{yy} = -\int_{-t/2}^{t/2} \sigma_{yy}\, z\, dz = Et^3/12(1-v^2)\left[\frac{\partial^2 w}{\partial y^2} + v\frac{\partial^2 w}{\partial x^2}\right] \tag{9.80}$$

$$M_{xy} = -\int_{-t/2}^{t/2} \sigma_{xy}\, z\, dz = Et^3/12(1+v)\left[\frac{\partial^2 w}{\partial x \partial y}\right]$$

The distribution of the stress σ_{xx}, σ_{yy} and σ_{xy} is shown in Figure 9.19 *b*, and the forces and moments are shown in Figure 9.20.

Figure 9.20 *Forces and moments acting on rectangular plate*

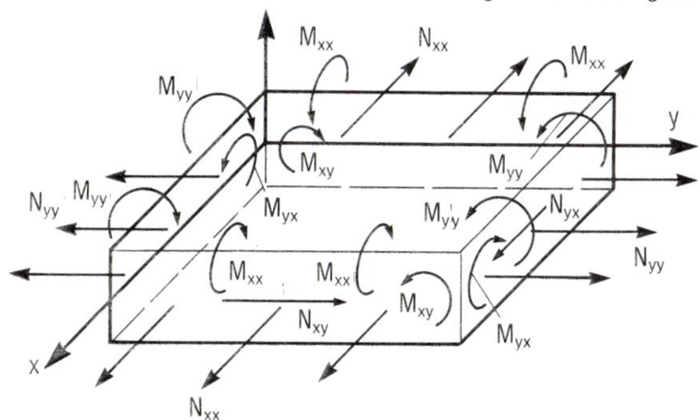

2 Orthotropic composite plates

The relationship between in-plane stresses and strains along the principal axes 1 and 2 for orthotropic plates is defined by

$$\begin{bmatrix} \sigma_{11} \\ \sigma_{22} \\ \sigma_{12} \end{bmatrix} = \begin{bmatrix} Q_{11} & Q_{12} & 0 \\ Q_{21} & Q_{22} & 0 \\ 0 & 0 & Q_{33} \end{bmatrix} \begin{bmatrix} \varepsilon_{11} \\ \varepsilon_{22} \\ \varepsilon_{12} \end{bmatrix} \tag{9.81}$$

where
$$Q_{11} = \frac{E_{11}}{1 - \nu_{21}\nu_{12}}$$

$$Q_{22} = \frac{E_{22}}{1 - \nu_{21}\nu_{12}}$$

$$Q_{12} = \frac{\nu_{12}E_{22}}{1 - \nu_{12}\nu_{21}} = Q_{21} = \frac{\nu_{21}E_{11}}{1 - \nu_{21}\nu_{12}}$$

$$Q_{33} = G_{12}$$

If the stresses and strains are defined about the arbitrary axis (x, y) there will be coupling between the normal and shear effects, giving the relationship

$$\begin{bmatrix} \sigma_{xx} \\ \sigma_{yy} \\ \sigma_{xy} \end{bmatrix} = \begin{bmatrix} \bar{Q}_{11} & \bar{Q}_{12} & \bar{Q}_{13} \\ \bar{Q}_{21} & \bar{Q}_{22} & \bar{Q}_{23} \\ \bar{Q}_{31} & \bar{Q}_{32} & \bar{Q}_{33} \end{bmatrix} \begin{bmatrix} \varepsilon_{xx} \\ \varepsilon_{yy} \\ \varepsilon_{xy} \end{bmatrix} \tag{9.82}$$

or
$$[\boldsymbol{\sigma}] = [\bar{Q}][\boldsymbol{\varepsilon}] \tag{9.83a}$$

$$= [\bar{Q}](\varepsilon_0 - z[\chi]) \tag{9.83b}$$

(where the components of $[\bar{Q}]$ are of similar form to Equation 9.24) from which it can be shown that the in-plane forces and out-of-plane moments acting along a rectangular plate per unit width can be expressed as

$$[N] = [\bar{Q}]t[\varepsilon_0]$$

and
$$[M] = [\bar{Q}](t^3/12)[\chi] \tag{9.84}$$

If the composite is manufactured from laminae having different properties but with a symmetric cross-section about the centre line of the section, as shown in Figure 9.21a, then the in-plane and out-of-plane behaviour will be uncoupled and the equations will be similar to those for the orthotropic plate just considered. The form of the equations, however, will be more complicated.

The in-plane forces and out-of-plane moments per unit width acting on a rectangular plate consisting of three orthotropic laminae can be expressed as:

$$N = ([\bar{Q}]_A 2t_0 + [\bar{Q}]_B t_I)[\varepsilon_0] \tag{9.85a}$$

$$M = \left([\bar{Q}]_A \left(\left(\frac{2t_0 + t_I}{12}\right)^3 - \frac{t_I^3}{12} \right) + [\bar{Q}]_B \frac{t_I^3}{12} \right) [\chi] \tag{9.85b}$$

Again, if a plate is made from laminae having different properties and an asymmetric cross-section about the centre line of the section, as shown in

Figure 9.21 *Composites manufactured from lamina (a) Symmetric laminae (exploded view), (b) Asymmetric laminae (exploded view)*

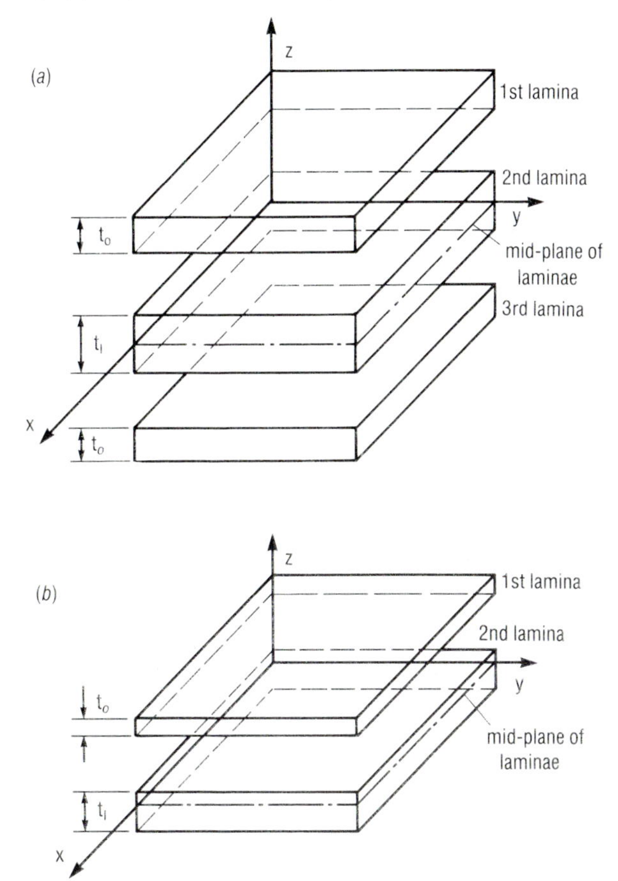

Figure 9.21*b*, the in-plane and out-of-plane behaviour will be coupled and the resulting equations will be more complex than those of the two previous cases.

The in-plane and out-of-plane moments for unit width acting in a rectangular plate consisting of two asymmetric orthotropic laminae will be expressed as:

$$[\boldsymbol{N}] = ([\bar{Q}]_A t_O + [\bar{Q}]_B t_I)[\varepsilon_0] + ([\bar{Q}]_A - [\bar{Q}]_B)t_I t_O/2[\chi] \quad (9.86a)$$

$$[\boldsymbol{M}] = ([\bar{Q}]_A - [\bar{Q}]_B)t_I t_O/2[\varepsilon_0] + ([\bar{Q}]_A t_O[t_O^2 + 3t_I^2] + [\bar{Q}]_B t_I[3t_O^2 + t_I^2])[\chi]/12 \quad (9.86b)$$

The examples which have been given above have been based upon a simple laminate stacking procedure. In practice, a composite consists of a large number of laminae and consequently it is recommended that further reading should be undertaken (Jones 1975) to gain a greater understanding of the development of the stiffness properties of multilayered composites. The work has shown that, generally, for symmetric and asymmetric laminates the relationship between the in-plane and flexural stress resultants and the corresponding strains may be expressed as

$$\begin{bmatrix} N \\ M \end{bmatrix} = \begin{bmatrix} R_{11} & R_{12} \\ R_{21} & R_{22} \end{bmatrix} \begin{bmatrix} \varepsilon_0 \\ \chi \end{bmatrix} \tag{9.87}$$

where $R_{12} = R_{21}$ is the coupling matrix.

9.7 DESIGN OF COMPOSITE COMPONENTS

The design of composites is an integrated procedure incorporating simultaneously solutions to the following items:

(i) The analysis of the structural system: This analysis will provide information on the magnitudes of stress and strain which the composite has to sustain; this information will enable a decision to be made on the fibre orientation and thence whether the final laminate will be isotropic, orthotropic or a combination of both.

(ii) The manufacturing techniques of the composite: These techniques will influence the final fibre volume fraction and the degree of compaction of the component parts. These parameters will reflect the final strength of the composite and its durability.

(iii) Testing of representative component parts of the manufactured composite to provide mechanical properties and failure strengths: The properties which will be required are stiffness in tension, compression, flexure and shear, together with a knowledge of how these vary with time and temperature. The theories which have been derived to predict stiffness and strength have been discussed in Section 9.4. In addition, it is necessary to specify whether the structural component is to have an extended life, in which case the structural properties utilised would need to be related to long-term tests to ensure that an adequate factor of safety exists at the end of the life of the component. It should also be remembered that composites are generally orthotropic materials and therefore the stiffness and strength properties will be direction-dependent. Figure 9.22 is a diagrammatic representation of the fibre orientation dependence of the modulus of elasticity of composite materials. It is now possible to obtain tailor-made composites with pre-determined fibre orientation distributions so that the elastic and strength properties can be designed to meet specific requirements.

Figure 9.22 *Typical orientation dependence of modulus of elasticity*

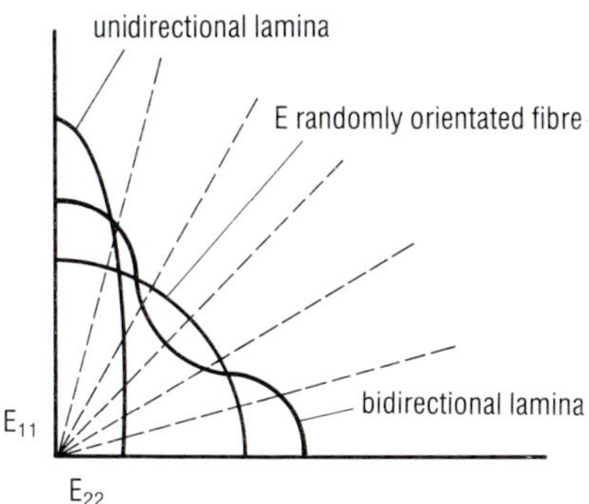

The relationship between the stresses in the fibre, the matrix and the composite has been derived from the basic equation in Section 9.5.

The basic issue in designing with composites is to understand and to be able to use the directionally dependent properties of the material and to realise that the number of stiffness constants will be increased to four for the stiffness and at least five for the strength of an on-axis unidirectionally aligned ply compared with the two constants that are required for isotropic materials.

The framework in which composites may be designed is shown in Figure 9.23. This diagrammatic representation must be expanded for composites to include the effects of the material variables on the functional relationships of the composite. This has been done in Figure 9.24.

Figure 9.23 *Diagrammatic representation of an integrated framework*

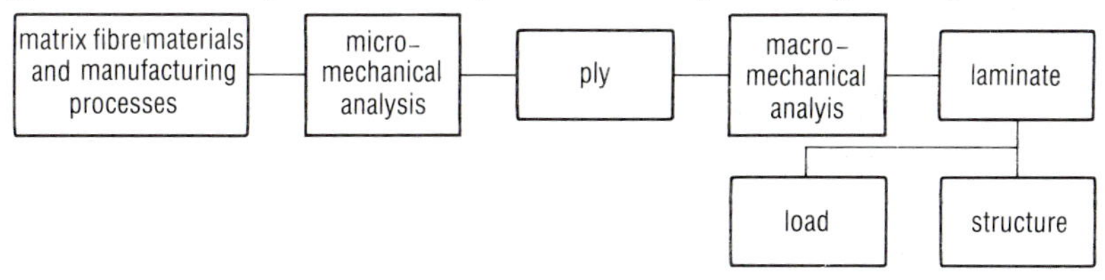

Figure 9.24 *Diagrammatic representation showing the principal variables of composite materials*

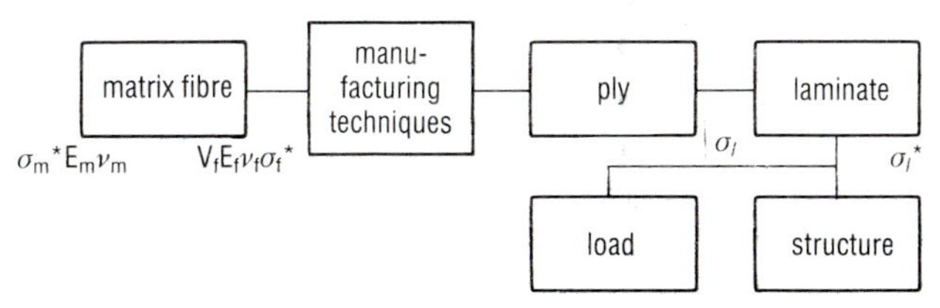

There are many fibre configurations which can be formed to produce composite materials with specific properties. The design approach can be divided into three broad sections:

(i) To optimise material configurations, it is essential to use the micromechanical formulae so that the contribution of the fibre and matrix can be related to the stiffness and strength of the laminate.
(ii) To optimise the stiffness properties, laminate theory would be used to provide a selection of composites.
(iii) To optimise the strength properties one of the approaches outlined in Section 9.4.3 should be used. The design would be verified at the limit and ultimate strengths of the composites.

Owing to the large number of material constants involved in the design of composites, computers must play an important role in this process. Spreadsheets, database and integrated graphics capabilities of computers are utilised to facilitate the design procedure.

9.8 SANDWICH MEMBERS, BEAMS AND PANELS

Sandwich members consist of a low-density core material which is sandwiched between two stiff-face or skin materials. The members so formed are efficient structural components for supporting axial or flexural loads. Under flexural loads, the sandwich beam or panel behaves in a similar way to a plate girder where the flange (face) material supports the moment in the section, and the core material, which has a lower modulus, supports the shear force at the section. The sandwich beams or panels can be divided into two types:

(i) a low-shear-modulus core material, such that the sandwich is subjected to significant shear deformation under load
(ii) a relatively high-shear-modulus core material, such that the sandwich is stiff enough for shear effects to be small.

The sandwich beam consists of two thin, stiff and strong sheets of relatively dense material, which in the present case would be a composite; these sheets are separated by a thick layer of a lower-density material which, in the case of item (i) above, would be a foam polymer or honeycomb material and in the case of item (ii) would be a slightly lower-modulus composite material than the faces. The bending stiffness of these arrangements is much greater than that of a single solid plate of the same total weight made of the same material as the faces. Figure 9.25 shows typical methods of forming sandwich systems.

Figure 9.25 *Various sandwich configurations*

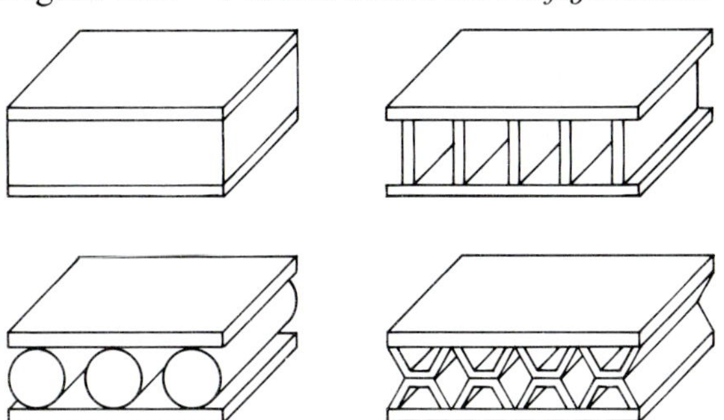

The core material must be stiff enough in the direction perpendicular to the faces to ensure that they remain the correct distance apart. It must also be stiff enough in shear to prevent the faces from sliding over each other when the panel is bent. In addition, the core must be stiff enough to prevent local buckling under compressive forces in its own plane. The degree of stiffness of the core will influence the analysis and design procedures of the panel; the lower the density of the core, the smaller will be the contribution to bending, and this will lead to a simplification of the analysis of stresses and deflection.

9.8.1 Low-density rigid-foam materials for sandwich systems

A cellular polymer is here defined as a plastics material, the apparent density of which is decreased substantially by the presence of numerous cells dispersed throughout its mass. Expanded plastics, foamed plastics and rigid foams are two-phase systems of a gas dispersed in solid plastics; in most cases the solid plastics represent only a minor proportion of the total volume but contribute greatly to the properties and utility of the foam.

The gas phase in a cellular polymer is usually distributed in voids or pockets called cells. If these cells are interconnected, the material is termed open-celled. If the cells are discrete and the gas phase of each is independent of that of the other cells, the material is called closed-cell. Foamed plastics are classified as rigid or flexible. The rigid plastics foams are the ones used in sandwich construction.

Honeycombs consist of very thin sheets attached to form connected cells which closely resemble bees' honeycomb found in nature. Development of the modern honeycomb commenced at the beginning of the Second World War when it was used in aircraft construction. The hexagonal core is one of the most common forms, and for structural applications it is attached to the composite face materials by adhesive bonding.

Honeycomb cores can be made from non-metallic materials such as glass-reinforced polymers, Nomex and Kraft paper. The adhesives come in film, paste and liquid forms and are composed of modified epoxies, phenolics, polyimides and urethanes.

9.8.2 Structural sandwich systems

Structural composite sandwich panels for the aircraft industry would probably incorporate a composite of carbon fibre/epoxy resin with polymer honeycomb or corrugated cores.

Panels for the construction industry would be manufactured from randomly orientated polyester composite face materials and a foam core. These units would be used in a structural or semi-structural way and would probably carry only relatively small loads over fairly long spans. Building panels should be light and relatively cheap. In the building industry new materials and new combinations of old ones are constantly being proposed and used in sandwich construction.

In addition to the structural aspects, sandwich systems can also provide good technical insulation properties, but problems may arise concerning acoustic insulation, vapour transmission and fire resistance. These difficulties can be overcome but the solutions are outside the scope of this chapter.

1 Sandwich beam flexural theory – low shear modulus core material

In this theory it is assumed that the faces are much stronger and stiffer than the core material, which consists of a low-shear-modulus, low-density material, such that the sandwich is subjected to significant shear deformations under transverse loads.

The flexural rigidity D of a sandwich beam (width b), a section of which is shown in Figure 9.26, is the sum of the flexural rigidities of the two separate parts of the composite, namely the core and the faces, measured about the centroidal axis of the whole section. From Figure 9.23 it is clear that

$$D = E_f bt^3/6 + E_f btd^2/2 + E_c bc^3/12 \tag{9.88}$$

where E_f and E_c are the modulus of elasticity of the face and core, respectively.

The first term on the right-hand side of the equation represents the local stiffness D_f of the faces bending about their own centroidal axes. The third term represents the bending stiffness of the core. In practice the second term is the dominant one, and provided the first and third terms are less than 1% of the second they can be discounted.

Figure 9.26 *Sandwich beam dimensions and sign convention*
The sign convention adopted for the stress resultants is shown above. The positive direction of the external forces required to balance the stress resultants at the cut section is derived from a right-handed system of co-ordinates x, y and z. Consequently, the relationship between the curvature and bending moment is:

$$M = -EI\,\frac{d^2z}{dx^2} \qquad \textit{defln} = w \qquad \textit{moment } M = -D\,\frac{d^2w}{dx^2}$$

$$\textit{slope} = \frac{dw}{dx} \qquad \textit{shear force } Q = -D\,\frac{d^3w}{dx^3}$$

$$Q = \frac{dM}{dx} \qquad \textit{curvature} = \frac{d^2w}{dx^2} \qquad \textit{distributed load } q = D\,\frac{d^4w}{dx^4}$$

$$q = -\frac{dQ}{dx}$$

(*a*) dimensions of sandwich beam for flexural rigidity equations

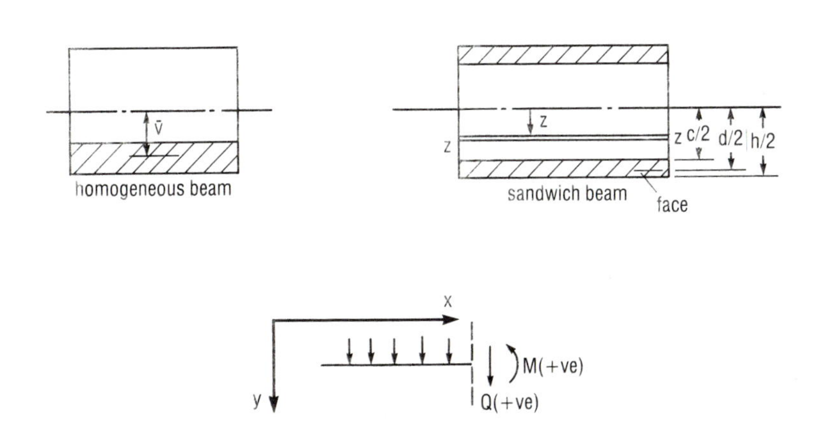

(*b*) dimensions for shear stress equations

Therefore, provided

$$[E_f bt^3/6]/[E_f btd^2/2] < \frac{1}{100}$$

and

$$[E_c bc^3/12]/[E_f btd^2/2] < \frac{1}{100}$$

ie

$$\frac{d}{t} > 5.77 \quad \text{and} \quad \frac{E_f}{E_c}\frac{t}{c}\left(\frac{d}{c}\right)^2 > \frac{100}{6} \tag{9.89}$$

$$D = E_f btd^2/12 \tag{9.90}$$

Under these conditions, the stresses in the faces may be determined by the use of ordinary bending theory. As the sections remain plane and perpendicular to the longitudinal axis, the strains at any point distance z below the centroidal axis are Mz/D. The sign convention for bending of beams is given in the caption to Figure 9.26.

The stress on the faces $[Mz/D]E_f$ is

$$(\sigma_f)_{max} = \pm MhE_f/2D \tag{9.91}$$

Similarly, the stress in the core = $[Mz/D]E_c$

$$(\sigma_c)_{max} = \pm McE_c/2D \tag{9.92}$$

Sandwich beams can be categorised broadly into three groups, related to the bending stiffness of the beams, in such a way that:

(i) for a very thin-face beam, in which the stiffness in bending about its own axis is taken as zero and is sufficiently thin to assume d is equal to c, the ratio $d/t > 100$
(ii) for a thin-face beam, in which the stiffness in bending about its own axis is assumed to be zero, but the thickness of the face is assumed to have a finite value so that d is not equal to c, and the relationship $100 > d/t > 5.77$ is satisfied
(iii) for a thick-face beam, in which the stiffness in bending about its own axis is significant and the thickness of the faces is assumed to have a finite value so that d is not equal to c, and the relationship $d/t > 5.77$ is satisfied.

It will be seen, therefore, that the bending stiffness for a thin-face sandwich beam, as given by Equation 9.90, conforms to the condition (ii) above and that from condition (iii) the stiffness of a thick-face sandwich beam is

$$D = E_f btd^2/2 + E_f bt^3/6 \tag{9.93}$$

2 Shear stress in core

The assumptions in the ordinary bending theory lead to the shear stress distribution in a homogeneous beam at a depth z below the neutral axis as

$$\tau = Qaz/b \cdot I \tag{9.94}$$

where Q = shear force at section
a = area of the section of the beam above (or below) position under consideration
z = distance of centroid of area a to centroidal axis of beam
b = width of beam at section under consideration
I = second moment of area of whole section of beam about neutral axis.

For a composite sandwich beam the shear stress in the core is

$$\tau = Q \sum (a_i z_i E_i)/Db \tag{9.95}$$

where D = the flexural rigidity of entire section
E_i = the modulus of elasticity of the various component sections above (or below) section under consideration.

Considering Equation 9.95 and referring to Figure 9.26, the product within the summation bracket becomes:

$$\sum a_i z_i E_i = [E_f btd/2] + [c/2 - z][c/2 + z](b/2)E_c$$
$$= E_f btd/2 + [c^2/4 - z^2]bE_c/2$$

Therefore

$$\tau = Q/D[E_f td/2 + E_c/2\{c^2/4 - z^2\} \tag{9.96}$$

The maximum core shear stress will occur at $z = 0$ and the minimum core shear stress will occur at $z = \pm c/2$.

If the ratio of the maximum core shear stress to the minimum core shear stress is within 1% of unity,

ie

$$\left[\frac{E_c}{E_f}\frac{c}{t}\frac{c}{d}\frac{1}{4}\right] \leqslant \frac{1}{100}$$

where

$$4\frac{E_f}{E_c}\frac{t}{c}\frac{d}{c} > 100$$

it may be assumed that the core is too weak to contribute to the shear stress which may be considered constant across the core; the last term in the right-hand side of Equation 9.96 is ignored and the constant shear stress in the core is

$$\tau = (Q/D)(E_f td/2) \tag{9.97}$$

Figure 9.27 *Shear stress distribution for sandwich beam*

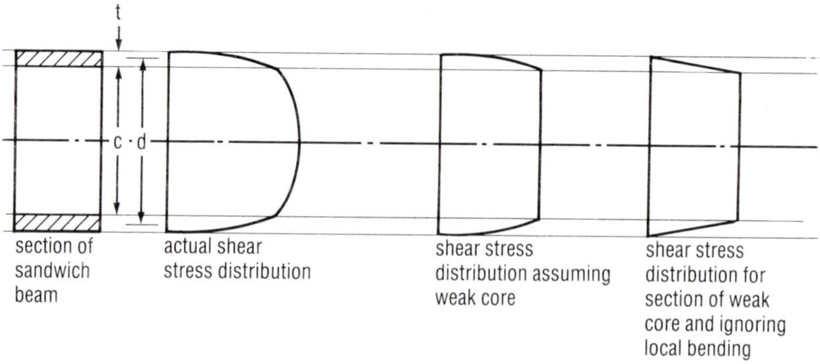

t

c · d

section of sandwich beam

actual shear stress distribution

shear stress distribution assuming weak core

shear stress distribution for section of weak core and ignoring local bending

If, in addition to the above, the condition associated with Equation 9.90 is also satisfied, then the shear stress in the core is

$$\tau = Q/bd \tag{9.98}$$

Figure 9.27 shows the shear stress distribution in the sandwich beam.

In using Equations 9.91, 9.92 and 9.98 for the stresses in bending and shear, the beam is defined as a thin-face one and the following assumptions have been made:

(i) A core with a low modulus resulting from a cell structure is considered to be a continuous elastic material.

(ii) In sandwich construction of composite faces and foam cores, the faces are much stronger, stiffer and denser than the core. Therefore, the contribution of the core to the overall stiffness of the sandwich construction is small and may be ignored. Consequently, the direct stresses which arise in the core under bending are small and may also be neglected; the shear stresses in planes perpendicular to the faces are uniform across the thickness of the core.

(iii) The core is assumed to be sufficiently stiff in the direction normal to the faces to maintain them at the correct distance apart; this assumes stiffening members are placed under concentrated loads.

The criterion for design of sandwich beams is taken as the limiting deflection, unlike that for ductile materials which would be limiting stress. The total deflection w of a beam can be regarded as the sum of the primary deflections (bending deflections) w_1 and the secondary deflections (shear deflections) w_2. The primary deflection may be calculated by the ordinary theory of bending and the secondary deflection also by this theory, but based on the combined flexural rigidity of the faces as they bend locally about their own separate centroidal axis.

The total deflection due to bending and shear for a sandwich beam carrying, for instance, a single point load at the centre of the beam under simply supported conditions is

total deflection = primary deflection + secondary deflection

$$w = w_1 + w_2$$

$$w = WL^3/48D + WL/4G_c A \qquad (9.99)$$

where D = the flexural stiffness
L = span of beam
W = point load on beam
G_c = modulus of rigidity of core material
$A = bd^2/c$

3 Buckling of sandwich struts with thin faces
The maximum axial compressive force which a pin-ended elastic strut can support before it becomes unstable is equal to the Euler load given by

$$P_E = \pi^2 D/L^2 \qquad (9.100)$$

In a sandwich strut where the core material is of low modulus, it is likely that shear deformations within the core will occur; these reduce the stiffness of the strut member.

For a thin face and an antiplane core, the flexural rigidity D_1 is given by Equation 9.90. As the strut deflects, the shear stress distribution is similar to that given in Figure 9.24 and the total displacement is given by Equation 9.99.

The buckled strut is shown in Figure 9.28 and at any cross-section the moment is equal to

$$P(w_1 + w_2) = -D_1 w_1''$$

Figure 9.28 *Deformations of buckled strut*

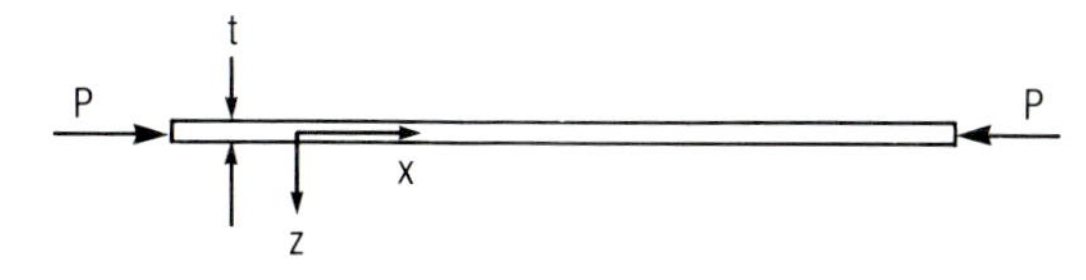

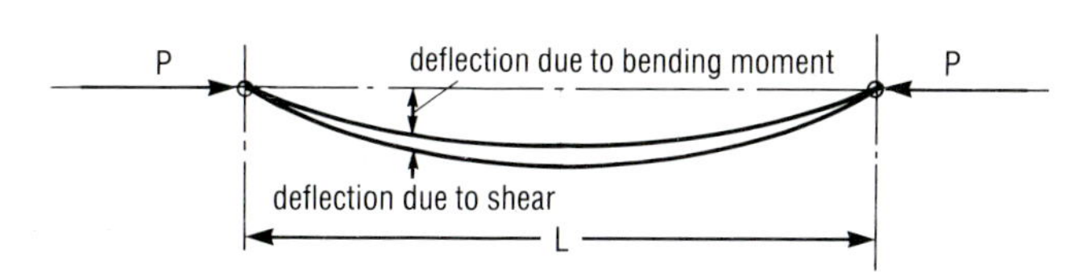

This equation yields a differential equation in w_1:

$$w_1''' + \alpha^2 w_1' = 0$$

The solution of this equation with the boundary conditions of

$$(w_1 + w_2 = 0 \text{ at } x = 0 \text{ and at } x = 1)$$

leads to the critical load of the sandwich strut as

$$P = P_E/(1 + P_E/AG_c) \tag{9.101}$$

where $\quad P_E = \pi^2 D_1/L^2$

4 Buckling of sandwich struts with thick faces

The critical buckling load of a thick-face sandwich beam has been given (Allen 1969) as

$$P_{cr} = P_E\left[\frac{1 + (P_{Ef}/P_c) - (P_{Ef}/P_c)(P_{Ef}/P_E)}{1 + (P_E/P_c) - (P_{Ef}/P_c)}\right] \tag{9.102}$$

where $P_E = \pi^2 D/L^2$ = Euler load of strut when there are no shear strains in the core

$P_{Ef} = \pi^2 D_f/L^2$ = sum of the Euler loads of the faces, considered as two independent struts

P_c = critical load for shearing instability.

Figure 9.29 *Sandwich panel with equal faces*

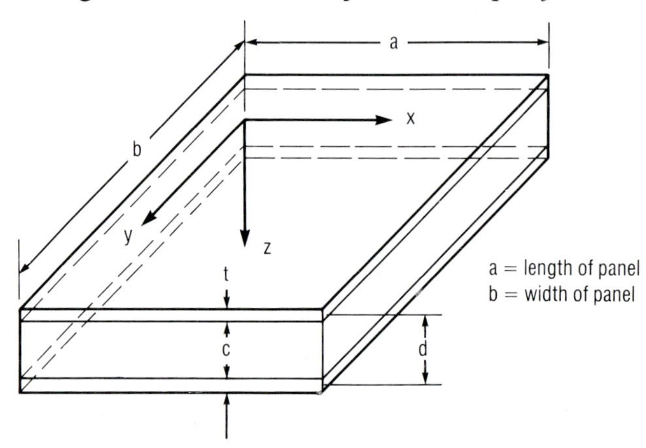

5 Buckling of sandwich panels

The essential equations for determining the critical edge force per unit length of sandwich panels, as illustrated in Figure 9.29 are:

$$P_{cr} = (\pi^2 D_2/b^2)K \qquad \text{(this load causes buckling in the } m, n \text{ mode)} \tag{9.103}$$

where $D_2 = (E_f td^2)/(2(1 - \nu_f^2))$ = flexural rigidity of the panel per unit length, assuming cylindrical bending and neglecting the local bending stiffness of the faces

ν_f = Poisson's ratio of face material

b = width of panel

K = dimensionless coefficient

$= \{mb/a + n^2a/mb)\}^2/\{1 + \rho[(m^2b^2/a^2) + n^2]\}$

a = length of panel

$\rho = (\pi^2/b^2)[E_f d^2 t/2(1 - \nu_f^2)][c/G_c d^2]$

$= [\pi^2/2(1 - \nu_f^2)][E_f/G_c][tc/b^2]$

$G_c d^2/c$ = shear rigidity of the panel per unit length

or $\rho = [\pi^2/b^2]$ times ratio of flexural rigidity to shear rigidity.

6 Buckling of thick-face sandwich panels

The critical buckling edge load per unit length in the x-direction for a thick-face sandwich is

$$P_{cr} = [\pi^2 D_2/b]K_t \tag{9.104}$$

where P, a, b and D_2 are defined in Equation 9.103, and

$$K_t = K + [mb/a + n^2a/mb]^2[t^2/3d^2]$$

To facilitate the use of these equations, the families of curves for ρ as K varies with a/b and for ρ as K_t varies with a/b have been given (Timoshenko and Woinowsky-Krieger 1959, and Allen 1969). It is recommended that, for a greater understanding of this topic, reference to these publications should be made.

7 The relatively high shear modulus sandwich beam and panel – axial and flexural theory

This section will discuss the analysis and design methods for sandwich beams and panels in which the core material is stiff enough for the shear effects to be small.

A symmetric material made from three composites fabricated into a composite structural unit is shown in Figure 9.30. It is assumed that the transverse strains arising from the differences in Poisson's ratio are negligible. The tensile modulus of the sandwich material E_s is given by the rule of mixtures (cf Section 9.5.1).

The force in the sandwich composite is equal to the summation of the forces in the core and face materials,

$$F_s = F_f + F_c \tag{9.105}$$

Figure 9.30 *Sandwich beam with high shear modulus core*

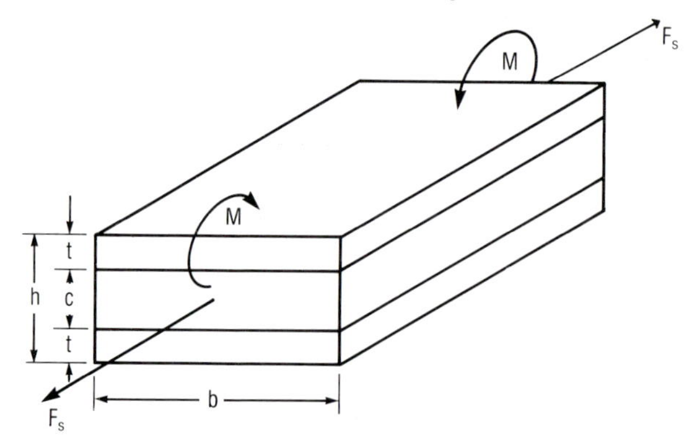

where the suffixes, s, f and c refer to the sandwich beam, face and core materials, respectively;

$$bh\sigma_s = 2t_f b\sigma_f + cb\sigma_c$$

$$h\varepsilon_s E_s = 2t_f \varepsilon_f E_f + c\varepsilon_c E_c$$

now the strains at any section in the face and core materials are equal; therefore $\varepsilon_s = \varepsilon_f = \varepsilon_c$

and $$E_s = 2t_f E_f/h + cE_c/h$$

or $$E_s = E_f(1 - c/h) + E_c(c/h) \tag{9.106}$$

It is clear, therefore, that the face and core materials contribute to the axial stiffness in proportion to their thickness.

The flexural rigidity D of the sandwich beam is derived as follows (cf Equation 9.93):

$$D = E_s I = E_s \frac{bh^3}{12} = E_f\{bh^3/12 - bc^3/12\} + [bc^3/12]E_c$$

Therefore $$E_s = E_f\{1 - c^3/h^3\} + (c^3/h^3)E_c \tag{9.107}$$

hence $$D_s = \{E_f[1 - c^3/h^3] + (c^3/h^3)E_c\}bh^3/12 \tag{9.108}$$

The flexural rigidity of a panel is

$$D_p = Eh^3/12(1 - \nu^2) \text{ per unit width} \tag{9.109}$$

Therefore the flexural rigidity D_{sp} of a sandwich panel is

$$D_{sp} = \{E_f[1 - (c/h)^3] + (c/h)^3 E_c\}h^3/12(1 - \nu^2) \text{ per unit width} \tag{9.110}$$

It can be seen from Equations 9.106 and 9.107 that the sandwich tensile and flexural modulus values are different and that the latter is larger than the former, provided that the modulus of elasticity of the face material is greater than that of the core.

The bending failure stress σ_f^* in the face materials of a beam is given by

$$\sigma_f^* = M_f h E_f / 2D_s \tag{9.111}$$

where D_s = stiffness of sandwich given by Equation 9.108.

It is also necessary to check the failure stress σ_c^* in the core material, and this is given by

$$\sigma_c^* = M_f c E_c / 2D_s \tag{9.112}$$

For the maximum bending failure stress of a panel, the Equation 9.111 would be used, where the value D_{sp} from Equation 9.110 would be substituted for the value D_s in Equation 9.111.

The maximum deflection w of relatively high shear modulus sandwich panel may be determined from the equation

$$w = \alpha P a^2 / 12 D_{sp} \tag{9.113}$$

where D_{sp} = value given by equation 9.110
P = total transverse load on plate
a = a typical dimension
α = a plate stiffness value which may be obtained from Timoshenko and Woinowsky-Krieger (1959).

The main advantage of a sandwich construction is that the core material enables the panel thickness h to be increased in value for the same overall weight. Although the modulus of elasticity of the material decreases with increase in core thickness, the flexural rigidity is increased due to the increased depth of the panel.

In the stiffness-limited case, Johnson and Sims (1987) have given a formula for the choice of sandwich geometry for a least weight panel as

$$\frac{c}{h} = \left[\frac{1 - \rho_c/\rho_f}{1 + E_c/E_f}\right]^{1/2} \tag{9.114}$$

where ρ_c and ρ_f are the densities of the core and face materials, respectively.

Using this ratio of c/h the modulus of elasticity of the beam or panel may be determined from Equation 9.107, and the least weight panel, for a given flexural rigidity, has a value of 'h' given by equation 9.110.

9.9 COMPUTER PROGRAMS

It is not easy to describe mathematically the complex nature of composites, and designers are using computer software systems to perform structural analysis of systems manufactured from anisotropic (laminated) composite materials. There are more than 600 finite-element programs available to the structural engineer, but when the features are specifically addressed to the applications on structural analysis of composites the number of suitable software systems decreases to about ten.

It is not the intention here to discuss the various packages which are available, because it is a matter of personal choice whether the designer wishes to use one system rather than another, and indeed only the designer knows what he requires of the system and whether it is capable of this requirement. An article has been written on selected software[1] which discusses some of the packages available, and in this section their likely specific requirement is discussed.

One of the main problems with commercial packages is that various programs are designed to run on specific computers. In addition, software programs vary in the quality of their computing capability. It is possible to have a mesh generation option as well as a range of coloured, shaped and illuminated options to display the model and applied loads. It is also possible to obtain software which has static, dynamic and thermal capabilities and to be able to analyse hundreds of laminae with many different ply materials in the composite. Some software can evaluate the laminate stiffness and compliance matrices and can calculate the laminate equivalent elastic and physical engineering constants.

9.10 THE FACTOR OF SAFETY

The factor of safety which should be applied to composite materials is dependent upon its component materials, upon the fabrication technique and upon the service life of the composite. The concept of a finite permissible working stress as applied to conventional materials; steel for instance, has no parallel in composite design. Some guidance, however, may be obtained from BS 4994 'Vessels and tanks in reinforced plastics'. In this Standard the allowable tensile stress, for instance, is derived from the ultimate tensile strength by the introduction of a number of factors which depend upon the manufacturing techniques, long-term behaviour of the material, design temperature, cyclic loading etc, and the product of these factors is then divided into the ultimate strength to obtain the allowable tensile strength. A factor of safety of 6 for glass-reinforced polyester used in the construction industry would be appropriate for a correctly manufactured and fully cured component in normal environmental conditions.

Head and Templeman (1986) have developed a limit state design principle for fibre-reinforced polymers. They have suggested that all or part of the composite structure or structural component can be described with reference to a limited set of limit states beyond which the design requirements are no longer satisfied. They have divided the limit states into three categories:

(i) The ultimate limit states corresponding to the maximum load-carrying capacity
(ii) The serviceability limit states relating to the criteria governing normal use
(iii) The conditional limit states, corresponding to an infrequent major random event, eg fire. Conditional limit states are frequently placed in one of the other categories.

In the design of any structure or component it would be necessary to consider all the relevant limit states before making a final selection.

NOTE

1 Selecting the software. *Advanced Composites Engineering*. September 1987

10 Damage and Fracture of Fibre Composites

Peter W R Beaumont DPhil

Lecturer, Department of Engineering
University of Cambridge

10.1 INTRODUCTION

There are more than 50,000 materials from which the engineer can select. For simplicity they can be divided into six broad classes: metals, polymers, elastomers, ceramics, glasses and composites. Within a class, there is some commonality in properties and behaviour. Ceramics, for instance, have high moduli, and polymers low moduli; metals have high toughness whereas glasses are brittle; elastomers are weak and composites strong.

To the physicist, these classes are distinguished by the nature of the bonds which hold their atoms together and the regular or non-regular pattern of the atomic structure. Ultimately, the strength and directionality of the bonding determine the properties of strength and stiffness which are so important to the engineer, and which set a fundamental upper limit to the value of any material. The property of toughness is more subtle, and does not correlate in any simple way to the bulk properties of strength and stiffness; it is determined by localised microscopic events of deformation and fracture that take place at the tip of an inherent flaw or crack in the material.

10.1.1 Toughness enhancement of brittle materials

There are two general categories of toughening mechanism. The first includes processes (typical of composites) that occur along the crack plane in the crack wake, such as crack bridging. Good examples are fibre pull-out in a cracked matrix and rubber particle stretching in a rubber-toughened epoxy. These mechanisms exert a direct influence on the crack tip stress intensity K_I and on the local crack propagation resistance of the material.

The second category influences toughness by means of events occurring in a process zone, such as phase transformation in, for example, ZrO_2/Al_2O_3 composites and crazing in rubber-toughened polystyrene. Mechanisms of this type result in a toughness that typically scales with the width of the process zone. Process zone effects are multiplicative with crack wake mechanisms.

In composite design, it is generally felt that toughening is achieved by increasing the inherent microstructural resistance, for example, by increasing fibre content, changing bond strength etc – a process referred to as intrinsic toughening. However, in any composite system, particularly one with a brittle matrix, the actual source of toughness is distinctly different, arising from mechanisms of crack-tip shielding (category one mechanisms), where toughness is achieved (and crack extension is impeded) by mechanical, microstructural and environmental factors which locally reduce the crack-driving force. This process is known as extrinsic toughening.

An understanding of the micromechanisms of fracture and their dominance under service conditions of the composite is important in engineering design because, for a particular application, they determine the ductility of the component and its design allowables, the toughness of the composite and its resistance to impact damage, and the crack advance (or damage) accumulation rate, and they effect residual strength and stiffness in cyclic loading (fatigue) situations. These factors are of particular significance in adverse environments and at extremes of temperature.

This chapter considers damage and fracture at two levels: at the microscopic level, where the origins of toughness are found and explained in terms of the microstructural features of the composite; and at the macroscopic level, where the maximum design (working) stress of a component can be determined from a knowledge of the crack (notch) tip damage zone of the composite and the size and geometry of any cut-out or hole in the part. But first, a generalised approach to fracture.

10.2 A GENERALISED APPROACH TO FRACTURE

Roughly speaking, the stress to pull atoms apart in a perfect crystal, causing it to separate on a plane normal to the stress axis, is

$$\sigma_{\text{ideal}} = \left(\frac{2E\gamma_s}{\pi b}\right)^{1/2} \approx \frac{E}{10}$$

γ_s is the surface free energy, E is Young's modulus and b is the atomic size.

Nearly all crystalline solids fail in this way if the temperature is sufficiently low, but they do so at a much lower stress. This is because any microcrack or growth defect in the brittle material may propagate at a nominal applied stress given by

$$\sigma_f \approx \left(\frac{EG_c}{\pi a}\right)^{1/2} \tag{10.1}$$

where $2a$ is the size of the pre-existing crack and G_c is the 'toughness'. Put another way, this sort of cleavage with no plasticity will occur when the stress at the crack tip is sufficient to pull atoms apart; that is, when

$$\sigma(r) = \frac{K_I}{\sqrt{2\pi r}} > \frac{E}{10} \tag{10.2}$$

at a distance $r = b$ from the tip. We find, when $\sigma = \sigma_f$,

$$K_c \approx \frac{E}{10}\sqrt{2\pi r} \tag{10.3}$$

where K_c is the 'fracture toughness', which is related to G_c in the following way:

$$K_c^2 = EG_c \tag{10.4}$$

This is the basis of linear elastic fracture mechanics (LEFM).

Fracture mechanics has been used successfully in the development of structural materials in two complementary ways: first, by relating K_c to microstructure; second, by identifying the origins of cracking. The strength of a component can then be maximised by optimising the microstructure through careful control of manufacturing conditions to reduce flaw size and increase K_c. This requires an understanding of the dependence of K_c (or G_c) on microstructure.

In general, materials which show some ductility can fail in one of two competing ways, depending on the size of any crack or notch they may contain. A material containing a short notch fails by general damage, in which ductile fracture occurs at a net section stress which is independent of notch length. Alternatively, the localised stresses at the tip of a long notch or crack may be large enough to cause crack propagation while deformation in the bulk of the material remains elastic. In this case, fracture occurs at a stress which is proportional to $1/\sqrt{a}$ (Equation 10.1). We call this the regime of single crack propagation.

The overall behaviour is summarised in Figure 10.1; roughly speaking, the material fails at the lower of the two stresses.

Figure 10.1 *Schematic diagram showing the variation of fracture strength with crack length for a ductile material*

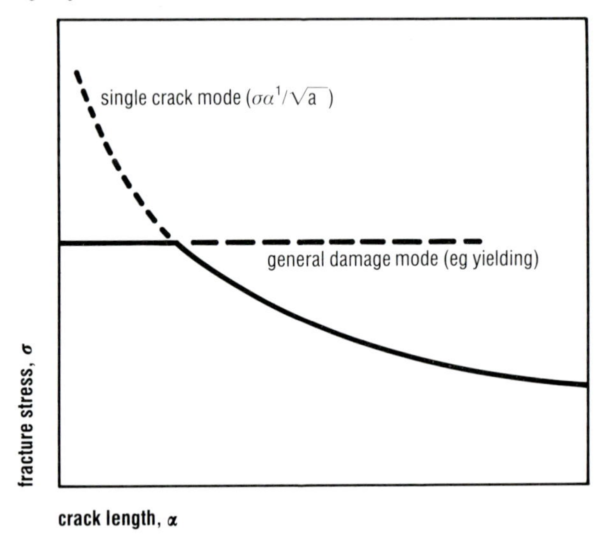

Source Wells and Beaumont 1981

Similar regimes may be identified in composite materials. When notches are absent or short, general damage can occur as dispersed, small cracks in the matrix, as fibre breakage or as decohesion at the fibre-matrix interface, which leads to degradation and failure of the entire section. However, when a long notch is present, it may propagate as a single crack, preceded by splitting and delamination processes at the notch tip, followed by fibre fracture and pull-out, with little or no deformation in the material remote from the notch tip damage zone.

10.2.1 Order of magnitude of K_c

A notched laminate under monotonic loading will exhibit localised damage close to the notch tip. Such damage will take the form of delamination, matrix cracking, splitting, fibre fracture etc (Figure 10.2). It is reasonable to suspect that fast fracture of the laminate will occur when the size of the delaminations, or the length of the crack tip split, or size of the damaged zone, exceeds some critical value.

For an elastic isotropic plate containing a centre crack of length $2a$ (Figure 10.3), the local crack tip tensile stress $\sigma^0_{yy}(x, 0)$ close to the tip is given by

$$\sigma^0_{yy}/K_I = [2\pi(x - a)]^{-1/2} \tag{10.5}$$

Figure 10.2 *Composites characterised by three material failure modes*

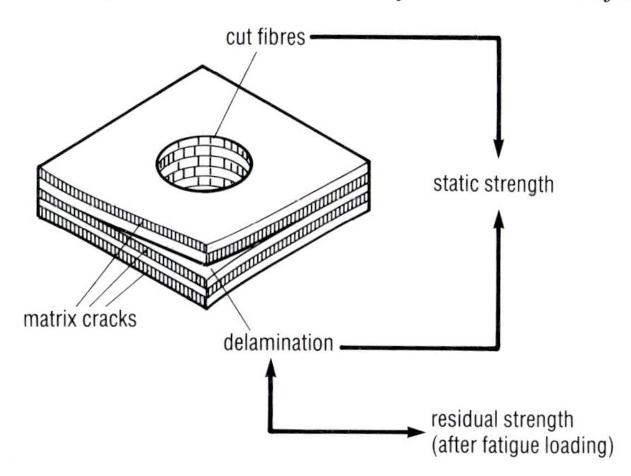

Crack propagation can occur when the local stress σ_{yy}^{0} exceeds the inherent strength σ_0 at some distance $x - a \geq r_c$ in front of the crack tip (Figure 10.3). In fact, Figure 10.3 shows that the region of material represented by the cross-hatch has already failed locally, and the stress should therefore be σ_0 in that damaged zone. Including the locally failed

Figure 10.3 (*a*) *Conditions for fraction of a material volume ra^3 in the neighbourhood of a crack tip under uniform tensile stress in the y-direction, and* (*b*) *Dependence of strength on flaw size*

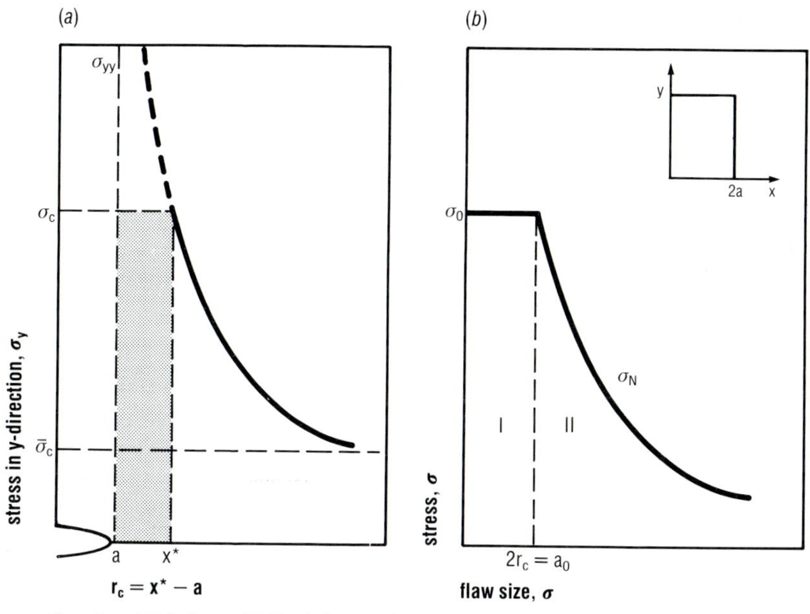

Source Caprino, Halpin and Nicolais 1979

damage zone, we can consider the crack to have an equivalent half-length of $a + a_0$. Thus we can write:

$$\frac{\sigma_{yy}^0}{K_I} \to \frac{\sigma_0}{K_c} \text{ at } \frac{1}{\sqrt{\pi a_0}} = \frac{1}{\sqrt{\pi r_c}} \tag{10.6}$$

where
$$K_c \to \sigma_N \sqrt{\pi(a + r_c)} = \sigma_N \sqrt{\pi(a + a_0)} \tag{10.7}$$

Here $a_0 = r_c$ and σ_N is the strength of the notched material. As the crack half-length approaches zero, then σ_N approaches σ_0, and the unnotched strength of the material is given by

$$\sigma_0 = \frac{K_c}{\sqrt{\pi a_0}} \tag{10.8}$$

The ratio of the unnotched strength σ_0 to the notched strength σ_N is equal to the stress concentration factor

$$\frac{\sigma_0}{\sigma_N} = K_t = [(a_0 + a)/a_0]^{1/2} \tag{10.9}$$

and the critical damage zone a_0 can be calculated using

$$a_0 = r_c = \frac{a}{(\sigma_0/\sigma_N)^2 - 1} \tag{10.10}$$

We can now evaluate the damage size a_0, using a circular hole of radius r as the discontinuity in the solid. A symmetrical damage zone of size a_0 emanates from the edge of the hole. Here

$$K_I = \sigma_c^\infty \sqrt{\pi a_0} f(a_0/r) \tag{10.11}$$

where σ_c^∞ is the applied composite stress far from the crack plane. At fracture,

$$\sigma_c^\infty = \frac{K_c}{\sqrt{\pi a_0} f(a_0/r)} \tag{10.12}$$

where $f(a_0/r)$ is a function which has been tabulated by Bowie (1956). If there were no hole, then the fracture stress σ_0 would be

$$\sigma_0 \equiv \sigma_c^\infty |_{c/r \to \infty} = \frac{K_c}{\sqrt{\pi a_0}(1.00)} \tag{10.13}$$

Combining Equations 10.11 and 10.12, we have

$$\frac{\sigma_0}{\sigma_c^\infty} = f(a_0/r) \tag{10.14}$$

Table 10.1 Small hole data summary (Waddoups *et al.* 1971)

Specimen*	Static strength (MPa)	(lbf/in^2)
control	463	67,135
1.6 mm (0.06 inch)-dia. hole	310	44,950
0.8 mm (0.03 inch)-dia. hole	355	51,475
0.4 mm (0.02 inch)-dia. hole	420	60,900

* Coupon specimens were 25 mm (0.98 inch) in width and 255 mm (10.04 inch) in length

We can, therefore, take a specimen and measure its flawed and unflawed strength and evaluate $f(a_0/r)$. From a table of $f(a_0/r)$ we can obtain a value of a_0/r and then determine a_0, accordingly.

10.2.2 Numerical examples

For example, let us consider the strength data of a $[0/\pm 45]_{2s}$ carbon-epoxy laminate (Table 10.1) containing a small circular hole. Inserting values of $\sigma_0 = 463$ MPa (67,135 lbf/in^2) and $\sigma_c^\infty = 310$ MPa (44,950 lbf/in^2) into Equation 10.14 gives

$$\frac{\sigma_0}{\sigma_c^\infty} = f(a_0/r) = 1.495$$

From tables of $f(a_0/r)$, we find $a_0/r = 0.89$ and therefore $a_0 = 0.7$ mm (0.28 inch). We now calculate K_c using the expression (from Equation 10.11):

$$\begin{aligned} K_c &= \sigma_c^\infty (\pi a_0)^{1/2} f(a_0/r) \\ &\approx 20 \text{ MN/m}^{3/2} \text{ (18,200 lbf/in}^{3/2}\text{)} \end{aligned}$$

Equation 10.11, together with a value of $K_c = 20$ MN/m$^{3/2}$ can be used to determine σ_c^∞ for specimens containing a hole of diameter 0.8 mm (0.03 inch) and 0.38 mm (0.01 inch). Values of σ_c^∞ calculated in this way are 360 MPa (52,200 lbf/in^2) and 418 MPa (60,610 lbf/in^2), respectively, which compare favourably with experimental measurements of 355 MPa (51,475 lbf/in^2) and 420 MPa (60,900 lbf/in^2) (Table 10.1).

Similarly, we can use the data presented in Table 10.2 to estimate the notched strength:

$$f(a_0/r) = \frac{\sigma_0}{\sigma_c^\infty} = \frac{524}{192} = 2.72$$

This corresponds to an a_0/r value of 0.100, and therefore $a_0 = 1.27$ mm (0.05 inch). Using Equation 10.11, then $K_c = 31$ MN/m$^{3/2}$ (28,210 lbf/in$^{3/2}$).

Table 10.2 Large hole data summary (Waddoups *et al.* 1971)

Specimen†	Static strength			
	Actual		Corrected for finite width	
	(MPa)	(lbf/in^2)	(MPa)	(lbf/in^2)
control	524	75,980	524	75,980
25 mm (0.98 inch)-dia. hole	183	26,535	192	27,840
68 mm (2.68 inch)-dia. hole	110	15,950	157	22,765
75 mm (2.95 inch)-dia. hole	91	13,195	158	22,910

† Coupon specimens were 125 mm (4.92 inch) in width and 965 mm (37.99 inch) in length. An isotropic width correction was used

We use this value, together with Equation 10.12 to obtain a value of σ_c^∞ for a 63 mm (2.48 inch)-diameter hole:

$$a_0/r = 0.04$$

$$f(a_0/r) = 3.13$$

$$\therefore \quad \sigma_c^\infty = 166 \text{ MPa } (24{,}070 \text{ lbf/in}^2)$$

which is in good agreement with a measured value of 157 MPa (22,765 lbf/in^2). For a plate containing an internal slit of length $2l$, we can use Equation 10.9, together with the strength data in Table 10.3 for a $[0/90]_s$ carbon/epoxy laminate, to determine σ_c^∞:

$$a_0 = \frac{l}{(\sigma_0/\sigma_c^\infty)^2 - 1}$$

where l = (hole diameter + slit)/2 = 1.80 mm (0.07 inch), $\sigma_c^\infty = 401$ MPa (58,145 lbf/in^2), $\sigma_0 = 575$ MPa (83,375 lbf/in^2) and therefore $a_0 = 1.7$ mm

Table 10.3 $[0/90]_s$ Carbon fibre-epoxy in fatigue, room temperature, $R = 0.1$ (Waddoups *et al.* 1971)

Specimen*	Average static strength		Average residual strength after 5 × 10^6 cycles	
	(MPa)	(lbf/in^2)	(MPa)	(lbf/in^2)
no stress concentration	575	83,375	497	72,065
1.6 mm (0.06 inch)-dia. hole	475	68,875	535	77,575
1.6 mm-dia. hole + 2 mm (0.008 inch) notch	497	72,065	512	74,240
1.6 mm-dia. hole + 2 mm notch	402	58,290	547	79,315

* Coupon specimens were 25 mm (0.98 inches) in width and 225 mm (8.86 inches) in length

(0.07 inch). Using the expression

$$K_c = \sigma_0 \sqrt{\pi a_0} \tag{10.15}$$

we estimate $K_c = 39.2\ \mathrm{MN/m^{3/2}}$ ($35{,}673\ \mathrm{lbf/in^{3/2}}$)

$$\sigma_c^\infty = \frac{K_c}{\sqrt{\pi(a_0 + a)}} \tag{10.16}$$

together with a value of $(a + a_0) = 2.7$ mm, gives a value of $\sigma_c = 460$ MPa ($66{,}700\ \mathrm{lbf/in^2}$), which compares well with a measured value of 497 MPa ($72{,}065\ \mathrm{lbf/in^2}$) (Table 10.3).

10.3 POINT STRESS AND AVERAGE STRESS FAILURE CRITERIA

Whitney (Nuismer and Whitney 1975) has extended the Waddoups model (Waddoups, Eisenmann and Kaminski 1971) explicitly to include the anisotropy of the material. In Whitney's analysis, orthotropic symmetry is assumed. For a plate containing a circular hole of radius R, the ratio of notched strength to the unnotched strength is given by

$$\frac{\sigma_N^\infty}{\sigma_0} = 2/[2 + \xi_1^2 + 3\xi_1^4 - (K_T - 3)(5 - \xi_1^6 - 7\xi_1^8)] \tag{10.17}$$

where $$\xi_1 = R/(R + a_0)$$

and $$K_T = 1 + 2\{2[(\bar{E}_{11}/\bar{E}_{22})^{1/2} - \bar{\nu}_{12}] + (\bar{E}_{11}/\bar{E}_{22})\}^{1/2} \tag{10.18}$$

$\bar{E}_{11}$, $\bar{E}_{12}$ are the effective elastic moduli of the composite or laminate. For large holes, $\xi_1 \to 1$ and $\sigma_N/\sigma_0 = (K_T)^{-1}$; for small holes, $\xi_1 \to 0$ and $\sigma_N/\sigma_0 \to 1$.

As in the Waddoups analysis, the assumption is that failure occurs when the point stress at r_c reaches a critical value. We can, in an alternative model, assume failure to occur when the average value of σ_{yy} over a distance d_0 in front of the hole exceeds the unnotched tensile strength of the material.

In this case,

$$\frac{\sigma_N^\infty}{\sigma_0} = 2(1 - \xi_2)/[2 - \xi_2^2 - \xi_2^4 + (K_T^\infty - 3)(\xi_2^6 + \xi_2^8)] \tag{10.19}$$

where $$\xi_2 = R/(R + d_0)$$

For an infinite anisotropic plate containing a centre crack of length $2a$, subjected to a uniform uniaxial tensile load, the point stress failure criterion gives

$$\frac{\sigma_N^\infty}{\sigma_0} = (1 - \xi_3^2)^{1/2} \tag{10.20}$$

The average stress failure criterion gives

$$\frac{\sigma_N^\infty}{\sigma_0} = [(1 - \xi_4)/(1 + \xi_4)]^{1/2} \tag{10.21}$$

Here ξ_3 and ξ_4 are given by

$$\xi_3 = a/(a + a_0),\ \xi_4 = a/(a + d_0)$$

We can rewrite Equations 10.20 and 10.21 in terms of the fracture toughness parameter K_c:

$$K_c = \sigma_0[\pi a(1 - \xi_3^2]^{1/2} \tag{10.22}$$

and

$$K_c = \sigma_0[\pi a(1 - \xi_4^2)/(1 + \xi_4)]^{1/2} \tag{10.23}$$

The expected limit of $K_c = 0$ for vanishingly small cracks is reached, while for large cracks K_c asymptotically approaches a constant value. This asymptotic value is given by

$$K_c = \sigma_0(2\pi a_0)^{1/2} \quad \text{(point stress failure criterion)} \tag{10.24}$$

and

$$K_c = \sigma_0(\pi d_0/2)^{1/2} \quad \text{(average stress failure criterion)} \tag{10.25}$$

Inherent to these models is the idea that a_0 and d_0 are perhaps material constants. The Whitney models predict a decrease in strength of the laminate with increasing hole size and an apparent increase in fracture toughness with increasing crack size.

An example of a comparison between theory and experimental data is shown in Figure 10.4 for centre-notched $[0°/\pm45°/90°]_{2s}$ laminates made from high-tensile-strength (HTS) carbon/epoxy. Here σ_N/σ_0 is plotted against notch length a. The theoretical curves of σ_N/σ_0 against notch size are computed using Equations 10.17–10.21. Values of $d_0 = 1$ mm (0.04 inch) and $a_0 = 3.8$ (0.15 inch) provided the best fit to these and other data for laminates with both circular holes and centre notches. The uniformly good fit lends credence to the postulate that a_0 (or d_0) is a material constant. The finite plate stress concentration factor K_T was computed to be 5.11 for the $[0/90]_{4s}$ HTS carbon/epoxy plate.

For comparative purposes, Whitney's data of K_c and a_0 are listed in Table 10.4. Earlier data of Caprino, Halpin and Nicolais (1979) for carbon, boron and glass fibre laminates are listed in Table 10.5. These data were obtained using the fracture mechanics concepts outlined earlier

Figure 10.4 *Comparison of predicted and experimental failure stresses for centre cracks in* $[0/\pm 45/90]_{2s}$ *HTS carbon fibre/epoxy*

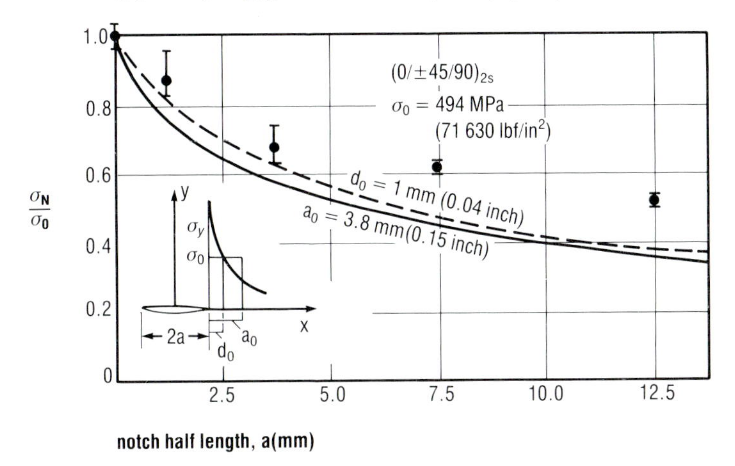

Source Nuismer and Whitney 1975

(Waddoups, Eisenmann and Kaminski 1971). The average characteristic dimensions are $a_0 = 2.1$ mm (0.08 inch) and $d_0 = 1.05$ mm (0.04 inch).

Table 10.6 lists the predicted strength (using laminated plate theory) and an estimate of fracture toughness, K_c, using the expression

$$\frac{K_c}{\sigma_0} = \sqrt{2\pi a_0} \tag{10.26}$$

for $a_0 = 2$ mm (0.08 inch), for each material system and laminate geometry presented. The theoretical projections correlate with the trends reported in Tables 10.3 and 10.5.

Table 10.4 Comparison of experimental estimate of the tensile strength and fracture toughness parameters (Caprino *et al.* 1979)

Laminate construction angle	Tensile strength (σ_0) (MPa)	Tensile strength (σ_0) (lbf/in²) [000]	Fracture toughness (K_c) (MPa $\sqrt{\text{m}}$)	Fracture toughness (K_c) (lbf/in$^{3/2}$) [000]	Damage dimension (a_0) (mm)	Damage dimension (a_0) (inches)
HTS carbon/epoxy:						
$[0/90]_{4s}$	635	92	42.1	38	2.03	0.08
$[0/\pm 45]_{4s}$	540	78	39.8	36	2.03	0.08
$[0/\pm 45/90]_{2s}$	465–495	67–72	33.2	30	2.03	0.08
E-glass/epoxy:						
$[0/90]_{4s}$	425	62	30.7	28	2.03	0.08
$[0/\pm 45/90]_{2s}$	320	46	24.3	22	2.03	0.08

Table 10.5 Comparison of experimental estimates of the fracture toughness parameters (Caprino *et al.* 1979)

Angular content θ	Tensile strength σ_0 (MPa)	(lbf/in^2) [000]	Fracture toughness K_c (MPa$\sqrt{m}$)	(lbf/in$^{3/2}$) [000]	Damage zone dimension a_0 (mm)	(inches)
HTS carbon/epoxy:						
0	1,045	151	36.5	33	—	
$[0/90]_{2s}$	530	77	42.1	38	1.78	0.07
$[0/\pm45]_{2s}$	460	67	33.2	30	1.02	0.04
$[0/\pm45/90]_{2s}$	340	49	24.3	22	—	
with round hole	270	39	30.1	27	2.79	0.11
with slit	275	40	28.4	26	2.29	0.09
90	35	5	1.7	1.5	—	
Boron epoxy						
0	1,325	192	—		—	
$[0_2/\pm45]_{2s}$	700	10	62.0–74.1	56—67	2.54–3.56	0.10–0.14
$[0_3/\pm45/90]_{2s}$	690	100	59.8	54	2.54	0.10
$[0/\pm45]_{2s}$	610	88	49.8	45	2.54	0.10
$[0/\pm45/90]_{2s}$	420	6	38.7	35	2.79	0.11
E-glass/epoxy:						
0	1,035	150	—		—	
$[0/90]_{2s}$	540	78	—		—	
$[0_2/\pm45]_{2s}$	600	87	—		—	
$[0/\pm45/90]_{2s}$	350	51	—		—	
with hole	195	28	37.7	34	2.54	0.01

The dimensions of a_0 and d_0 appear to be sensitive to variations in laminate construction and possibly for different fibrous reinforcements.

10.3.1 Point (average) stress criteria against LEFM

Over the past decade or so, a vast amount of experimental data on the notched tensile strength of a variety of laminates have been generated. In general, fracture from the root of the notch involves splitting and delamination, followed by fibre fracture and pull-out, and the final fracture surface shows a complex crack path which depends on the fibre type and the laminate configuration. Some of these data, shown in Figure 10.5, are presented in the form of the normalised apparent fracture toughness K_c/σ_0, given by

$$\frac{K_c}{\sigma_0} = \left(\frac{\sigma_N}{\sigma_0}\right)\sqrt{\pi a} \tag{10.27}$$

Table 10.6 Comparison of predicted strengths and fracture toughness for different materials and laminates (Caprino *et al.* 1979)

Laminate construction angle	π/n	Tensile strength (MPa)	Tensile strength (lbf/in²) [000]	Fracture toughness (MPa $\sqrt{m}$)	Fracture toughness (lbf/in$^{3/2}$) [000]
HTS carbon/epoxy (specific manufacturers):					
0	$\pi/1$	1,310	190	105.2	96
0/90	$\pi/2$	645	94	51.8	47
0/±45	—	520	75	42.0	38
0/±60	$\pi/3$	470	68	38.0	35
0/±45/90	$\pi/4$	470	68	38.0	35
0/±36/±72	$\pi/5$	470	68	38.0	35
0/±30/±60/90	$\pi/6$	470	68	38.0	35
HTS carbon/epoxy (generic average):					
0	$\pi/1$	1,045	152	83.7	76
0/90	$\pi/2$	530	77	42.4	39
0/±45	—	440	64	35.7	32
0/45/90	$\pi/4$	370	54	29.9	27
E-glass epoxy:					
0	$\pi/1$	1,035	150	83.0	76
0_2/±45	—	620	90	49.6	45
0/90	$\pi/2$	545	79	43.7	40
0/±45/90	$\pi/4$	370	54	29.9	27
Boron/epoxy:					
0	$\pi/1$	1,325	192	106.3	97
0_2/±45	—	725	105	58.1	53
0_2/±45/90	$\pi/4$	485	70	38.8	35

The ratio of $K_c/\sigma_0 = 0.045$ m$^{1/2}$ (0.082 ft$^{1/2}$) and $K_c/\sigma_0 = 0.09$ m$^{1/2}$ (0.163 ft$^{1/2}$) appear as two horizontal lines that encompass much of the data for long notches. For notches of length less than about 2 mm (0.08 inch), the data show a rising trend consistent with failure by general damage at a constant stress. These data lie between two curves representing the ratio $\sigma_N/\sigma_0 = 1.0$ and $\sigma_N/\sigma_0 = 0.65$. Figure 10.5 also shows the predictions of the 'point stress' and 'average stress' failure criteria of Whitney, for the values of the parameters a_0 and d_0 shown. The main feature of these criteria is that by choosing a fixed critical distance (a_0 or d_0) they lead to an apparent fracture toughness which falls when the notch length is shorter than some critical distance. An alternative interpretation is that the reduction is due to a change from a single-crack mode of failure to one associated with general damage (see Section 10.2).

Figure 10.5 *The variation of normalised fracture toughness, K_{IC}/σ_0 for carbon, glass and Kevlar 49 composites. The predictions of the point stress and average stress criteria are also shown*

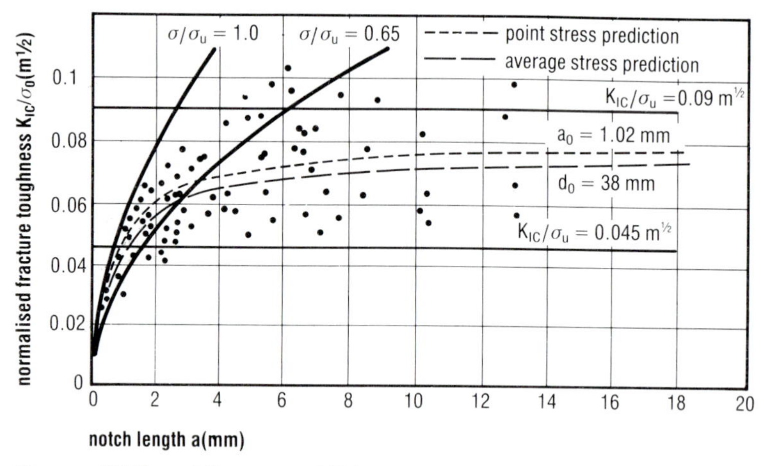

Source Wells and Beaumont 1981

In addition, the results displayed in Figure 10.5 for a variety of fibres and configurations show a remarkable consistency in notched strength properties. For the lay-ups and fibre types considered here, the normalised apparent fracture toughness lies in the range (Wells and Beaumont 1981):

$$\frac{K_c}{\sigma_0} = 0.07 \pm 0.02 \text{ m}^{1/2} \ (0.13 \pm 0.04 \text{ ft}^{1/2}) \tag{10.28}$$

This is the more remarkable when one considers the wide variation in fibre properties and lamina toughness (see Section 10.5.2). However, detailed differences between them are observed using the parameter K_c/σ_0 for each individual class of material as a measure of notch sensitivity, and their fracture behaviour may be summarised (Wells 1982) as follows:

(i) Where there is a high proportion of 0° fibres, eg $(0/90)_s$, $(0/\pm45)_s$ and $(0/\pm45/0)_s$, the laminate is notch-sensitive and LEFM seems to predict notched strength reasonably accurately.

(ii) Where there is a lower percentage of 0° fibres, eg $(0/\pm45/90)_s$, the laminate shows intermediate notch sensitivity, ie K_c/σ_0 is higher than in (1) and varies with notch length; LEFM should be treated with caution.

(iii) Pure angle-ply laminates, eg $(\pm45)_s$ are commonly notch-insensitive, and the methods of LEFM are invalid.

(iv) Notch sensitivity will be affected by factors that include interlaminar shear strength and lamina toughness, which will blur the distinctions between these classes of materials.

A word of caution: It will not have escaped the notice of even the most casual of readers that unlike many, perhaps all materials, Equation 10.28 indicates that the stronger the fibre composite the tougher it becomes. While this most certainly appears to be the case in practice, such a simple relationship begs the question whether or not methods of toughness measurement based on LEFM have any significance.

Furthermore, for most notched laminates, subcritical localised damage (where splitting and delamination dominates), is extensive enough to alter the original notch tip stress distribution substantially. Since most notched strength models are based on this original stress distribution in the notch tip region, the validity of these models, including LEFM, should be questioned. An alternative, new model based on the actual state of stress in the notch tip vicinity is described in Section 10.4.

10.3.2 Comparison of damage zone size and the point stress parameter a_0

The strength of laminate containing a central notch of length $2a$ can be predicted using the 'point stress failure criterion':

$$\sigma_N = \sigma_0 \sqrt{1 - \xi_3^2} \tag{10.29}$$

where

$$\xi_3 = \frac{a}{a + a_0}$$

a_0 is, of course, the critical distance ahead of the notch tip at which the localised tensile stress exceeds the un-notched strength of the material.

Alternatively, using LEFM

$$\sigma_N = \left(\frac{E^* G_c}{\pi(a + r_c)} \right)^{1/2} \tag{10.30}$$

where E^* is an apparent modulus, G_c is toughness and r_c is the crack advancement (damage zone size) at which fibre fracture occurs (Wells 1982):

$$r_c = \pi \left(\frac{E_c \bar{\sigma} l_d}{8K_c E_f} \right)^2 \tag{10.31}$$

$\bar{\sigma}$ is the average stress in the fibre over its debond length l_d when it snaps, E_c is longitudinal laminate modulus, and E_f is fibre modulus. Equating the squares of Equations 10.29 and 10.30,

$$\frac{E^* G_c}{\pi(a + r_c)} \approx \sigma_0^2(1 - \xi_3) \tag{10.32}$$

Rearranging,

$$a_0 \approx a\left[\left(1 - \frac{E^* G_c}{\pi\sigma_0^2(a + r_c)}\right)^{1/2} - 1\right] \tag{10.33}$$

This expression has been evaluated for a variety of CFRP (carbon fibre-reinforced plastics) laminates and compared with experimental data (Figure 10.6). The parameter a_0 which best fits the deduced values is then plotted against damage zone size r_c. While the agreement with the simple model is reasonable, it still does not provide a physical justification of the 'point stress' criterion (Wells, 1982).

Figure 10.6 *Comparison between the predicted and observed variation of a_0 with damage zone size and material type*

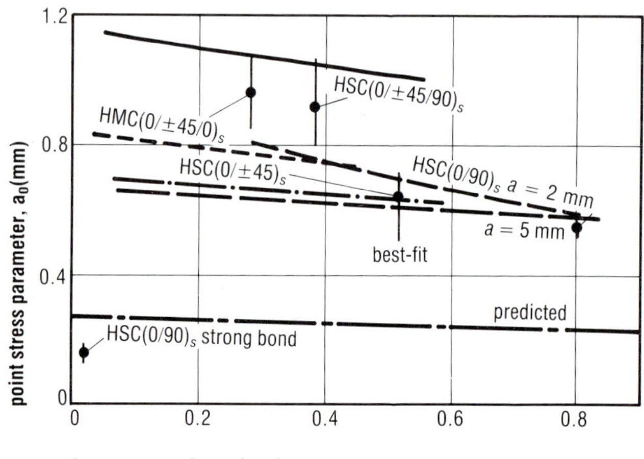

Source Wells 1982

10.4 DAMAGE MECHANICS OF STATIC FRACTURE

To understand what makes for a crack-resistant, damage-tolerant composite laminate we must first identify the underlying physical mechanisms that bring about its ultimate failure. This approach has led to a formulation of the damage mechanics of composite fracture which, although still incomplete, is now a foundation on which to build. To begin with, we will formulate a model for the mechanisms of subcritical damage growth at a notch tip to predict the fracture stress. This is based on the processes of splitting and delamination in front of the notch (see Figure 10.2). We will apply the model to see the effect of notch size on fracture stress. The basic premise of this conceptual approach to fracture is that there is some aspect of the terminal notch tip stress distribution which is identical to all specimens, regardless of lay-up, notch size, terminal damage state etc. The stress distribution clearly must depend on the shape and extent of the notch tip damage, and it should be possible to derive a unique relationship between damage accumulation and applied stress (Kortschot 1988).

10.4.1 A damage-based strength model

In a monotonic tensile test of a notched cross-ply laminate, early signs of damage include splitting of the 0° plies and the formation of a triangular-shaped delamination zone between the 0° and 90° plies. Subsequent damage includes transverse ply cracking of the 90° plies leading to fibre fracture within the 0° plies. The two CFRP specimens of $(90°/0°)_{2s}$ and $(90^\circ_2/0^\circ_2)_s$ configuration which display quite different amounts of notch tip damage also have completely different tensile strengths. This is despite the identical in-plane stress distributions for these two samples where the number of 0° and 90° plies they contain is the same. Most macro-models of fracture, eg Whitney's 'point stress' and 'average stress' criteria would require new empirical parameters to be evaluated for each lay-up. Our interpretation is that the extent of damage in the notch tip zone has affected the critical stress distribution ahead of the notch tip at the point of fast fracture. Assuming that the amount of energy dissipated by delamination is far greater than that of splitting ($G_d \sim 400$ J/m^2 (8.88 cal/ft^2) compared with $G_s \sim 160$ J/m^2 (3.55 cal/ft^2), and the split area is much smaller than the size of the delaminations), split growth-rate is determined by the delamination energy

$$\frac{l}{a} \propto \frac{\sigma_\infty^2}{G_d} \qquad (10.34)$$

This parabolic relation is a good one to describe the experimental data of split growth l/a shown in Figure 10.7. The theoretical solid curve is for a delamination energy G_d of 400 J/m^2 (8.88 cal/ft^2), which is reasonable for mixed modes I and II failure.

10.4.2 Notch size effect

Based on a finite-element analysis, we know that the notch stress concentration factor K_t is a function of l/a (for a given specimen geometry)

Figure 10.7 *Split growth as a function of applied stress*

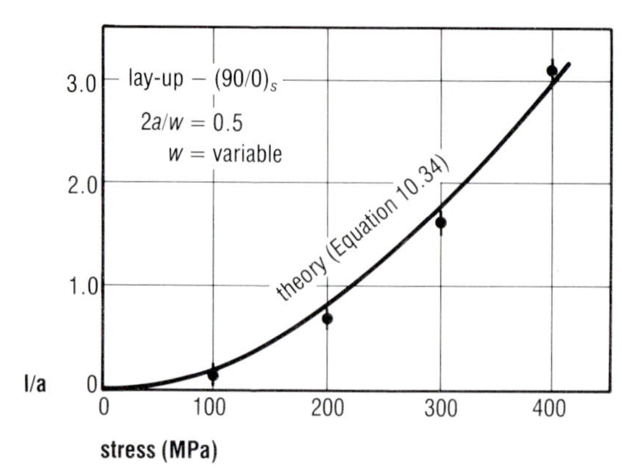

Source Kortschot 1988

(Kortschot 1988). Since l/a depends only on the applied stress (Equation 10.34), it follows that K_t is a function of the applied stress also. Furthermore, the peak stress in the 0° ply is a function of K_t and the applied stress σ_∞:

$$K_t \propto (l/a)^{-0.28} \tag{10.35}$$

$$\sigma_{(0°)p} = K_t \sigma_\infty \tag{10.36}$$

From a knowledge of K_t and σ_∞, combined with an X-ray measurement of split length, we can determine the peak stress in the 0° ply as a function of σ_∞ (Figure 10.8). The theoretical curve through the data is based on the theoretical split growth curve (Equation 10.34) together with the equations for K_t and $\sigma_{(0°)f}$, where

$$\sigma_{(0°)p} \propto \sigma^\infty \left[\frac{1}{\left(1 - \left(\frac{2a}{w}\right)^2\right)\left(\frac{l}{a}\right)^{0.28}} \right] \tag{10.37}$$

This equation can be used to estimate the strength of any $(90°/0°)_s$ specimen. Failure will occur when the peak stress in the 0° ply exceeds its strength. Provided the strength of the 0° ply is known, the failure stress of the laminate can be read directly from Figure 10.8.

An important observation is that small variations in the strength of the 0° ply can lead to large variations in laminate fracture stress. This explains the large scatter of results commonly found in a batch of tensile tests. Small variations in the strength of the 0° ply are magnified by the progressive notch-blunting mechanisms. The slope of the line in Figure 10.8 determines the magnification factor.

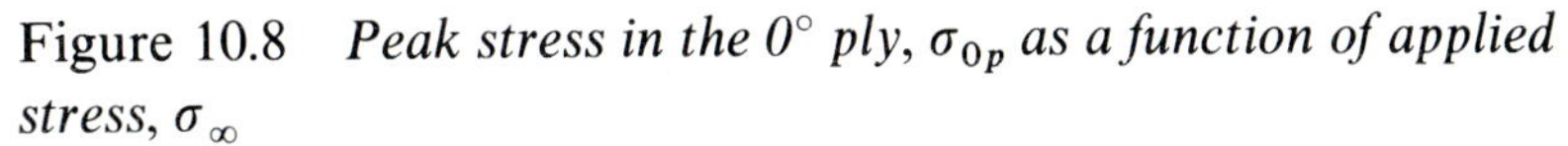

Figure 10.8 *Peak stress in the 0° ply, σ_{0p} as a function of applied stress, σ_∞*

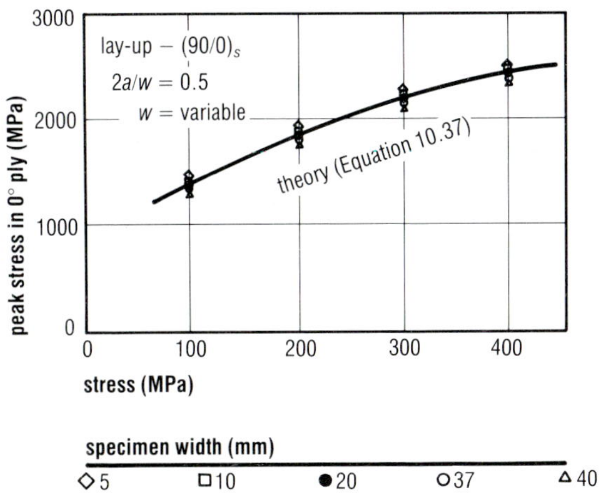

Source Kortschot 1988

To model the notch size effect requires a knowledge of the strength of the 0° ply for each specimen size, and the fracture stress of the laminate is determined using Equation 10.37. The strength of the 0° ply within a $(90°/0°)_s$ laminate (Kortschot 1988) is given by

$$\sigma_{(0°)f} = C_3[(l/a)(a^2)]^{-1/m} \tag{10.38}$$

where the Weibull modulus m is 20 for cross-ply CFRP laminates. This relationship is based on a Weibull model which suggests that the strength of the 0° ply is affected by the volume of the specimen which influences the terminal damage state at the notch front.

Since our failure criterion is

$$\sigma_f^\infty = \sigma_{(0°)f}/K_t \tag{10.39}$$

then combining the above equations

$$\sigma_f^\infty = C_3[(l/a)a^2]^{-1/20}\left[\frac{(l/a)^{0.28}}{C_2}\right] \tag{10.40}$$

$$\sigma_f^\infty = C_4[(\sigma_f^\infty)^2 a^2]^{-1/20}\,[(\sigma_f^\infty)^2]^{0.28} \tag{10.41}$$

$$(\sigma_f^\infty)^{(1-2(0.28)+0.1)} = C_5\,a^{-0.1} \tag{10.42}$$

$$\therefore \quad \sigma_f^\infty = C_5\,a^{-0.19} \tag{10.43}$$

where C_1, C_2, C_3, C_4, C_5 are constants. The results of this analysis, with the constant C_5 evaluated, is shown in Figure 10.9. The model slightly

Figure 10.9 *Strength against crack (notch) length for $(90/0)_s$ DEN specimens*

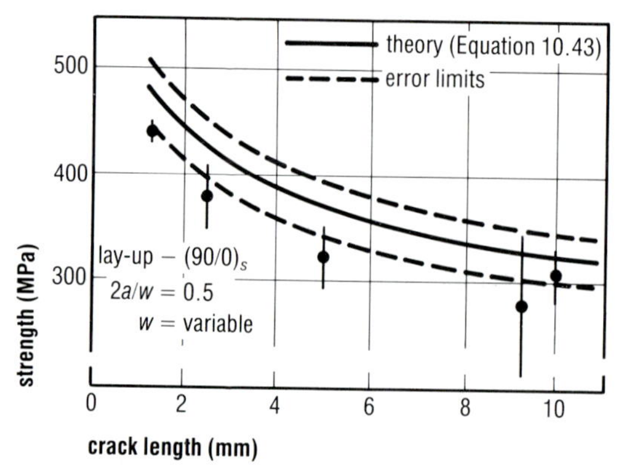

Source Kortschot 1988

overestimates the strength but the trend is accurately predicted without the need for a fitting parameter. This represents a significant departure from the previous strength models described in Section 10.3. One likely cause of the discrepancy is the effect of transverse-ply (matrix) cracking of the 90° ply, which leads to localised fibre breakage in the 0° ply.

10.5 FAILURE MECHANISMS AND TOUGHNESS MAPS

It is useful to have a way of summarising, for a given fibre composite, information about how microstructure affects each mechanism of cracking and fracture, and the resulting toughness. One way of doing this is shown in Figure 10.10. This is a diagram with axes of matrix (epoxy) toughness and interface (glass/epoxy) toughness. The diagram or map is divided into fields separated by a boundary which show the regions of matrix and interface toughness over which each of the three failure mechanisms labelled is dominant, ie fibre/matrix decohesion (debonding), fibre fracture and fibre pull-out. Superimposed into the map are contour lines of toughness where toughness can be equated to the maximum dissipation of energy by these three mechanisms during crack growth normal to aligned fibres, ie transverse crack growth of a single ply (Wells 1982).

A word of warning: one must be careful not to attribute too much precision to these maps; they are far from perfect or complete. As we shall see, they have been constructed from a simple model of a partially debonded fibre spanning a matrix crack which ultimately fractures within the resin and pulls out as the matrix crack surfaces open. The maps are

Figure 10.10 *A toughness map for E-glass fibre/epoxy ($V_f = 0.60$), where the axes are matrix toughness and interface toughness. The small circles denote the boundary between one dominant mechanism (field) and another. The mechanisms are:*
PO = pull-out
INT = interfacial debonding
EL = fibre fracture.
The triangle defines a point on the map for a typical E-glass composite

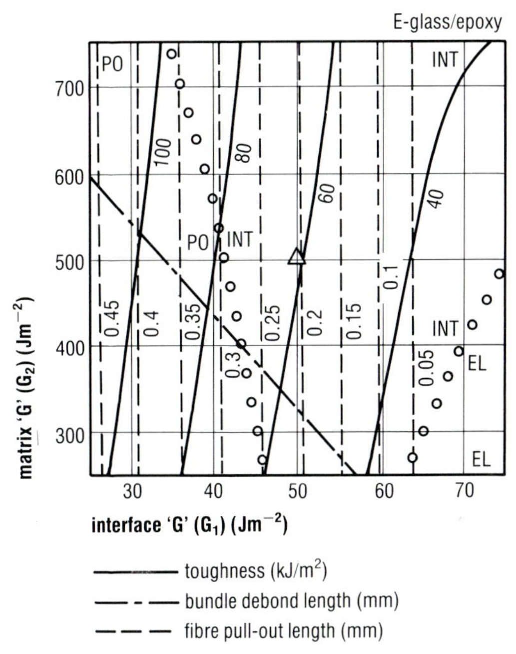

Source Wells and Beaumont 1985

no better (and no worse) than the model and equations, combined with the material data used to construct them.

10.5.1 Method of construction

The aim is to derive for each fibre-dominated mechanism an energy absorption equation to predict toughness, based on the physically sound microscopic process. Our starting point assumes a matrix crack spanned by an unbroken brittle fibre that has partially debonded (Figure 10.11). We select an equation which describes the build-up of a non-uniform stress along the length of that portion of debonded fibre (Figure 10.11). Based on this equation (Figure 10.11), we can derive an expression for each failure mechanism in terms of the energy dissipated during debond-

Figure 10.11 *A single, partially debonded, fibre spanning a matrix crack. Load builds up along debonded length of fibre by frictional (sliding) forces*

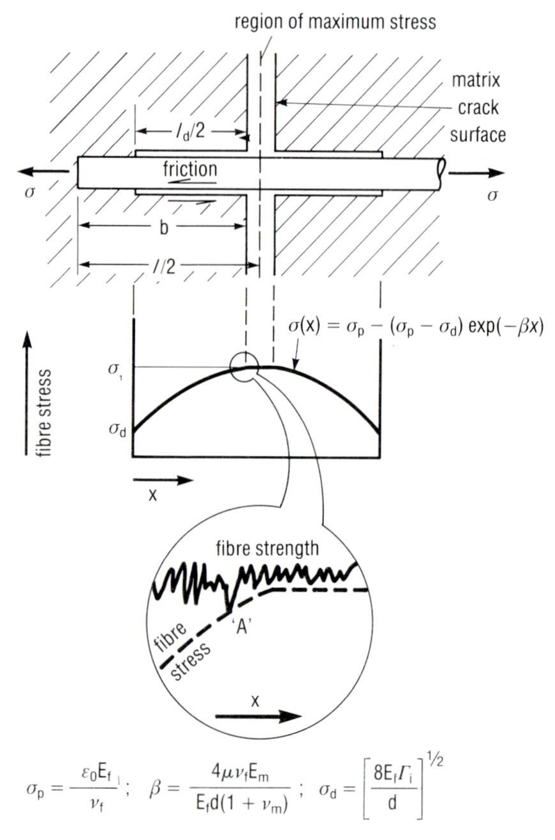

Source Wells 1982

ing, fibre fracture and pull-out. The failure mechanism which dominates (meaning it contributes most to the total work to fracture the composite) will depend on the property values which we insert into the equations.

1 Toughening by interfacial debonding

While the fracture energy of an interface is small, the total surface area of cylindrical cracks along fibre-matrix interfaces can be extremely high. It follows that the contribution this mechanism makes to the enhancement of toughness of the composite is in proportion to the total surface area:

$$\Delta G_i = 8 l_d \gamma_i V_f / d \qquad (b > l_d/2) \tag{10.44}$$

V_f is fibre volume fraction, d is fibre diameter and l_d can be determined. In a composite containing short fibres, where one end of a fibre is close to the matrix crack, the maximum length of debond crack, l_d, will be unattainable if the fibre pulls out (Figure 10.11). The model, therefore, has to be modified for the case where $b < l_d/2$, where b is defined in Figure 10.11:

$$\Delta G_i = 8b\gamma_i V_f/d \qquad (b < l_d/2) \tag{10.45}$$

2 Toughening by fibre fracture

A loaded debonded fibre dissipates energy when it snaps. An estimation of this energy can be made from a knowledge of the states of stress in the fibre and matrix immediately before and after fibre fracture.

The enhancement of composite toughness per unit fracture area due to fibre fracture is:

$$\Delta G_f = \frac{V_f}{E_f}\left[\frac{\sigma_p^2 l_d}{2} - \frac{(\sigma_p - \sigma_d)^2 (\exp(-\beta l_d) - 1)}{2\beta} + \frac{2(\sigma_p - \sigma_d)(\exp(-\beta l_d/2) - 1)}{\beta}\right] \qquad (b > l_d/2) \tag{10.46}$$

Likewise, the model has to be modified for short fibre composites when $b < l_d/2$, by replacing l_d with b.

3 Toughening by fibre pull-out

If the average pull-out length is $\langle l_p \rangle$, then, to a good approximation,

$$\Delta\langle G_p \rangle = V_f \sigma_p \left[\langle l_p \rangle + \frac{(\exp(-\beta\langle l_p \rangle) - 1)}{\beta}\right] \tag{10.47}$$

First, data for the properties are gathered: fibre strength and modulus, fibre diameter, matrix modulus and toughness, interfacial bond strength etc. Second, we predict (or measure) the fibre debond length and average fibre pull-out length; also, we estimate the frictional stress distribution parameters σ_p and β from these material properties (Figure 10.11). Values of these parameters, together with the appropriate equations, are then used to predict the maximum energy dissipated for each mechanism. Furthermore, a computer is used to construct a diagram or map by allowing any two of the many material properties to vary in turn (the remaining parameters are held constant), and for the two variables to be plotted against one another. These two properties form the two axes of the map. Contours of constant total toughness are then superimposed on the map. Finally, predictions of fibre debond length and fibre pull-out length can be displayed as contours on the map also. They may be useful in a failure analysis of a broken component.

Examples of other maps are shown in Figures 10.12 and 10.13. They all show toughness increasing rapidly with fibre strength, and decreasing slowly with increasing fibre stiffness. Likewise, toughness increases with

Figure 10.12 *Toughness map of fibre strength against fibre modulus for HMS carbon fibre/epoxy*

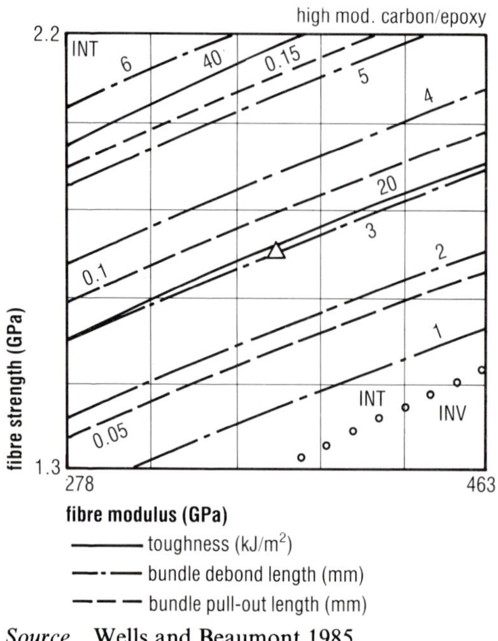

Source Wells and Beaumont 1985

Figure 10.13 *Toughness map of fibre strength against fibre modulus for Kevlar fibre/epoxy*

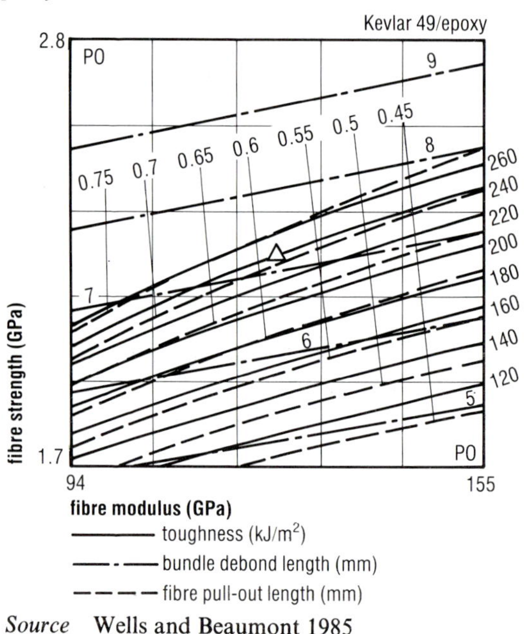

Source Wells and Beaumont 1985

Table 10.7 Summary of fracture property dependence on constitutive properties of the composite (Wells and Beaumont 1985a)

Composite parameter increasing	Glass and Kevlar fibre composites: Effect on l_{pf}	l_{db}	G	Note	Carbon fibre composites: Effect on l_{pb}	l_{db}	G
σ_f	↑	↑	↑	direct effect on l_p and l_d	↑	↑	↑
E_f	↓	↓	↓	increases debond stress	↓	↓	↓
σ_m	—	—	—	negligible effect on strength of bundle	—	—	—
E_m	↓	↓	↓	residual stress for a given misfit strain affected	↓	↓	↓
r_f	↑	—	↑	fibre debond stress and friction build-up affected	—	—	—
r_b	—	↑	↑	reduced bundle debond stress and friction build-up	↑	↑	↑
ε_0	↓	—	↓	faster friction stress build-up	—	—	—
ε_b	—	↓	↓	faster friction stress build-up	↓	↓	↓
G_1	↓	↓	↓	debond stress and interface energy	↓	↓	↓
G_2	—	↓	↓	debond stress and interface energy	↓	↓	↑
V_f	—	↑	↑	increased number of fibres; increased bundle strength	↑	↑	↑

increasing fibre (and fibre bundle) diameter, decreasing matrix modulus and decreasing fibre-matrix bond strength. Composite toughness also increases with increasing matrix toughness, but the dependence is a weak one (see Figure 10.10). The gradient of the toughness contours and their spacing indicates the sensitivity of composite toughness on a particular material property.

The models predict reasonable estimates of composite toughness, ranking the Kevlar fibre composite the toughest and the high-modulus carbon fibre composite the least tough. Table 10.7 summarises the effects of all material properties on the toughness of glass, Kevlar and carbon fibre composites.

10.5.2 Hygrothermal ageing effects

A map for the fracture of a cross-ply glass fibre epoxy laminate is shown in Figure 10.14. The axes of the map are fibre misfit strain ε_0 and fibre

Figure 10.14 *Toughness map of fibre-matrix misfit strain against fibre strength for a cross-ply E-glass fibre/epoxy. The arrows show how the measured toughness changes with exposure to steam for a given length of time (in hours)*

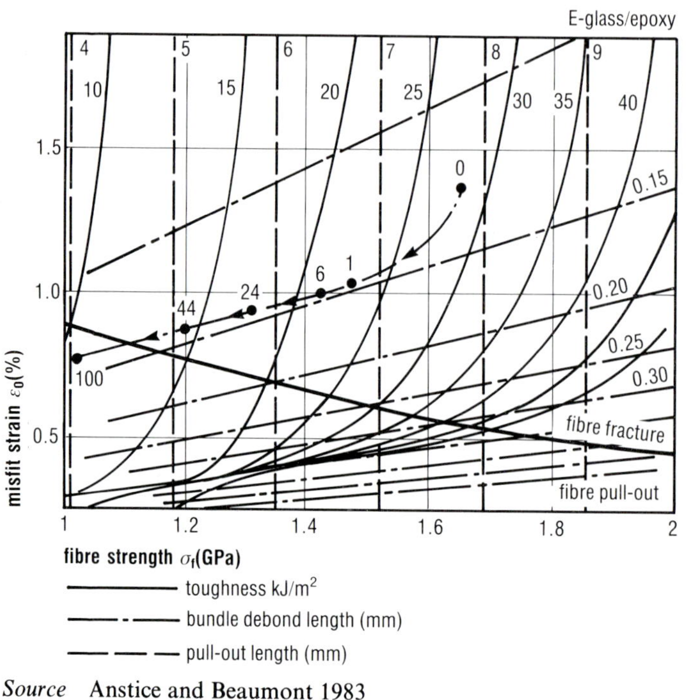

Source Anstice and Beaumont 1983

strength σ_f. A positive value of misfit strain indicates that the matrix exerts a compressive stress over the surface of the fibre; a negative value would indicate a tensile stress across the fibre-matrix interface. The map also displays data of toughness collected on samples exposed to hot-wet conditions for up to 100 hours. The data are fitted on the map by comparing measurement and prediction of toughness and measurement and prediction of fibre debond length.

The map suggests that hygrothermal degradation (the toughness has fallen from nearly 30 kJ/m^2 (666 cal/ft^2) to about 10 kJ/m^2 (222 cal/ft^2) over 100 hours) is brought about by a moisture weakening effect on the glass fibre, combined with a swelling of the epoxy (the misfit-strain falls). The implication is that glass is attacked by water and the diffusion of water into the resin has caused the matrix to expand and therefore it grips the fibre less tightly. There may be other degradation mechanisms in addition to these.

10.6 FATIGUE DAMAGE MECHANICS AND LIFETIME PREDICTION

The concept of damage-tolerant design requires that a structure retains adequate strength (and stiffness) after sustaining damage such as impact or fatigue. We can assess the effect of damage in several ways. One approach is to measure residual strength with load-cycling. Then, an S/N curve can be drawn through the focus of data points where the residual strength curves equal the applied stress. If the application is critical, a Weibull analysis could be used to describe the variations in strengths and probability curves drawn through the vast collection of data. This is an empirical method, 'macroscopic' in nature, in which no account is taken of the irregularities in microstructure brought about by the damaging mechanisms.

An alternative method is to consider microstructural changes during damage growth and the effect they have on the residual properties. This approach is preferred since a prediction of damage growth and lifetime includes the actual failure processes. The current level of damage could also be used to determine the remaining lifetime of a structure.

10.6.1 Development of a new damage model

We know that cyclic loading of composite laminates causes damage, eg splitting and delamination, matrix cracking and eventually fibre breakage, and that such damage accumulates with time. It is reasonable to assume, then, that failure will occur when some critical level of damage in the composite is exceeded. The physical nature of this damage is likely to depend on load history.

Damage grows with cycling until either the net section stress (allowing for the loss of section caused by the damage) exceeds the ultimate strength, or until a critical crack forms by the aggregation of damage. If the damage growth rate depends on the cyclic stress range $\Delta\sigma$, the load ratio R and the current 'value' of damage D, then, provided all other variables (temperature, frequency etc) do not change, then

$$\frac{dD}{dN} = f(\Delta\sigma, R, D) \qquad (10.48)$$

It follows that the fatigue lifetime is simply the number of load cycles it takes to raise the initial damage state D_i (usually equal to zero unless there is some prior damage, like impact, caused during manufacture or installation) to the final or critical level of damage D_f.

$$N_f = \int_{D_i}^{D_f} \frac{dD}{f(\Delta\sigma, R, D)} \tag{10.49}$$

The problem, of course, is that we do not know the function f. However, if damage affects the stiffness E of the laminate, and it usually does, we can write

$$E = E_0\, g(D) \tag{10.50}$$

where E_0 is the undamaged modulus. By differentiating Equation 10.50 and combining with Equation 10.48, we have

$$\frac{1}{E_0}\frac{dE}{dN} = g'\left[g^{-1}\left(\frac{E}{E_0}\right)\right] f\left[\Delta\sigma, R, g^{-1}\left(\frac{E}{E_0}\right)\right]$$

where g^{-1} is the inverse of g:

$$D = g^{-1}\left(\frac{E}{E_0}\right)$$

To obtain the function f we must first establish the function $g(D)$. This can be done experimentally; it will depend on the laminate properties. Having gathered data, then for E/E_0 with load cycles N, and knowing $g(D)$, we can evaluate f:

$$f(\Delta\sigma, R, D) = \frac{1}{g'\left[g^{-1}\left(\frac{E}{E_0}\right)\right]}\left(\frac{1}{E_0}\right)\frac{dE}{dN} \tag{10.51}$$

To do so, the right-hand side of the equation is evaluated for a range of $\Delta\sigma$ values at constant E/E_0 and R, for a range of R at constant $\Delta\sigma$ and E/E_0 and for a range of E/E_0 at constant $\Delta\sigma$ and R (Poursartip 1983).

10.6.2 Application to CFRP

A quasi-isotropic carbon fibre-epoxy laminate (CFRP) in tensile cyclic loading fails by matrix cracking of the 90° ply followed by the 45° ply, delamination between the 45° and 90° and 90°/−45° ply interfaces and finally by fibre fracture. In high-cycle fatigue (long fatigue lives), the dominant mode of failure is by delamination (Poursartip 1983).

Ignoring other forms of damage, we can define a damage parameter D as the normalised delamination area A/A_0, where A is the actual delamination area and A_0 is the total area available for delamination. A simple model for the loss of modulus with delaminations is

$$E = E_0 + (E^* - E_0)A/A_0 \tag{10.52}$$

Figure 10.15 *Comparison between observed damage-growth rate and stress range for a quasi-isotropic CFRP and theoretical prediction*

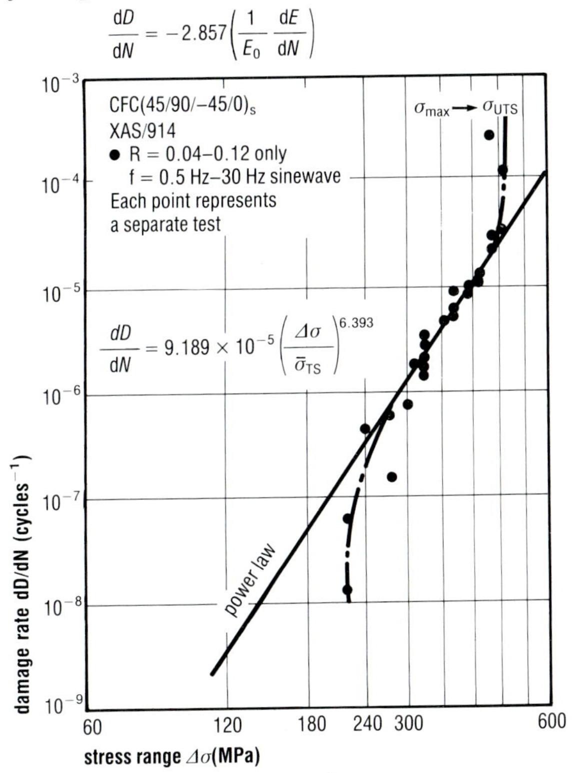

Source Poursartip, Ashby and Beaumont 1986

where E^* is the modulus which corresponds to total delamination of the laminate, ie where $A/A_0 = 1$ and $E/E_0 = 0.65$. It follows that

$$D = 2.857(1 - E/E_0) \tag{10.53}$$

Thus,

$$\frac{dD}{dN} = f(\Delta\sigma, R, D) = -2.857\left(\frac{1}{E_0}\frac{dE}{dN}\right) \tag{10.54}$$

The right-hand side of Equation 10.54 can easily be evaluated experimentally.

Figure 10.15 shows the damage rate dD/dN as a function of $\Delta\sigma$ ($R = 0.1$). Clearly, there are three regimes of fatigue behaviour:

(i) There is a power-law relation between dD/dN and $\Delta\sigma$ over most of the stress range.

(ii) The damage rate is much higher than expected from a power law at stresses approaching the ultimate strength; essentially we have a static failure.

(iii) There is an apparent threshold stress of 250 MPa (36,250 lbf/in^2) below which no damage occurs.

It is interesting to note that this is the stress which corresponds to the first cracking stress (the 'design ultimate') observed in a monotonic tensile test.

1 The terminal damage
Finally, to predict fatigue life we need to know the critical terminal value of D. We can determine D_f as follows: in a monotonic tensile test, assuming no stiffness reduction,

$$\varepsilon_c = \frac{\sigma_{TS}}{E_0} \tag{10.55}$$

After cycling at some maximum stress σ_{max} the increased strain is

$$\varepsilon = \frac{\sigma_{max}}{E} \tag{10.56}$$

Since failure occurs when the maximum strain during the fatigue cycle equals ε_c, by equating Equations 10.55 and 10.56 and substituting for D_f from Equation 10.53, we have, at failure,

$$D_f = 2.857\left(1 - \frac{E_f}{E_0}\right) = 2.857\left(1 - \frac{\sigma_{max}}{\sigma_{TS}}\right) \tag{10.57}$$

For power-law damage growth (from Figure 10.14),

$$\left(\frac{dD}{dN}\right) = 9.2 \times 10^{-5}\left(\frac{\Delta\sigma}{\sigma_{TS}}\right)^{6.4} \quad (\text{for } R = 0.1) \tag{10.58}$$

The effect of mean stress is such that

$$\left(\frac{dD}{dN}\right)_{R>0.1} = \left(\frac{dD}{dN}\right)_{R=0.1} \times \left\{\frac{\sigma_m(R > 0.1)}{\sigma_m(R = 0.1)}\right\}^P \tag{10.59}$$

where $P \approx 2$, depending on the value of $\Delta\sigma$. Substituting for $f(\Delta\sigma, R, D)$ from Equations 10.58 and 10.59 into Equations 10.48 and 10.49, we have

$$N_f = \int_{D_i}^{D_f} 1.1 \times 10^4\left(\frac{\Delta\sigma}{\sigma_{TS}}\right)^{-6.4}\left\{\frac{\bar{\sigma}_m(R = 0.1)}{\bar{\sigma}_m(R > 0.1)}\right\}^P dD \tag{10.60}$$

D_f is determined by Equation 10.53. We can substitute for σ_{max} and σ_m in terms of $\Delta\sigma$ and R, and by integrating Equation 10.60:

$$N_f = 3.1 \times 10^4 \left(\frac{\Delta\sigma}{\sigma_{TS}}\right)^{-6.4} \left(1.22\,\frac{1-R}{1+R}\right)^P \left(1 - \frac{\Delta\sigma}{(1-R)\sigma_{TS}}\right) \quad (10.61)$$

Figure 10.16 predicts the S/N curve for the laminate ($R = 0.1$). The predicted life (Equation 10.61) is plotted for three values of σ_{TS}. Clearly, there is consistency with this treatment and experimental observations. At low stress, high-cycle fatigue the effect of variations in tensile strength is small, whereas in low-cycle fatigue, small changes in σ_{TS} are significant (Poursartip and Beaumont 1986).

Figure 10.16 *Comparison between measured lifetime and theoretical prediction of a quasi-isotropic CFRP laminate*

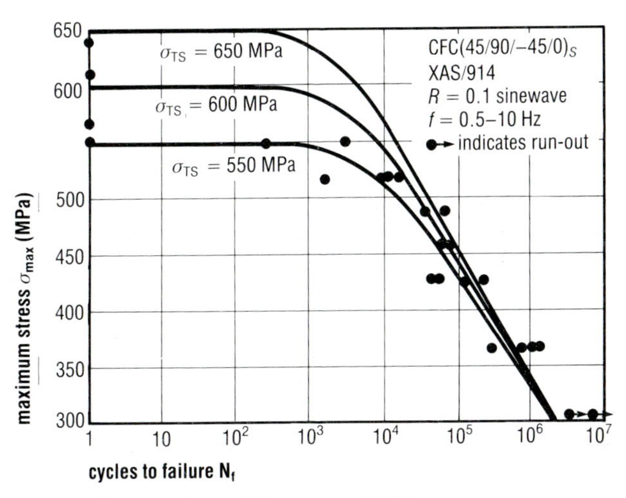

Source Poursartip and Beaumont 1986

2 Variable-amplitude loading effects

To predict fatigue lives under variable-amplitude loading we must be able to predict the average damage growth-rate $(dD/dN)_{av}$ and the critical damage level to cause failure, D_f. In a variable-amplitude loading sequence we would expect the critical damage to be determined by the cycle with the highest maximum stress.

Let us consider an experiment (Poursartip 1983) in which the following 2-block loading sequence was run:

$\Delta\sigma_1$ = 328 MPa (47,560 lbf/in^2)	$\Delta\sigma_2$ = 395 MPa (57,275 lbf/in^2)
σ_{max1} = 365 MPa (52,925 lbf/in^2)	σ_{max2} = 500 MPa (72,500 lbf/in^2)
N_1 = 1,000–2,000 cycles	N_2 = 1–100 cycles

Since the low stress was maintained for the greater number of cycles, it is not surprising that the observed damage rates were found to lie close to that expected of type 1 cycling on its own. The average measured value of D_f was 0.64. Using Equation 10.53 and assuming $\sigma_{TS} \sim 600$–650 MPa (87,000–94,250 lbf/in^2), then

$$0.48 < D_f < 0.66$$

if the damage at failure is to be determined by the highest maximum stress seen, and

$$1.12 < D_f < 1.25$$

if the damage at failure is to be determined by the most frequent maximum stress seen. The observed D_f agrees best with the highest maximum stress criterion.

Let us now consider the simple case of two blocks of duration N_1 and N_2, with corresponding damage rates $(dD/dN)_1$, and $(dD/dN)_2$. Provided there are no load interaction effects (and this may not be true under all loading conditions), then we may assume a linear sum of the rates to represent the average damage rate:

$$\left(\frac{dD}{dN}\right)_{av} = \frac{N_1}{N_1 + N_2}\left(\frac{dD}{dN}\right)_1 + \frac{N_2}{N_1 + N_2}\left(\frac{dD}{dN}\right)_2 \qquad (10.62)$$

We can check this as follows: in a simple test where $N_1 = 1{,}000$, $\Delta\sigma_1 =$ 395 MPa (57,275 lbf/in^2), $R = 0.2$, followed by $N_2 = 2{,}000$, $\Delta\sigma_2 = 328$ MPa (47,560 lbf/in^2), $R = 0.1$, repeated for about 40,000 cycles, we found that $(dD/dN)_{av} = 3.6 \times 10^{-6}$.

Using damage rates determined by Equation 10.58 for the two separate loading sequences, inserted into Equation 10.62, we obtain

$$\frac{dD}{dN} = \frac{1}{3}(7.4 + 10^{-6}) + \frac{2}{3}(2.25 \times 10^{-6})$$

$$= 4 \times 10^{-6}$$

which compares well with the observed rate (Poursartip and Beaumont 1986).

10.6.3 Application to GFRP

This approach in which the development of a damage mechanics analysis is based on a dominant failure mechanism has also been adapted successfully for a glass fibre/epoxy laminate in which transverse ply (matrix) cracking is recognised as the principal cause of damage (Ogin *et al.* 1985).

As before, the difficulty is in finding the appropriate form of the damage function f (Equation 10.48). One way of solving the problem with respect to the cross-ply glass fibre/epoxy laminate is to consider the stored elastic strain energy U between two adjacent matrix cracks in the transverse (90°) ply:

$$U = \frac{K^2 \sigma_{max}^2 4sdW}{2E_2} \tag{10.63}$$

where $K\sigma_{max}$ is that fraction of the applied stress on the laminate carried by the uncracked transverse ply, $2s$ is the average crack spacing, d is ply thickness and W is the laminate width. E_2 is the modulus of the 90° ply.

The postulate is made (which is supported by experimental observation) that the total length a of all those matrix cracks increases with successive load cycles, and that the total crack growth rate obeys a power-law function of the stored elastic strain energy:

$$\frac{da}{dN} \propto (\sigma_{max}^2 2s)^n \tag{10.64}$$

This is analogous to the Paris law of single-crack mechanics of isotropic materials.

Since the crack density D $(= 1/2s)$ is proportional to this total crack length a, it follows, therefore, that

$$\frac{dD}{dN} \propto \left[\frac{\sigma_{max}^2}{D}\right]^n \tag{10.65}$$

Alternatively, in terms of the damaged modulus E,

$$E = E_0(1 - cD) \tag{10.66}$$

where c is a material constant, Equation 10.65 can be rewritten as

$$-\frac{1}{E_0}\frac{dE}{dN} \propto \left[\frac{\sigma_{max}^2}{E_0^2(1 - E/E_0)}\right]^n \tag{10.67}$$

where the introduction of the term E_0^2 makes the right-hand side of the equation dimensionless. We now have a form of the damage function f. The modulus reduction rate (LHS of Equation 10.67) can be determined experimentally at any value of E/E_0. Figure 10.17 shows experimental data of reduced modulus over a range of maximum applied stress. The data fit a power-law relation. It follows that the relationship between the current (damaged) modulus E, the number of load cycles N and the maximum applied stress σ_{max} is

$$\frac{E}{E_0} = 1 - \left[25\left(\frac{\sigma_{max}}{E_0}\right)^{1.5} N^{0.25}\right] \tag{10.68}$$

Figure 10.17 *Comparison between observed modulus change and maximum applied cyclic stress and prediction for a cross-ply GFRP laminate*

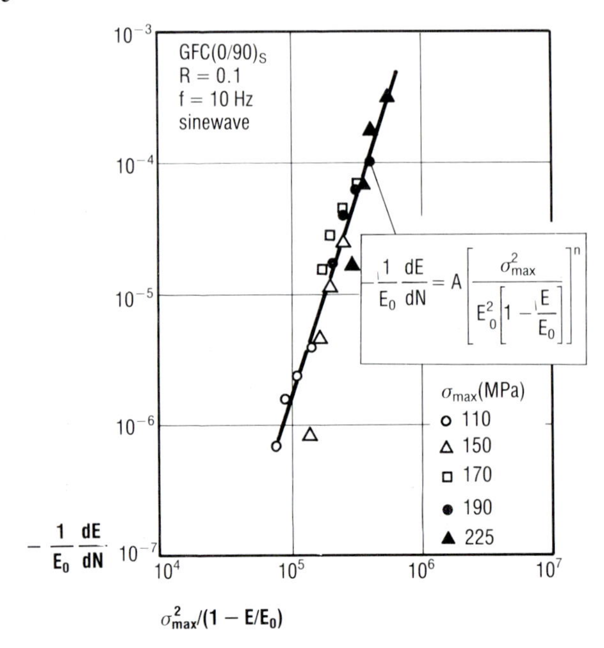

Source Ogin, Smith and Beaumont 1985a

Figure 10.18 *Prediction of S-N curves for various values of E/E_0 for a cross-ply GFRP laminate*

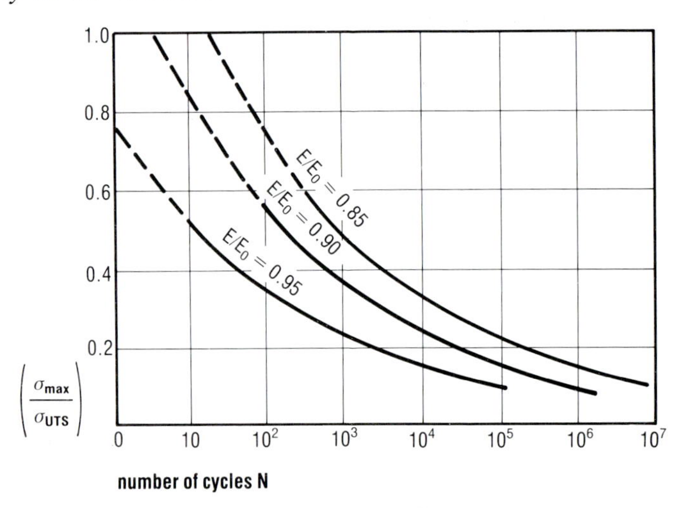

Source Ogin, Smith and Beaumont 1985a

We can, therefore, determine the state of damage and the effect on residual modulus at any stage in the lifetime of the material (Figure 10.18).

10.7 A FINAL WORD

The failure models presented in this chapter provide a quantitative basis for assessing the fracture stress and toughness enhancement of fibre composites in terms of the microstructure, laminate and specimen geometry, and modes of failure at a notch tip. Likewise, a knowledge of the mechanisms of fatigue combined with damage models that relate failure modes to property changes can be used to predict the lifetime for variable stress history. Built into these damage mechanics analyses is the ability to predict the residual strength (or stiffness) and the remaining lifetime of a damaged composite. Order of magnitude calculations of K_c and G_c allow a mechanism to be associated with toughness enhancement. Similarly, approximate estimates of fatigue damage growth rate give insight into the role of the microstructure and micromechanisms in damage accumulation.

ACKNOWLEDGEMENTS

Much of the work on which this chapter is based was carried out in collaboration with my past post-graduate research students of the University Engineering Department Composites Group, Cambridge. It is a special pleasure to acknowledge the work of Dr J K Wells, Dr A P Poursartip, Dr P A Smith, Dr S Ogin and Dr M T Kortschot. In particular, this work has benefited greatly from the contributions of Professor M F Ashby.

References

1 Introduction

Atkinson, R J (1948) *Derivation of test factors and permissible design values.* RAE Technical Note, Structures 15. Farnborough (May).

Bagg, G E *et al.* (1970) Paper no. 6. *Proceedings of BPF 7th International Reinforced Plastics Conference*, Brighton, UK.

Bowers, B (1971) *R E B Crompton, Pioneer Electrical Engineer.* Science Museum Booklet. London, HMSO.

Clark, R W (1977) *Edison.* London, Macdonald & James, p 91.

Dietz, A G H (1963) Fibrous composite materials. *International Science and Technology* (August).

Ferguson, J (1988) Of spiders and spinning. *Chemistry in Britain*, 24(1):39.

Fitzer, E (1986) *Carbon Fibres and their Composites.* Berlin, Springer-Verlag.

Gordon, J E (1978) *Structures.* Harmondsworth, Penguin Books, p 246.

Gordon, J E and Evans, M E N (1964) Silicon nitride and other whiskers. *Proc Roy Soc* A282, p 26.

Holister, G S and Thomas, C (1966) *Introduction to Fibre Reinforced Materials.* London, Elsevier.

Kaufmann, M (1967) *Giant Molecules.* London, Aldus Books. Also published (1964) as Science Museum Booklet. London, HMSO.

Kelly, A (1973) *Strong Solids.* 2nd edition. Oxford, Clarendon Press, p 245.

Ladizesky, N H and Ward, I M (1986) High modulus fibres and their composites. Paper 29/1, *Proceedings of 3rd Plastics and Rubber Institute International Conference, London* (October 1985). Published (1986) as *Carbon Fibres.* NJ, USA, Noyes Publications, p 200.

Lovell, D (1986) A comparison of available carbon fibres. *Proceedings of 3rd Plastics and Rubber Institute International Conference: Carbon Fibres III, ibid.*

McCrum, N G (1971) *A Review of the Science of Fibre Reinforced Plastics.* London, HMSO.

Parker, B M (1974) Boron fibre and boron fibre composites. *Composites* (Jan), pp 7–15.

Phillips, L N (1972) *Composites.* Chap. 26 in A D Jenkins (ed) *Polymer Science*, Amsterdam, North-Holland Publishing Co.

Ridding, A (1964) *S Z de Ferranti – Pioneer of Electric Power.* London, HMSO.

Schwartz, M M (1984) *Composite Materials Handbook.* New York, McGraw-Hill, Section 2.6.

Sprecher, N (1983) *Comparative Data (S2 Glass Fibre).* Owens Corning Fibreglass, Europe (Brussels).

Watt, W (1970) Production and properties of High-Modulus Carbon Fibres. *Proc Roy Soc London.* A319:5–15.

Watt, W and Johnson, W (1969) The effect of length changes during the oxidation of polyacrylonitrile on the Young's modulus of carbon fibres. Applied Polymer Symposia. No 9. London. pp 215–227.

Watt, W and Johnson, W (1975) Mechanism of oxidation of polyacrylonitrile fibres. *Nature* 257(5523):210–212.
Watt, W, Phillips, L N and Johnson, W (1966) High-strength high-modulus carbon fibres. *The Engineer* 221:815.
Welling, M S (1981) *Processing and Uses of CFRP*. (Translation.) Dusseldorf, VDI Verlag.
Winter, J (1970) *Industrial Architecture*. London, Studio Vista, p 58.

2 Fabrication

Bader, M G (1983) In Kelly and Mileiko (eds) *Handbook of Composites*. Amsterdam, North Holland Publishing Co. p 177.
Ballance, B (1986) Progress and prospects from CFRTPs. *Proceedings of 3rd International Plastics and Rubber Institute Conference: Carbon Fibres III*. NJ, USA, Noyes Publications, p 72.
Charnock, G F (1942) *Mechanical Technology*. 2nd edition. London, Constable & Co.
Childs, R (1980) Some aspects of the autoclave fabrication process. *Proceedings of Salford Symposium on Fabrication Techniques for Advanced Reinforced Plastics*. Guildford, Surrey, IPC Science & Technology Press. p 1.
Cook, T (1980) Production of precision filament-wound tubes. *Ibid*. p 48.
Iddon, K J and Blundell, C (1980) The pultrusion process. *Ibid*. p 58.
Johnson, F C (1980) Compression moulding of composite aero engine components with elevated thermal stability. *Ibid*. p 10.
Jones, W R and Johnson, J W (1980) A resin injection technique for the fabrication of aero engine composite components. *Ibid*. p 40.
Kelly, A and Mileiko, S T (1983) *Handbook of Composites*. Amsterdam, North-Holland Publishing Co, vol 4, p 45.
Mohr, J G, Oleesky, S M, Shook, G D and Meyer, L S (1973) *SPI – Handbook of Technology*. 2nd edition. New York, Van Nostrand Reinhold, p 27.
Newling, D O (1968) *AWRE Report no 081/68*. London, HMSO.
Parker, B and Waghorne, R M (1982) Surface treatment of CFRP for adhesive bonding. *Composites* 13(3): 280.
Phillips, L N (1953) *The Moulding Behaviour of Durestos under Vacuum Conditions*. RAE Technical Report no Chem 496. London, Ministry of Supply.
Phillips, L N (1955) Structural plastics for aircraft use in theory and practice. *Trans. Plast. Inst. (London)* (October): 325.
Phillips, L N (1969) Forming processes for carbon fibre and resin. *Composites* 1(2):101.
Phillips, L N (1980) Fabrication of reinforced thermoplastics by means of the film-stacking technique. *Proceedings of Salford Symposium on Fabrication Techniques for Advanced Reinforced Plastics*. Guildford, Surrey, IPC Science & Technology Press, p 101.
Phillips, L N and Parker, D B V (1964) *Polyurethanes*. Plastics Institute Monograph. London, Iliffe Books, p 58.
Phillips and Murphy (1980) The properties of carbon fibre reinforced thermoplastics, moulded by the film-stacking method. *Proceedings of Salford Symposium on Fabrication Techniques for Advanced Reinforced Plastics*. IPC Science & Technology Press.
Rogers, K F, Kingston-Lee, D M and Phillips, L N (1980) The development of a lightweight all-plastics apparatus for vacuum moulding composites. *Proceedings of Salford Symposium on Fabrication Techniques for Advanced Reinforced Plastics*. Guildford, Surrey, IPC Science & Technology Press, p 27.
Rosato, D V and Grove, C S (1964) *Filament Winding*. New York, Interscience and John Wiley.

Ruegg, C (1981) *Processing and Uses of CFRP*. Dusseldorf, VDI-Verlag GmbH, p 188.
Tarnopol'ski, Yu M (1983) In Kelly and Mileiko, *op. cit.*
Taylor, P (1980) The development and use of a tape-laying machine. *Proceedings of Salford Symposium on Fabrication Techniques for Advanced Composite Materials.* IPC Science & Technology Press.
Trewin, E, Turner, R and Cluley, A (1980) Carbon fibre reinforced thermoplastics. In A Bunsell (ed) *Proceedings of ICCM3, Paris*, Pergamon Press. vol 2, p 1796.
Walker, N J (1986) Papers, a new dimension in carbon fibre materials. *Proceedings of 3rd Plastics & Rubber Institute International Conference: Carbon Fibres III.* NJ, USA, Noyes Publications. p 55.
Weaver, J and Taylor, E (1986) Hybrid systems – best of both worlds? *Ibid.*

3 Properties of thermoset polymer composites and design of pultrusions

Adamson, M J (1983) Long term behaviour of composites. In T K O'Brien (ed) *ASTM STP*, p 813.
Anderson, J J (1987) *Creep Properties of Composite Rods.* British Ropes internal report.
Birley, A L, Dawkins, J V, and Strauss H E (1984) Blistering in glass reinforced polyester laminates. *Proceedings of BPF Congress.*
Blackford, R W (1985) Improved durability airline paint schemes. *Proceedings of 6th International Conference SAMPE, Scheveningen.* Amsterdam, Elsevier.
Caddock, B D, Evans, K E and Hull, D (1986) The role of diffusion in the micromechanisms of stress corrosion cracking of E glass/polyester composites. *Proceedings of IMechE 2nd International Conference on Fibre Reinforced Composites.*
Cochram, D and Scrimshaw, G F (1980) Acid corrosion of GRP. *Proceedings of 35th Conference RP/CI SPI.*
Crump, S (1986) A study of blister formation in gel coated laminates. *Proceedings of 41st Conference RP/CI SPI.*
Duthie, A C (1986) The preliminary design of composite structures. *Proceedings of BPF RPG Congress.*
Forsdyke, K L (1984) Phenolic resin composites for fire and high temperature applications. *Proceedings of International Conference FRC 84 PRI.*
Hogg, P. J, Price, J N and Hull, D (1984) Stress corrosion of GRP. *Proceedings of International Conference: Fibre Reinforced Composites.*
Holmes, M and Just, D J (1983) *GRP in Structural Engineering.* Barking, Essex, Elsevier Applied Science.
Imuro, H and Yoshida, N (1986) Differences between Technora and PPTA aramid. *Proceedings of 25th International Man-Made Fibres Conference*, Austria.
Jones, L A (1985) In-service environmental effects on carbon fibre composite materials. *Proceedings of 6th International Conference SAMPE, Scheveningen.* Amsterdam, Elsevier.
Krenchel, H (1964) *Fibre orientation.* Copenhagen, Akademisk Forlag.
Lubin G (1982) *Handbook of Fibreglass and Advanced Plastics Composites.* New York, Van Nostrand Reinhold.
McKague, E L, Halkias, J E and Reynolds, J D (1975) *Journal of Composite Materials* 9 (Jan).
Mandell, J F and Meier, U (1975) Fatigue crack propagation in E-glass/epoxy composites. *ASTM STP.* 569.
Payne, K G, Burrows, B J and Jones, C R (1987) Practical aspects of applying lightning protection to aircraft and space vehicles. *Proceedings of 8th International Conference SAMPE, La Baule.*

Ramani, S V and Nelson, H G *Effect of Moisture on the Fatigue Behavior of Graphite/Epoxy Composite Laminates.* NASA Report TM 78548, no date.

Schaefgen, J R and Wardle, M W (1984) Aramid fibres: Structure vs properties and composite applications. *Fibre Producer* (Aug).

Steard, P A and Jones, F R (1986) The effects of the environment on stress corrosion of single E-glass filaments and the nucleation of damage in GRP. *Proceedings of IMechE 2nd International Conference FRC.*

Steiner, B A, Blair, M, Gong, G, McKillen, J and Browne, J (1985) Development of a toughened bismaleimide resin prepreg for carbon fibre composites. *Proceedings of 6th International Conference SAMPE, Scheveningen.* Amsterdam, Elsevier.

Tsai, S W and Hahn, W T (1980) *Introduction to Composite Materials.* Lancaster, Philadelphia, Technomic Publishing Co.

Tuhru Hiramatsu, Tomitake Higuchi, and Junichi Matsui (1987) Torayca T1000 ultra high strength fibre and its composites properties. *Proceedings of 8th International Conference SAMPE, La Baule.*

Walton, J M and Yeung, Y C T (1987) Flexible tension members from composite materials. *Proceedings of 6th Arctic Eng Symposium.* vol III. American Soc of Mech Eng.

4 Resin matrices

Boenig, H V (1964) *Unsaturated Polyesters: Structure and Properties.* Amsterdam, Elsevier.

Bowles, K J and Vannucci, R D (1986) *SAMPE Quarterly* 17(2):12–18.

Bruin, P F (1976) *Unsaturated Polyester Technology.* New York, Gordon & Breach.

Cassidy, P E (1980) *Thermally Stable Polymers.* New York, Marcel Dekker.

Chaudhari, M A, Cobuzzi, C A and King, J J (1984) *Proceedings of the 16th National SAMPE Technical Conference.* pp 565–576.

Chaudhari, M A, Galvin, T and King, J J (1985) *SAMPE Journal* 21(4):17–21.

Clemans, S R, Handermann, A C and Western, E D (1987). In J de Bossu, G Briens and P Lissac (eds) *Looking Ahead for Materials and Processes.* Amsterdam, Elsevier.

Di Guilio, C, Gautier, M and Jasse, B (1984) *Journal of Applied Polymer Science* 29:1771–779.

Fischer, M, Lohse, F and Schmid, R (1980) In *Makromolekulare Chemie* 181:1251–287.

Gaku, J, Suzuki, K and Nakamichi, K (1978) *US Patent* 4 110 394.

Gibbs, H H (1985) *SAMPE Journal* 21(4):22–27.

Grundschober, F (1970) *US Patent* 3 533 996.

Grundschober, F and Sambeth, J (1968) *US Patent* 3 380 964.

Haug, T (1980) In *Makromolekulare Chemie* 181:2025–047.

Hay, J N, Johnson, F, Lind, D, Owens, G A, Wells, J K and Wilson, D (1987) *SAMPE Journal* 23(3):35–41.

Hergenrother, P M (1986) *Die Angewandte Makromolekulare Chemie* 145/6:323–341.

Hummel, D O, Heinen, K U, Stenzenberger, H D and Seisler, H (1974) *Journal of Applied Polymer Science* 18:2015–024.

Jones, J S (1983) *SAMPE Journal* 19(2):10–23.

Knop, A and Scheib, W (1979) *Chemistry and Applications of Phenolic Resins.* Berlin, Heidelberg, Springer-Verlag.

Kubens, R, Schultheis, H, Wolf, R and Grigat, E (1968) *Kuntstoffe* 58:827–832.

Lee, H and Neville, K (1967) *Handbook of Epoxy Resins.* New York, McGraw-Hill.

Lohse, F (1987) *Makromolekulare Chemie – Macromolecular Symposia* 7:1–16.

May, C A and Tanaka, Y (1973) *Epoxy Resins.* New York, Marcel Dekker.

Ostberg, G M K and Seferis, J C (1987) *Journal of Applied Polymer Science* 33:29–39.

Renner, A and Zahir, S A (1978) *US Patent* 4 100 140.
Seferis, J C (1986) *Polymer Composites* 7(3): 158–169.
Shimp, D A (1987) *SAMPE Quarterly* 19(1): 41–46.
Siebert, A R (1985) *Proceedings of PRI International Conference on Toughening of Plastics II* 2nd–4th July: 7/1. London.
Tai-Shung Chung and McMahan, P E (1986) *Journal of Applied Polymer Science* 31: 965–977.
Yoshikawa, M, Yokoi, H, Sanui, K and Ogata, N (1984) *Journal of Polymer Science – Polymer Letters Edition* 22: 125–127.
Zahir, S A (1982) *Advances in Organic Coatings Science and Technology* 4: 83–102.

5 Joining of composites

Adams, R D (1986) Theoretical stress analysis of adhesively bonded joints. In F L Matthews (ed) *Joining Fibre-Reinforced Plastics*. Chap 5, pp 185–226. Barking, Essex, Elsevier Applied Science.
AGARD 1984a and b *Report 716: Composite Structure Repair* (February 1984); *Addendum* (August 1984), Advisory Group for Aerospace Research and Development, North Atlantic Treaty Organisation (AGARD/NATO).
Allen, K W, Chan, S Y T and Armstrong, K B (1982) Cold-setting adhesives for repair purposes using various surface preparation methods. *International Journal of Adhesion & Adhesives* 2(4): 239–247.
Armstrong, K B (1983) The design of bonded structure repairs. *Ibid*, 3(1): 37–52.
Collings, T A (1986) Experimentally determined strength of mechanically fastened joints. In F L Matthews (ed) *Joining Fibre-Reinforced Plastics* Chap 2, pp 9–64. Barking, Essex, Elsevier Applied Science.
Cotter, J L (1978) *Proceedings of Symposium on Jointing in Fibre-Reinforced Plastics* pp 87–94. Guildford, IPC Science & Technology Press.
Godwin, E W and Matthews, F L (1980) A review of the strength of joints in fibre-reinforced plastics, Pt 1: Mechanically fastened joints. *Composites* 11(3): 155–160.
Goland, M and Reissner, E (1944) The stresses in cemented joints. *Journal of Applied Mechanics* 11: A17–27.
Hart-Smith, L J (1973a) Adhesive-bonded double-lap joints. *Report CR-112235*, NASA Washington DC, USA.
Hart-Smith, L J (1973b) Adhesive-bonded scarf and stepped-lap joints. *Report CR-112237*, NASA Washington DC, USA.
Hart-Smith, L J (1973c) Adhesive-bonded single-lap joints. *Report CR-112236*, NASA Washington DC, USA.
Hart-Smith, L J (1986a) Design of adhesively bonded joints. In F L Matthews (ed) *Joining Fibre-Reinforced Plastics* Chap 7, pp 271–311. Barking, Essex, Elsevier Applied Science.
Hart-Smith, L J (1986b) Design and empirical analysis of bolted or riveted joints. *Ibid*, Chap 6, pp 227–270. Barking, Essex, Elsevier Applied Science.
Matthews, F L (1986) Theoretical stress analysis of mechanically fastened joints. *Ibid*, Chap 3, pp 65–103. Barking, Essex, Elsevier Applied Science.
Matthews, F L (1987) In *Proceedings of Conference on New Materials and their Applications, University of Warwick*. London, The Institute of Physics.
Matthews, F L (1989) Chap 9 in D H Middleton (ed) *Composite Materials in Aircraft Structures*. London and New York, Longman Scientific & Technical.
Matthews, F L, Kilty, P F and Godwin, E W (1982) A review of the strength of joints in fibre-reinforced plastics, Pt 2: Adhesively bonded joints. *Composites* 13(1): 29–37.

Matthews, F L and Tester, T T (1985) The influence of stacking sequence on the strength of bonded CFRP single-lap joints. *International Journal of Adhesion & Adhesives.* 5(1):13–18.

Matthews, F L, Wong, C M and Chryssafitis, S (1982) The stress distribution in bolted joints in fibre-reinforced plastics. *Composites* 13(3):316–322.

Parker, B M and Waghorne, R M (1982) Surface pretreatment of carbon fibre-reinforced composites for adhesive bonding. *Ibid*, 13(3):280–288.

Quinn, W J and Matthews, F L (1977) The effect of stacking sequence on the pin-bearing strength in glass fibre reinforced plastic. *Journal of Composite Materials* 11(2):139–145.

Rogers, K F, Kingston-Lee, D M and Phillips L N (1980) *Proceedings 3rd International Conference on Composite Materials (ICCM-III)*, pp 1390–407. Oxford, Pergamon Press.

Smith, P A (1985) PhD thesis, Chap 4, pp 79–89. Cambridge University.

Volkersen, O (1938) Die nietkraftverteilung in zugbeanspruchten nietverbidungen mit konstanten laschenquerschnitten. *Luftfahrtforschung* 15:4–47.

6 Specifications, test methods and quality control of advanced composites

Alberti, F P (1985) *Mantech Journal* 10(2):37–45.

Anderson, T and Messick, V B (1980) In G Pritchard (ed) *Developments in Reinforced Plastics* 1. Chap 2. Barking, Essex, Elsevier Applied Science.

Apicella, A (1986) *Ibid*, 5. Chap 5.

Aronhime, M T and Gillham, J K (1984) *Journal of Coatings Technology* 56(718):35–47.

Aronhime, M T and Gillham, J K (1986) Time temperature transformation (TTT) cure diagram of thermosetting polymeric systems. In K Dusek (ed) *Advances in Polymer Science* vol 78: Epoxy resins and composites III. New York, Springer.

Ashton, J E, Halpin, J C and Petit, P H (1969) *Primer on Composite Materials.* Stanford, CT, USA, Technomic Publishing.

Bacon, R and Moses, C T (1986) Carbon fibres from light bulbs to outer space. In R B Seymour and G S Kirshenbaum (eds) *High Performance Polymers – Their Origin and Development.* Amsterdam, Elsevier. pp 341–345.

Batch, G L and Macosko, C W (1987) *Proceedings of 42nd SPI/RP Conference.* Paper 12-B.

Bosshard, A W and Schlumph, H P (1987) Chap 9 in R Gachter and H Muller (eds) *Plastics Additives.* 2nd edition. Munich, Hanser.

Broyer, E and Macosko, C W (1978) *Polymer Engineering and Science* 18(5):382–387.

Buckley, L and Roylance, D (1982) *Ibid* 22(3):154–159.

Chamis, C C (1980) *Proceedings of 35th SPI/RP Conference.* Paper 12-A.

Curtis, T P (1984) CRAG test methods for the measurement of engineering properties of fibre reinforced plastics. *RAE Technical Report 84012* (October).

Duke, J C (1983) Chap 6 in C E Browning (ed) *Composite Materials: Quality Assurance and Processing.* ASTM STP 797.

Elias, H-G and Vohwinkel, F (1986) Chap 10 in *New Commercial Polymers 2.* London, Gordon & Breach.

Frost and Sullivan (1986) *Interpals Business News* (September/October).

Galiotis, C, Young, R J, Yeung, P H J and Batchelder, D N (1984) *Journal of Materials Science* 19(11):3640–648.

Gutowski, T G, Morigaki, T and Cai, Z (1987) *Journal of Composite Materials* 21(2):172–187.

Hamstad, A E (1985) *Mantech Journal* 10(3):24–32.
Han, C D and Han, K-W (1983) *Journal of Applied Polymer Science* 28:3155–225.
Han, C D, Lee, D S and Chin, H B (1985) *Proceedings of 40th SPI/RP Conference.* Paper 2F.
Kabelka, J (1984) Chap 6 in G Pritchard (ed) *Developments in Reinforced Plastics 6.* Barking, Essex, Elsevier Applied Science.
Kardos, J L, Dudokovic, M P, McKague, E L and Lehman, M W (1983) Chap 7 in C E Browning (ed) *Composite Materials: Quality Assurance and Processing.* ASTM STP 797.
Lee, W I, Loos, A C and Springer, G S (1982) *Journal of Composite Materials* 16(11):510–520.
Liao, Y T and Koenig, J L (1984) Chap 2 in G Pritchard (ed) *Developments in Reinforced Plastics 4*, Barking, Essex, Elsevier Applied Science.
Loos, A C and Springer, G S (1983) *Journal of Composite Materials* 17(3):135–169.
Ma, C-C, Lee, K-Y, Lee, Y-D and Huang, J-S (1987) *SAMPE Journal* (September/October).
Margolis, R (1977) *Proceedings of 32nd SPI/RP Conference* Paper 14-F.
McLaughlin, P V, McAssey, E V and Dietrich, R C (1980) *NDT International* 13(2):56–62.
Miller, D R and Macosko, C W (1976) *Macromolecules* 19(12):199–211.
Nicolais, L (1975) *Polymer Engineering and Science* 15(3):137–147.
Nixon, J A and Hutchinson, J M (1985) *Plastics and Rubber Processing and Applications* 5(3):349–357.
Parnaby, J, Hassan, G A, Helmy, H A A and Ali, A (1981) *Ibid*, 1:303–315.
Pittman, J F T (1986) Chap 6 in A Whelan and J L Craft (eds) *Developments in Plastics Technology 3.* Barking, Essex, Elsevier Applied Science.
Pritchard, G (ed) (1982) *Developments in Reinforced Plastics 2.* Barking, Essex, Elsevier Applied Science.
Quinn, J A (1987) *Design Manual for Pultruded Products.* Runcorn, Cheshire, Fibreforce Composites Ltd.
Richter, E B and Macosko, C W (1978) *Polymer Engineering and Science* 18(13):1012–018.
Ryan, M E (1984) *Ibid*, 24(9):698.
Sayers, D R, Howard, R D and Holland, S J (1986) *Proceedings of 41st SPI/RP Conference*, Paper 1E.
Serabian, S (1985) *Mantech Journal* 10(3):11–23.
Silva-Nieto, R J, Fisher, B C and Birley, A W (1985) *Polymer Composites* 1:135–169.
Springer, G (1986) *SAMPE Journal* (September/October):22–26.
Sprouse, J F, Halpin, B M and Sacher, R (1978) *Cure Analysis of Epoxy Composites Using FTIR.* AMMRC TR78–45, Army Materials and Mechanics Research Center, Watertown, MA 02172, USA.
Stevenson, J F (1986) *SPE ANTEC* (October): 452–457.
Summerscales, J (1987a) Chap 22 in I H Marshall (ed) *Composite Structures 4*, vol 2. Barking, Essex, Elsevier Applied Science.
Summerscales, J (ed) (1987b) *Non-Destructive Testing of Fibre-Reinforced Composites*, vol 1. Barking, Essex, Elsevier Applied Science.

7 Applications

Ault, G M and Fresche, J C (1979) Composites emerging for aeropropulsion applications. *Proc Am Inst Aeronautics & Astronautics* (October):48.

Beardmore, P, Harwood, J J and Horton, J (1980) Design and fabrication of a GRFRP concept automobile. *Proceedings of ICCM3 Conference, Paris.* Oxford, Pergamon Press. vol 1, p 11.

Christensen, R M and Wu, E M (1978) *Proceedings of ICCM2 Conference, Toronto.* New York, AIMMPE. p 718.

Clarke, G P (1985) The use of composite materials in racing car design. *Proceedings of 3rd Plastics & Rubber Institute International Conference Carbon Fibres III, London.* New Jersey, Noyes Publications, p 113.

Conen, H and Kaitatzidis, M (1981) Elevator unit for the Alpha-Jet, made from carbon-reinforced plastics. In *Processing and Uses of Carbon Fibre Reinforced Plastics.* Dusseldorf, VDI Verlag GmbH. p 151.

Cools, J J (1981) Composite rotor-blades for an experimental 330 kW wind turbine. *Proceedings of SAMPE Conference, Cannes. SAMPE Journal,* vol 17 (May/June).

Dharan, C K (1978) Design of automobile components with advanced composites. *Proceedings of ICCM2 Conference, Toronto.* New York, AIMMPE. p 1446.

Dunbar, D R, Robertson, A R and Kerrison, R (1980) Graphite/epoxy booms for the space shuttle remote manipulator. *Proceedings of ICCM3 Conference, Paris,* p 1360.

Eymard, M (1985) Ariane 4 and composite materials. *Proceedings of 6th International SAMPE Conference, Scheveningen* (May). Penultimate paper.

Franz, J and Laube, H (1985) Strength of carbon fibre composite/titanium bonded joints as used for SPAS-type structures. *Proceedings of ESA Noordwijk Symposium,* Netherlands (October), p 301.

Galipienso, G and Dell'Amico, S (1985) High stability telescope structures. *Ibid,* p 119, sp 243.

Garilotti, J F, Johnson, R and Cwiertny, A J (1980) Material and structural approaches for large space structures. *Proceedings of 31st International Astronautical Federation Congress, Tokyo* (September). IAF paper 80–G–297.

Green, A K and Phillips, L N (1978) Non-aerospace applications of carbon and other high performance materials and their hybrids. *Materials in Engineering Applications* 1 (December):59.

Grosser, M (1981) Building the Gossamer Albatross. *Technology Review* (April), p 52. Also *Ibid* (May).

Haines, D W and Chang, N (1975) Application of graphite composites in musical instruments. *Proceedings of ASME Design Conference* (April), p 13.

Hall, D J and Hall, B L (1984) A review of the design and materials Robson evaluation programme for the GRP/foam sandwich composite hull of the RAN minehunter. *Composites* 15(4):266. Guildford, Butterworth Scientific.

Hardy, R (1976) *Longbow: A Social Military History.* Cambridge, Patrick Stephens Ltd. p 14.

Jacoby, D and Shorrock, L D (1980) The production of ultra-thin carbon fibres for use as targets in laser produced plasma studies. *Journal of Physics, E: Scientific Instruments* 13:1272.

Johansen, B S, Lilholt, H and Lystrop, A (1980) Wingblades of glass-reinforced polyester for a 630W wind turbine: Design, fabrication and materials testing. *Proceedings of ICCM3, Paris.* Oxford, Pergamon Press, p 1353.

John, L K (1980) On the design and fabrication of a graphite/epoxy violin. *Proceedings of ICCM3 Conference, Paris,* p 1347. Oxford, Pergamon Press.

Jones, L A (1985) In-service environmental effects in carbon fibre composite material. *Proceedings of 6th International SAMPE Conference, Scheveningen.* Amsterdam, Elsevier.

Kawashima, T, Inoue, T and Seko, H (1985) Design and development of the graphite epoxy structure for the CS-3 satellite. *Proceedings of ESA Noordwijk Symposium*, Netherlands (October). p 267.

Kliger, H S and Yates, D N (1980) Design and material implications of composite driveshafts. *Proceedings of ICCM3 Conference, Paris.* Oxford, Pergamon Press. p 1335.

Krings, H and Scharring-hausen, J (1985) Fibre-reinforced plastic cases for solid propellant motors. *Proceedings of Workshop on Composite Design for Space Applications.* The Netherlands, ESA Noordwijk (October) p 295.

Lancaster, J L (1966) The effect of carbon-fibre reinforcement on the friction and wear of polymers. *RAE Technical Report no. 66378* (December).

Lapinleimu, R (1985) Sports/leisure – market opportunities. *Proceedings of 3rd Plastics & Rubber Institute International Conference: Carbon Fibres III.* New Jersey, Noyes Publications. p 126.

McCarthy, R (1985) Manufacture of composite propeller blades for commuter aircraft. *SAE Technical Papers no. 850875.* SAE Technical Meeting, Wichita, Kansas, USA (April).

McKibbin, B (ed) (1983) *Recent Advances in Orthopaedics*, Edinburgh, Churchill Livingstone, pp 129–203.

Miska, K H (1978) Record-setting aeroplane and catamaran share many common materials. *Materials Engineering* (March) p 14.

Mordoff, K F (1986) Around the world flight. *Aviation Week and Space Technology* (22 December), p 18. Also (1987) Around the world flight. *Ibid* (5 January), p 22.

Newton, A G (1980) The future of civil turbo-fan engines. *Rolls-Royce Magazine* 6 (September/November): 27.

Phillips L N (1969) Improving racing car bodies. *Composites* 1(1): 50.

Phillips, L N and Judd, N C (1969) Carbon fibre reinforcement in plastics for process equipment. *Process Engineering* (December).

Picquet, E (1981) Miracle string for crippled joints. *Readers Digest.* p 77.

Potter, K D (1985) Advances in the resin injection process for the reliable production of complex structural components. In *Advanced Materials and Processes.* Amsterdam, Elsevier Science Publisher BV, p 247.

Riley, B L (1986) *AV-8B GR Mark 5 Airframe Composite Applications.* 22nd John Player Lecture. IMechE.

Ruegg, C (1981) Carbon shafts made of carbon fibre reinforced plastics and mixed laminates. In *Processing and Uses of Carbon Fibre Reinforced Plastics.* Dusseldorf, VDI-Verlag. p 185.

Scheer, W (1981) Carbon fibre reinforced epoxy resin, a material for human implants. In *Processing and Uses of Carbon Fibre Reinforced Plastics.* Dusseldorf, VDI-Verlag. p 279.

Schwartz, M M (1984) Section 7.11: Fighters. In *Composite Materials Handbook.* New York, McGraw-Hill.

Shankar, P (1980) Satellite propulsion composite tanks for space shuttle applications. *Proceedings of ICCM3, Paris*, p 1439. Oxford, Pergamon Press.

Shorter, J J (1986) Carbon fibre uses and prospects. *Proceedings of 3rd Plastics & Rubber Institute International Conference: Carbon Fibres III.* New Jersey, Noyes Publications.

Stinton, D (1983a) Reciprocating engines. In *The Design of the Aeroplane*, p 271. London, Collins. Also (1983) Turbine engines. *Ibid.* Chap 8, p 321.

Stinton, D (1983b) *Ibid* p 495.

Stöffler, G and Wurlinger, H (1981) Tension-compression struts from carbon fibre reinforced plastics. *Processing and Uses of Carbon Fibre Reinforced Plastics.* Dusseldorf, VDI-Verlag. p 129.

Torres, M (1980) Les pales d'helicoptères en composites: conception, réalisation et comportement en operation. *Proceedings of ICCM3, Paris.* Oxford, Pergamon Press.

Worthington, P J (1978) The properties and requirements of P J filament-wound CFRP retaining rings, hoops and cylinders for rotating electrical machines. *Proceedings of ICCM2 Conference, Toronto.* New York, AIMMPE. p 719.

Yates, B (1977) The thermal expansion of carbon fibre et al reinforced plastics – Part 1: The influence of fibre type and orientation. *Journal of Materials Science* 12:718.

8 The quality control and non-destructive evaluation of composite materials and components

Burch, S F (1982) *Computer Assisted Real Time Radiography.* NDT Centre, AERE, Harwell, UK.

Cawley, P (1987) The sensitivity of the mechanical impedance method of non-destructive testing. *NDT International* 20(4).

Cielo, P, Bertrand, L, Rouset, G and Parkinson, E P (1987) Pulsed opto-thermal inspection of composite materials. *Canadian Aeronautics and Space Journal* 33(3).

Dean, G D and Lockett, F J (1973) Determination of the mechanical properties of fibre composites by ultrasonic techniques. *ASTM STP 521*, pp 326–346.

Dickson, J (1982) Inspection of composite materials using air-coupled ultrasonic transducers. *Proceedings of ATA Conference*, Long Beach, CA, USA.

Dickson, J (1988) *Recent Advances in Dry-Coupled Wheel Probe Technology.* Staveley NDT Technology Inc, Report tr 254/88 (March).

Drummond, W, *et al.* (1980) *Enhanced X-ray Stereoscopic NDT of Composites.* AFWAL-TR-80-3053 (June).

9 Design of composites

Agarwal, B D and Broutman, L J (1980) *Analysis and Performance of Fiber Composites.* New York, John Wiley.

Allen, H G (1969) *Analysis and Design of Structural Sandwich Panels.* Oxford, Pergamon Press.

Ashton, J E, Halpin, J C and Petit, P H (1969) *Primer on Composite Materials: Analysis.* Stanford, CT, USA, Technomic Publishing.

Benjamin, B S (1982) *Structural Design with Plastics.* 2nd edition. New York, Van Nostrand Reinhold.

Broutman, L J and Krock, R H (1967) *Modern Composite Material.* Reading, MA, USA, Addison-Wesley.

Ewins, P D and Ham, A C (1973) *The Nature of Compressive Failure in Unidirectional Carbon Fibre Reinforced Plastics.* Royal Aircraft Technical Report 73057.

Head, P R and Templeman, R B (1986) The application of limit state design principles to fibre reinforced plastics. *The British Plastics Federation Congress*, Nottingham, UK, 17–19 September.

Hill, R (1960) *The Mathematical Theory of Plasticity.* London, Oxford University Press.

Hollaway, L (1978) *Glass Reinforced Plastics in Construction.* Surrey University Press/International Textbook Company.

Holmes, M and Just, D J (1983) *GRP in Structural Engineering.* Barking, Essex, Elsevier Applied Science.

Hull, D (1985) *An Introduction to Composite Materials.* London, Cambridge University Press.
Johnson, A F and Sims, G D (1987) *Design Procedures for Plastics Panels.* An NPL consortium project on engineering design in plastics.
Jones, R M (1971) Buckling of stiffened multi-layered circular cylindrical shells with different orthotropic moduli in tension and compression. *AIAA Journal* (May):917–923.
Jones, R M (1975) *Mechanics of Composite Materials.* Kogakusha, Tokyo, McGraw-Hill.
Ogorkiewicz, R M (1974) *Thermoplastics: Properties and Design.* Chichester, Sussex, Wiley.
Piggott, M R (1980) *Load Bearing Fibre Composites.* Oxford, Pergamon Press.
Timoshenko, S P and Woinowsky-Krieger, S (1959) *Theory of Plates and Shells.* 2nd edition. New York, McGraw-Hill.
Tsai S W (1968) Strength theories of filamentry structures. In R T Schwartz and H S Schwartz (eds) *Fundamental Aspects of Fibre Reinforced Plastic Composites.* New York, Wiley International Science.
Tsai, S W and Hahn, W T (1986) *Introduction to Composite Materials.* Lancaster, Philadelphia, Technomic Publishing Co.
Williams, J G (1980) *Stress Analysis of Polymers.* 2nd edition. London, Longman.

10 Damage and fracture of fibre composites

Anstice, P D and Beaumont P W R (1983) *Journal of Materials Science,* 18:3404.
Bowie, O (1956) *Journal of Mathematics and Physics* 35:60.
Caprino, G, Halpin, J C and Nicolais L (1979) *Composites* (October):223–227.
Kortschot, M T (1988) PhD thesis, University of Cambridge.
Nuismer, R J and Whitney, J M (1975) *Fracture Mechanics of Composites.* ASTM STP 593:117–142.
Ogin, S L, Smith, P A and Beaumont, P W R (1985a) *Composite Science and Technology* 24:47.
Ogin, S L, Smith, P A and Beaumont, P W R (1985b) *Ibid* 22:23.
Poursartip, A P (1983) PhD thesis, University of Cambridge.
Poursartip, A P and Beaumont, P W R (1986) *Ibid* 25:283.
Poursartip, A P, Ashby, M F and Beaumont, P W R (1986) *Ibid* 25:193.
Waddoups, M E, Eisenmann, J R and Kaminski, B E (1971) *Journal of Composite Materials* 5:446–454.
Wells, J K (1982) PhD thesis, University of Cambridge.
Wells, J K and Beaumont, P W R (1981) *Scripta Met.* 16:99.
Wells, J K and Beaumont, P W R (1985a) *Journal of Materials Science* 20:1735.
Wells, J K and Beaumont, P W R (1985b) *Ibid* 20:1275.

General Index

Index to Specifications